Marine Biodiversity and Ecosystem Functioning

Marine Biodiversity and Ecosystem Functioning

Frameworks, Methodologies, and Integration

EDITED BY

Martin Solan

University of Southampton, UK

Rebecca J. Aspden

University of St Andrews, UK

and

David M. Paterson

University of St Andrews, UK

OXFORD
UNIVERSITY PRESS

OXFORD

UNIVERSITY PRESS

Great Clarendon Street, Oxford, OX2 6DP,
United Kingdom

Oxford University Press is a department of the University of Oxford.
It furthers the University's objective of excellence in research, scholarship,
and education by publishing worldwide. Oxford is a registered trade mark of
Oxford University Press in the UK and in certain other countries

© Oxford University Press 2012

The moral rights of the authors have been asserted

First Edition published in 2012
Impression: 1

British Library Cataloguing in Publication Data

Data available

Library of Congress Cataloging in Publication Data
Library of Congress Control Number: 2012934408

ISBN 978–0–19–964225–0 (hbk.)
ISBN 978–0–19–964226–7 (pbk.)

Printed and bound by
CPI Group (UK) Ltd, Croydon, CR0 4YY

Contents

List of Contributors

Rebecca J. Aspden, University of St Andrews, Scottish Oceans Institute, East Sands, St Andrews, KY16 8LB, UK
rja4@st-andrews.ac.uk

Lisandro Benedetti-Cecchi, Dipartimento di Biologia, University of Pisa, CoNISMa, Via Derna 1, I-56126 Pisa, Italy
lbenedetti@biologia.unipi.it

Matthew E. S. Bracken, Northeastern University, Marine Science Center, 430 Nahant Road, Nahant, MA 01908, USA
m.bracken@neu.edu

John F. Bruno, Marine Sciences Department, University of North Carolina, Chapel Hill, North Carolina 27599-3300, USA
john_bruno@unc.edu

Matthew E. Clapham, Department of Earth and Planetary Sciences, University of California, Santa Cruz, 1156 High Street, Santa Cruz, CA 95064, USA
mclapham@ucsc.edu

Tasman P. Crowe, School of Biology and Environmental Science, Science Centre West, University College Dublin, Belfield, Dublin 4, Ireland
tasman.crowe@ucd.ie

Roberto Danovaro, Department of Life and Environmental Sciences, Polytechnic University of Marche, Ancona, Italy
r.danovaro@univpm.it

Emma C. Defew, University of St Andrews, Scottish Oceans Institute, East Sands, St Andrews, KY16 8LB, UK
Ecd2@st andrews.ac.uk

Stephen Q. Dornbos, Department of Geosciences, University of Wisconsin-Milwaukee, P.O. Box 413, Milwaukee, WI 53201, USA
sdornbos@uwm.edu

J. Emmett Duffy, Virginia Institute of Marine Science, College of William and Mary, Gloucester Point, Virginia 23062-1346, USA
jeduffy@vims.edu

Nicholas K. Dulvy, Biological Sciences, 8888 University Drive, Simon Fraser University, Burnaby, BC V5A 1S6, Canada
nick_dulvy@sfu.ca

Anton Edwards, University of St Andrews, Scottish Oceans Institute, East Sands, St Andrews, KY16 8LB, UK
AntonEdwards@aol.com

Mark C. Emmerson, School of Biological Sciences, Queen's University Belfast, 97 Lisburn Road, Belfast, BT9 7BL, Northern Ireland, UK
m.emmerson@qub.ac.uk

Margaret L. Fraiser, Department of Geosciences, University of Wisconsin-Milwaukee, P.O. Box 413, Milwaukee, WI 53201, USA
mfraiser@uwm.edu

Alan M. Friedlander, US Geological Survey-Hawaii Cooperative Fishery Research Unit, University of Hawaii at Manoa, Hawaii, USA
alan.friedlander@hawaii.edu

Jasmin A. Godbold, Ocean and Earth Science, National Oceanography Centre, Southampton, University of Southampton, Waterfront Campus, European Way, Southampton, SO14 3ZH, UK
j.a.godbold@soton.ac.uk

Caroline Hattam, Plymouth Marine Laboratory, Prospect Place, West Hoe, Plymouth PL1 3DH, UK
Caro4@pml.ac.uk

Alison R. Holt, Department of Animal and Plant Sciences, University of Sheffield, Sheffield S10 2TN, UK
a.holt@sheffield.ac.uk

Julia Jabour, Institute for Marine and Antarctic Studies, University of Tasmania, Private Bag 129, Hobart, Tasmania 7001, Australia
julia.jabour@utas.edu.au

Marc Laflamme, Smithsonian Postdoctoral Fellow, Department of Paleobiology, National Museum of Natural History, PO Box 37012, MRC 121, Washington, DC 20013-7012, USA
marc.laflamme@yale.edu

Andrew M. Lohrer, National Institute of Water and Atmospheric Research, P.O. Box 11-115, Hillcrest, Hamilton, New Zealand
drew.lohrer@niwa.co.nz

Elena Maggi, Dipartimento di Biologia, University of Pisa, CoNISMa, Via Derna 1, I-56126 Pisa, Italy
emaggi@biologia.unipi.it

Anne E. Magurran, School of Biology, University of St Andrews, St Andrews, Fife KY16 8LB, Scotland, UK
aem1@st-andrews.ac.uk

Stephen Mangi, Plymouth Marine Laboratory, Prospect Place, West Hoe, Plymouth PL1 3DH, UK
stcma@pml.ac.uk

Scot Mathieson, Scottish Environmental Protection Agency, Erskine Court, The Castle Business Park, Stirling, FK9 4TR, UK
Scot.Mathieson@sepa.org.uk

Shahid Naeem, Department of Ecology, Evolution, and Environmental Biology, Columbia University, 1200 Amsterdam Ave., 10th Floor Schermerhorn Extension, MC5557, New York, NY, 10027, USA
sn2121@columbia.edu

Nessa E. O'Connor, School of Biological Sciences, Queen's University Belfast, 97 Lisburn Road, Belfast, BT9 7BL, Northern Ireland, UK
n.oconnor@qub.ac.uk

Ruth Parker, Cefas Lowestoft Laboratory, Pakefield Road, Lowestoft, Suffolk, NR33 0HT, UK
Ruth.parker@cefas.co.uk

David M. Paterson, University of St Andrews, Scottish Oceans Institute, East Sands, St Andrews, KY16 8LB, UK
dp1@st-andrews.ac.uk

David Raffaelli, Environment, University of York, Heslington, York, YO10 5DD, UK
Dr3@york.ac.uk

Finlay Scott, Cefas Lowestoft Laboratory, Pakefield Road, Lowestoft, Suffolk, NR33 0HT, UK
finlay.scott@cefas.co.uk

Martin Solan, Ocean and Earth Science, National Oceanography Centre, Southampton, University of Southampton, Waterfront Campus, European Way, Southampton, SO14 3ZH, UK
m.solan@soton.ac.uk

Paul J. Somerfield, Plymouth Marine Laboratory, Prospect Place, West Hoe, Plymouth, PL1 3DH, UK
pjso@pml.ac.uk

John J. Stachowicz, Bodega Marine Laboratory, University of California at Davis, P.O. Box 247, Bodega Bay, CA 94923, USA
jjstachowicz@ucdavis.edu

Simon F. Thrush, National Institute of Water and Atmospheric Research, P.O. Box 11-115, Hillcrest, Hamilton, New Zealand, and DipTeRis, Università di Genova, Corso Europa, 26, 16132 Genoa, Italy
simon.thrush@niwa.co.nz

Stephen Widdicombe, Plymouth Marine Laboratory, Prospect Place, West Hoe, Plymouth, PL1 3DH, UK
swi@pml.ac.uk

CHAPTER 1

Marine biodiversity: its past development, present status, and future threats

Stephen Widdicombe and Paul J. Somerfield

1.1 Introduction

'We do not associate the idea of antiquity with the ocean, nor wonder how it looked a thousand years ago, as we do of the land, for it was equally wild and unfathomable always. The Indians have left no traces on its surface, but it is the same to the civilized man and the savage. The aspect of the shore only has changed.'

—Henry David Thoreau; Cape Cod
(1855–1865), in The Writings of Henry David
Thoreau, vol. 4, p. 188, Houghton Mifflin (1906)

Humans are terrestrial creatures, and those with a deep knowledge of the sea tend to be those associated either with its exploitation or exploration. Our lack of knowledge about life in the marine realm, what lives there, where it lives, and what it does, should not be underestimated. Seawater covers around 70% of the Earth, and the vast majority of landscape beneath it has never been seen by humans, let alone sampled in any way. Sampling the oceans is costly and having to work from ships or boats puts many constraints on what can be done, where, and how often. The vast majority of marine biodiversity records are from shallow waters (< 200 m), or from the seabed, and the deep pelagic ocean, which represents the largest volume on Earth in which life can exist (> 10^9 km^3) is recognized as particularly under-represented in global databases (Webb *et al.* 2010). The total area of seabed surveyed by human eyes, in submersibles, by diving, or remotely using cameras, represents an infinitesimal fraction of the total. To compound this problem, the majority of marine life doesn't even live on the seabed where cameras can see it, but

lives either buried within it or floating above it in the water column. In addition, much of the metazoan diversity in the water column consists of extremely delicate forms which are destroyed by conventional sampling using nets. Consequently, our knowledge of marine biodiversity has been likened to what our knowledge of rainforest biodiversity might be like if our only way of assessing it was by means of a grappling hook deployed from a hot-air balloon. Recent discoveries have highlighted the depth of our ignorance. These include the discovery in 1977 of hydrothermal vent communities on the seabed, deriving their energy from chemoautotrophy rather than sunlight (Lonsdale 1977), the identification in 1986 of *Prochlorococcus* in the photic zone, a tiny photosynthetic cyanobacterium which is probably the most abundant organism on Earth and is estimated to account for 20% of the oxygen in the Earth's atmosphere (Chisholm *et al.* 1988), and in the deep pelagic ocean, the discovery of the very large, charismatic and widely distributed megamouth shark *Megachasma pelagios* in 1976 (Berra 1997). Finally, but perhaps most notably, are the findings generated over the past 10 years by the Census of Marine Life (<http://www.coml.org>). This census has brought together nearly 3000 scientists from over 80 nations to assess the diversity, distribution and abundance of marine life. Their results indicate that there are approximately 250 000 valid marine (non-microbial) species formally described in the scientific literature with as many as another 750 000 species still waiting to be described. They also estimate that there may be more than a billion types of microbes living in the oceans. These recent, concerted, international

Marine Biodiversity and Ecosystem Functioning. First Edition. Edited by Martin Solan, Rebecca J. Aspden, and David M. Paterson.
© Oxford University Press 2012. Published 2012 by Oxford University Press.

activities have started to reveal the staggering enormity of marine biodiversity.

1.2 What is biodiversity?

The term biodiversity is a contraction of 'biological diversity' and first appeared in print in 1988 (Wilson and Peters 1988), very quickly becoming widespread (Haila and Kouki 1994). The use of this term arose amid concern over the loss of the natural environment and, as a result, it has become part of the lexicon of science, the media, and government on a global scale (Gaston 1996). The ultimate confirmation of this was the signing, in 1992, of the UN Convention on Biological Diversity (<http://www.cbd.int>). According to the convention biodiversity means 'the variability among living organisms from all sources including, inter alia, terrestrial, marine and other aquatic ecosystems and the ecological complexes of which they are part; this includes diversity within species, between species and of ecosystems.' Gaston (1996) provides a number of alternative definitions which are essentially variations on this theme, and an interesting discussion of what biodiversity is, or might be. Essentially, biodiversity is a multivariate concept with 'the variety of life' distributed across a range of hierarchical scales and cannot, therefore, be boiled down to a single measure. For example, even with a defined biological category such as macrofauna (multicellular animals larger than 0.5 mm), diversity is not only a function of the number of species in any given area (known as richness), but also the way in which individuals are distributed within those species (evenness). There are even different measures of richness depending on whether you are interested in taxonomic or functional diversity. This provides a dilemma in situations where there is a desire to measure biodiversity and use it as an end point in rigorous empirical studies (Solbrig 1991; Harper and Hawksworth 1994). In practice, how biodiversity is measured will depend on the specific nature and purpose of the study in question. To date, much of the biodiversity research effort exploring its relevance to human governance, well-being, and survival has tended to focus on terrestrial systems, with the possible exception of fisheries.

1.3 Comparing marine and terrestrial biodiversity

Marine ecosystems are fundamentally different from terrestrial ecosystems (Steele 1985; Strathmann 1990). Biomass per unit area in marine systems may be several orders of magnitude less than in terrestrial or freshwater systems, but it must be remembered that terrestrial systems are essentially two-dimensional with most living material concentrated into a layer 10–100 m thick. The seas and oceans, on the other hand, have an average depth in excess of 3500 m, and two-thirds of the Earth's surface lie below more than 200 m of seawater. Where terrestrial primary producers tend to be large and sessile, marine primary producers tend to be small and mobile, and therefore subject to fluid transport processes and spatial mixing. Photosynthesis is limited to the upper sunlit parts of the ocean, which over vast areas are very much less productive than the land because growing phytoplankton strip nutrients from the surface waters which are removed as they sink and die. Essentially, therefore, much marine productivity is self-limiting. Production and consumption may be decoupled in the sea with fixed carbon being consumed many times before it is completely remineralized and, although available organic matter becomes increasingly scarce with depth, it supports diverse pelagic and benthic ecosystems. At the other end of the marine food-web, consumers, grazers, and carnivores, which may be several times larger than their terrestrial counterparts, have a much greater range of life-history characteristics than terrestrial ones. Many have planktonic and benthic stages, with different associated environmental linkages. Marine systems may be relatively 'open' with gene-flow between distant marine habitats facilitated by larval dispersal. Many marine predators have extremely high reproductive output especially among older and larger members of their populations, producing millions of offspring. This may help them to survive over-exploitation, at least for a while, but also makes populations highly variable, difficult to predict, and vulnerable to threshold effects.

Unsurprisingly, we do not actually know how many species currently live in Earth's seas and oceans although current estimates for multicellular organisms are in the order of 250 000 species (May 1988, Census of Marine Life). If correct, this is far less than the numbers estimated to live on land, but marine diversity, at least animal diversity, is organized very differently to its terrestrial counterpart. To put it glibly, to a first approximation, all species of land animals are insects. In contrast, marine biodiversity is more evenly distributed between the major animal groups. There are 35 different phyla that occur in the sea, with most of them being benthic (Briggs 1994, May 1988) and only 11 occurring in the water column (Angel 1993). From the total 35 phyla, 14 are known to occur only in the sea. This compares favourably with freshwater systems, containing only 14 phyla, and terrestrial systems that can only boast representatives from 11 phyla. Thus animal phyletic diversity is clearly at its highest in the marine realm.

As a consequence of these fundamental differences, biodiversity paradigms developed in terrestrial systems may not be transferable to marine systems and the response of these two systems to human perturbations will be markedly different (National Research Council 1995).

1.4   The rise of marine biodiversity

Until recently, our knowledge of the origins and evolution of life on Earth was primarily derived from the fossil record, but currently our estimates of the timing and nature of major events are being revised using novel molecular and chemical forms of evidence. Marine biodiversity is extremely ancient and life probably arose in a marine environment between 4500 and 3500 Mya (million years ago), with chemoautotrophic prokaryotes and photosynthetic cyanobacteria appearing by 3000 Mya, and eukaryotic cells around 2000 Mya (Battistuzzi *et al.* 2004, Cavalier-Smith 2006). Thus, for most of the time life has existed on Earth, biodiversity was entirely marine and microbial. Microbes lack the capacity to exploit behavioural or morphological diversity, have limited capacity to engineer environments and drive co-evolutionary change, and their extremely large populations are effectively immune from extinction and other ecological

processes driving evolutionary change and diversification. The consequence would have been extreme evolutionary stasis. The Proterozoic oceans were generally anoxic, sulphidic, turbid, and stratified, and largely physically driven. Life in the water column was dominated by cyanobacteria, the small size of which limited their rate of sinking and maintained a positive feedback whereby cyanobacterial dominance, turbidity, and anoxia were maintained (Butterfield 2010). Life on the seafloor was characterized by microbial mats (Bottjer 2005). Precisely when multicellular organisms evolved in the sea is unknown. Molecular evidence suggests that plants, animals and fungi diverged about 1500–1600 Mya, the basal animal phyla (Porifera, Cnidaria, Ctenophora) diverged about 1200–1500 Mya, nematodes diverged from the line leading to arthropods and chordates about 1200 Mya, and arthropods and deuterostomes around 1000 Mya (Wang *et al.* 1999; Pisani *et al.* 2004). These estimates indicate the presence of complex metazoan life long before the oldest currently known direct fossil evidence, which indicates the presence of sponges at least 635 Mya (Love *et al.* 2009). The earliest known fossils of complex multicellular organisms date from somewhere between 595 and 565 Mya (Narbonnel and Gehlingl 2003), forming the Ediacaran biota, a phylogenetically diverse assemblage of organisms, probably including animals, protists, and fungi with a simple ecology dominated by epibenthic osmotrophs, deposit feeders and grazers with few, if any, predators (Xiao and Laflamme 2008). The earliest evidence of motile animals is from 565 Mya (Liu *et al.* 2010). This biota flourished worldwide until disappearing at the onset of the Cambrian approximately 542 Mya.

Most major groups of complex animals first appeared as fossils sometime in the Cambrian, the so-called 'Cambrian explosion'. The apparent absence of fossils from before this time (in the absence of knowledge about the Ediacaran biota) and the apparently sudden appearance of so many types of life without forerunners greatly concerned Charles Darwin (1859), who saw these facts as the greatest objection to his theory of evolution by natural selection. The Ediacaran and early Cambrian are marked by some of the largest biogeochemical perturbations in Earth's history including major changes in carbon and sulphur cycling, iron geochemistry, phosphate deposition, and oceanic oxygenation (Butterfield 2010). Until recently, such observations have been used to explain the sudden appearance of diverse animal assemblages in the Cambrian in terms of changes in the Earth system providing conditions conducive to metazoan evolution, with a focus on atmospheric oxygenation (e.g. Knauth and Kennedy 2009; Payne *et al.* 2009). Recently this picture has changed, with the realization that animals have particular abilities and qualities that allowed them to fundamentally alter the Earth system, and to drive evolutionary change (Butterfield 2010; Meysman *et al.* 2006). The evolution of animal organization allowed organisms to achieve different sizes, large and small organisms could interact, so predation and competition became factors driving evolution which could then occur at a pace more rapid and at an order of magnitude greater than previously. Predation promoted the evolutionary advantage of defensive and strengthening structures such as biomineralized skeletons with bristles and spines, promoting co-evolution among predators and prey and, coincidentally, the probability of organisms becoming fossilized. Animal feeding and bioengineering activities facilitated and altered biogeochemical cycles. For example, animals feeding on plankton converted slow-sinking cells to rapidly sinking faecal pellets (Logan *et al.* 1995), providing a source of organic-rich sediment, and animals burrowing in sediment could exploit buried carbon as a food source, at the same time oxygenating sediments and altering their biogeochemistry (Meysman *et al.* 2006). Phytoplanktonic organisms increased in size and developed biomineralized structures, causing them to sink more rapidly and therefore to change export production from the surface ocean. Concentrating biomass into fewer but larger organisms increased the availability of free oxygen. The evolution of filter-feeding provided a mechanism for clearing microbial populations and providing a relatively non-turbid photic zone in which algal dominated primary production could take place. Thus the co-occurrence of a major shift in the Earth

system and marine biodiversity, from a turbid, stratified, anoxic, microbially dominated Proterozoic ocean to a clear, oxygenated, algal, and animal-dominated Phanerozoic ocean, may mark the effects of life on Earth, rather than the other way around.

Many studies have used the documented occurrences of taxa in the fossil record to analyse global trends in marine diversity through deep time. A much repeated curve generated from these studies (see Sepkoski 2002) shows a rapid increase in marine diversity through the Cambrian and Ordovician (to 446 Mya), a plateau lasting until the end of the Permian (254 Mya), and then an apparently exponential increase to the present. However, concerns about the extent to which this pattern is artefactual have led to studies redrawing the Phanerozoic diversity curve using standardized methods (Alroy *et al.* 2008). These new studies tend to show that, although diversity varied through time, the overall trend was a slow increase, with levels in the Neogene (23–7.2 Mya) only slightly higher than the levels seen in the Devonian and Permian. Sepkoski (1981) identified three major 'Evolutionary Faunas'—the Cambrian, the Palaeozoic, and the Modern—which sequentially replaced each other. Despite these gradual changes, the repeating motif of life in clear shallow seas throughout the Phanerozoic and Cenozoic is of diverse and abundant reef-building organisms (archaeocyathid sponges, rudist bivalves, corals), epifauna and infauna (molluscs, brachiopods, bryozoans, crustaceans, echinoderms), and iconic top predators (anomolacaridids, eurypterids, nothosaurs, ichthyosaurs, plesiosaurs, diverse types of fish, especially sharks).

1.5 The distribution of marine biodiversity

Most of the Earth's surface (65–70%, depending on how one defines it) is covered by deep seawater (> 200 m). Life in the deep sea is diverse and specialized with organisms living in constant darkness at high pressures and low temperatures. However, the number and variety of organisms per unit area or volume is generally low compared to shallow coastal systems, with diversity tending to decline with increasing depth. It is undoubtedly true that the proportion of the actual deep-sea fauna which is known to science is lower than for other marine systems, but enough is known to estimate that the total number of species in the deep ocean is unlikely to exceed the number in shelf seas. It should also be borne in mind that life in the deep sea is comparably young, as for most of Earth's history the deep sea has been inhospitable to life.

Despite covering less than 10% of the Earth's surface, coastal seas and oceans deliver approximately 30% of marine production and 90% of marine fisheries (Millennium Ecosystem Assessment 2005). Some temperate ecosystems are, or at least were, among the most productive on Earth (Suchanek 1994). They are the burial site for 80% of organic matter, and 90% of sedimentary mineralization and nutrient cycling takes place there. They also provide the sink for up to 90% of the suspended load in the world's rivers, along with many associated contaminants (Gattuso *et al.* 1998). Marine biodiversity is highest in the coastal and shelf seas supported by the presence of diverse sedimentary habitats, rocky and biogenic reefs, kelp forests, and seagrass beds. Many of these habitats are highly specialized. For example, estuaries are generally inhabited by a limited number of specialist species that can tolerate reduced salinities. Generally those species that live in estuaries are marine in origin, and there is a gradient of declining biodiversity from full salinity water at the estuary mouth towards the freshwater at the head of the estuary. Another specialized set of organisms inhabits the intertidal zone which ranges from rocky to sediment shores, and exposure to waves is a crucial determinant of the community inhabiting a particular shore. Again, the vast majority of species are marine in origin and there tends to be a gradient of declining biodiversity from permanently submerged shallow waters at the base of the shore to higher parts of the shore which are only inundated rarely, or are only wetted by spray. The increasing gradients in marine biodiversity from the land to the open coastal and shelf seas contrasts strongly with our knowledge about the species present, which is greatest in estuaries and on coasts, and declines towards the open sea (Lotze and Worm 2009).

1.6 Human impacts on marine biodiversity

Modern humans evolved in Africa around 200 000 years ago with the earliest evidence of humans at the coast being from the Eritrean coast of Africa 125 000 years ago (Walter *et al.* 2000). Using hand-axes and obsidian blades, these early people were harvesting and processing oysters and other marine resources (Bruggemann *et al.* 2004). A likely scenario for the geographical spread of early humans is that they followed the coasts, especially during periods of low sea levels when much of the continental shelves would have been exposed. By 70 000 years ago, humans had spread to north-west Africa, and by 50 000 years ago, humans had reached south-east Asia and Australia. Humans were present on the coasts of east Asia and Siberia around 30 000 years ago and reached southern Europe some 20 000 years ago. Humans then colonized North America about 15 000 years ago and South America 15–12 000 years ago. Thus, by around 10 000 years ago, humans were present on all the major continents (except Antarctica, unreachable and uninhabitable without appropriate technology), leaving only remote islands to be colonized. During roughly 200,000 years as hunter-gatherers the human population probably never exceeded 10 million (May 2010); however, the introduction of agriculture approximately 10 000 years ago allowed the population to increase rapidly before a number of factors such as infectious diseases slowed growth, and it was not until 1830 that numbers exceeded 1000 million. From there the population doubled to 1930 and passed 4000 million in 1970. Following the scientific and industrial revolution and the development of science-based medical understanding, the Earth's human population has reached 6700 million and continues to increase. To put these figures in context, for every human alive today, only 15 have lived before, and the human population has trebled in less than a lifetime (May 2010). Nearly 70% of all humans now live within 60 km of the sea, many of them in cities of > 10 million inhabitants, of which 75% are in the coastal zone. It is therefore unsurprising that the activities of human beings are likely to have a significant impact on coastal marine systems.

Following the upsurge of interest in biodiversity in the early 1990s, many reviews were published outlining the state of marine biodiversity, in the open ocean (Angel 1993), sediments (Snelgrove 1999), the coastal zone (Ray 1991), coral reefs (Jackson 1991), rocky shores (Thompson *et al.* 2002), estuaries (Levin *et al.* 2001), temperate coastal seas (Suchanek 1994), and the deep sea (Grassle 1991), along with more general reviews of marine biodiversity (Gray 1997, Ray and Grassle 1991, National Research Council 1995). These reviews described biodiversity patterns in marine systems and discussed the threats faced by marine ecosystems and the organisms they support. The general conclusion was that the most severe threats to marine biodiversity are in the coastal zone and are a direct result of human population and demographic trends (Gray 1997), a conclusion supported and expanded in the comprehensive reviews of current state and trends in the Millennium Ecosystem Assessment (2005). Although many of those pressures acting on estuarine and intertidal biodiversity, particularly habitat loss and habitat modification or squeeze, can have extreme local effects on marine biodiversity, these are fairly predictable, and many of the species affected are tolerant and many may actually benefit, such as those which can use man-made structures as habitat. Restoration in such systems is also often possible.

At the other end of the spectrum, the effects of man's activities in deep sea systems may be widespread and some, such as the removal of predators or the accumulation of toxins or garbage, may alter food-webs. Localized effects, such as commercial fishing of slow-growing and long-lived fish in geographically limited areas such as seamounts and continental shelves, with concomitant damage to seabed organisms, may be severe. Commercial and technological developments are pushing the limits at which activities such as oil and gas extraction, or seabed mining for minerals, are viable ever deeper. Despite all of this, the direct effects of man's activities on deep-sea biodiversity as a whole are unlikely to be as severe as in shallower waters.

It is in the intermediate shallow shelf and coastal waters, where marine biodiversity evolved and has flourished throughout most of the history of life,

that the effects of pressures from human activities are highest. It is no coincidence that recorded extinctions such as the Great Auk, Stellar's Seacow, the Caribbean Monk Seal, and ecological extinctions such as many large species of skates and rays and commercially fished populations of other fish, or major changes in ecosystems such as the demise of coral reefs and cod, are to be found in such waters. Species have not evolved mechanisms to cope with sudden change and the effects of pressures are relatively poorly understood. It used to be thought that the fluid nature of life in the sea, with widespread dispersal, could insure most marine organisms against total extinction. More recently it has become clear that there are many invisible barriers to dispersal in the sea, so species may be more vulnerable than was thought. Sadly we have no idea of the number of populations or species that may already have been lost or how many will be lost in the future.

The major threats to marine biodiversity are numerous: habitat degradation, fragmentation and loss, climate change and other large-scale atmospheric changes such as ozone depletion, direct and indirect effects of fishing and other forms of overexploitation, nutrient runoff, mineral extraction, pollution and marine litter, species relocations and invasions, watershed alteration and the physical alteration of coasts, tourism, limited knowledge and willingness to conserve marine biodiversity, and legal complexities in dealing with marine issues. Realizing that, even among professionals, knowledge about what marine biodiversity ought to be like in the absence of human intervention was generally missing (Pauly 1995, Jackson 2001), and in recognition that a research strategy based on short-term experiments or observations is unlikely to be helpful in addressing this issue, a number of studies have subsequently taken a multidisciplinary approach, incorporating elements of palaeontology, archaeology, history, and ecology, to describe long-term trends in marine biodiversity. Using a range of different data sources, a strategy based on the facts that different parts of the world were colonized by people at different times, and that in different places cultural development followed similar paths, a clear pattern of degradation has been shown in a diverse range of marine communities and ecosystems: coastal (Jackson *et al.* 2001), coral reefs (Pandolfi *et al.* 2003), estuaries and coastal seas (Lotze *et al.* 2006), large marine animals (Lotze and Worm 2008). These studies, summarized and expanded by Jackson (2008), point to the fact that even small human populations may have severe effects on marine biodiversity, and that although humans have been degrading marine systems for a very long time, the recent increase in human population and its consequences have led to a step-shift in the rate and extent of degradation.

A primary cause of biodiversity degradation is the removal of large animals and the consequent disruption of food-webs through trophic cascades. In the industrial age (circa 1750 to present day), these effects may be very rapid and widespread (Myers and Worm 2003). Large-scale exploitation of marine resources probably did not occur in Europe before AD 1000, and when it did it may have been a response to the rise of urbanization and the overexploitation and degradation of freshwater ecosystems (Barrett *et al.* 2004). Major exploitation of marine resources elsewhere in the world largely followed the arrival of Europeans. The consequent rapid decline of marine ecosystems is therefore remarkable with the majority of coastal ecosystems now being severely degraded compared to long-term historical baselines. Through human activities, complex ecosystems are being transformed into monotonous level bottoms, clear and productive seas into anoxic dead zones (Vaquer-Sunyer and Duarte 2008), complex food-webs with large apex predators into microbially dominated systems. These changes, along with similar but better-known changes in terrestrial and freshwater ecosystems, have led to some proposing a new term, the 'Anthropocene epoch', to separate the world as it is now from the world as it was until recently. As Jackson (2008) points out, we do not know what organisms will benefit from and come to dominate marine ecosystems in the near future, but there is already evidence that jellyfish may be among them (Richardson *et al.* 2009) along with microbes, including some that may have the potential to be toxic or cause disease (Jackson 2008). We are only beginning to realize the extent of the damage we have already

done, are doing, and are likely to do, and what the likely consequences will be (Worm *et al.* 2006). For the anthropogenic impacts that are performed at local scales—e.g. fishing, pollution, habitat destruction—these trends may still be reversible. However, the impacts associated with climate change are likely to be far more ubiquitous and long lasting.

1.7 The relationship between global climate and marine biodiversity

Close examination of the geological record provides chilling evidence that Planet Earth has historically been vulnerable to climate-induced biodiversity loss on a huge scale. Prior to large extinction events in the late Permian/early Triassic, early Jurassic, middle Cretaceous, and late Paleocene, major volcanic events occurred and the carbon-isotope record confirms that during these periods atmospheric CO_2 concentrations rose considerably and stayed high for hundreds of thousands of years. Although the current rise in atmospheric CO_2 levels is not volcanically driven, the consequences could essentially be similar. The largest of all climatically driven extinction events occurred 251 Mya at the end of Permian period. During this event, 80% of all marine species and 49% of all marine families went extinct (Raup and Sepkoski 1982; Stanley and Yang 1994). It is suggested that the environmental changes associated with the late-Permian extinction event were triggered by huge quantities of volcanically produced CO_2 and methane that drove rapid global warming. As the oceans warmed, they carried less oxygen and these conditions encouraged the growth of bottom-dwelling anaerobic bacteria. These bacteria produced large quantities of hydrogen sulphide (H_2S), and eventually the chemocline (the division between oxygen-rich surface water and H_2S-rich deep water) rose up into the photic zone allowing the proliferation of green and purple photosynthesizing sulphur bacteria while oxygen-breathing organisms suffocated. The H_2S also diffused into the air where it killed animals and plants on land. Once into the upper atmosphere, the H_2S destroyed the ozone layer and the Sun's ultraviolet radiation caused even more extinctions (Kump *et al.* 2005).

Whilst it is difficult to imagine that such a fate could befall modern Earth, it is worth reflecting on the atmospheric levels of CO_2 that correspond to some of these large extinction events. At the start of the late-Permian event, the concentration was around 3000 parts per millon (ppm), almost an order of magnitude higher than today's value of 385 ppm. However, the smaller yet still significant extinctions which occurred at the end of the Paleocene and at the end of the Triassic began when atmospheric CO_2 levels were much lower, around 1000 ppm. If current increases in atmospheric CO_2 levels continue as projected, we could be looking at levels approaching 1000 ppm by the end of the next century (IPCC 2007).

1.8 Could marine biodiversity be facing large-scale climate-induced extinction?

At the start of the industrial revolution (circa 1750), the atmospheric concentration of carbon dioxide (CO_2) in the atmosphere was around 280 ppm. In the subsequent 250 years, this has risen rapidly to a current value of more than 385 ppm (Feely *et al.* 2004). This increase has been driven primarily by the burning of fossil fuels, wide scale deforestation, changes in land use, and the manufacture of cement (IPCC, 2007). Once in the atmosphere CO_2, along with other 'greenhouse' gases such as methane and nitrous oxide, retains the Sun's heat and warms up the Earth's lower atmosphere. As 65–70% of the planet's surface is covered by water, much of this heat is being absorbed by the oceans, causing mean global sea-surface temperatures to have increased at a rate of 0.13 °C per decade over the past 30 years (IPCC). In addition to causing this warming, the process of radiative forcing (the entrapment of the Sun's energy in the lower atmosphere) is supplying the energy which is driving a number of other climate-change phenomena. These include increased average wind velocities and storm frequency, increased global average precipitation and riverine runoff, and enhanced sea-level rise through thermal expansion and melting of ice. All of these physical effects could have a significant indirect impact on the marine environment by altering ocean currents, changing the supply and cycling of key nutrients,

and by reducing seawater salinity. For a summary of these impacts on a range of marine ecosystems, see Brierley and Kingsford (2009). It must also be considered that climate change impacts will be geographically variable with some locations, such as the Arctic Ocean, experiencing much larger changes than the global average. An appreciation of the potential severity of these impacts has lead some authors to predict that climate change will be the main driver of global biodiversity loss over the next century (Thomas *et al.* 2005).

As organisms look to optimize the structural and kinetic coordination of molecular, cellular and systemic process, they are constrained to exist within a physiologically prescribed thermal window (Figure 1.1). At the edges of these thermal windows an organism will experience decreased performance with respect to many functions: growth, reproduction, immune response (Pörtner and Farrell 2008). At these edges, biological interactions, such as predation and competition, will determine which species are able to persist and therefore the structure and diversity of the resident community. Consequently, changes in seawater temperature have the potential to alter the performance of resident species, change the competitive balance between organisms, and increase the likelihood of invasive species entering an area. All of this will modify the strength and direction of the key biological interactions which underpin biodiversity. However, temperature is not the only environmental stressor involved in this process. Figure 1.1 also indicates that additional environmental stressors, such as hypercapnia, hypoxia, and salinity change, can narrow an organism's thermal window and therefore increase its susceptibility to extreme temperature events.

Climate change could also promote biodiversity loss through indirect impacts. In many marine systems the activities of individual species can be instrumental in maintaining high levels of biodiversity in the associated fauna (see review by Widdicombe and Austen 2005). This biological control can be exerted through reducing competitive exclusion *via* density dependent predation or consumption, i.e. 'keystone species' (Paine 1966) or through habitat modification and creation of

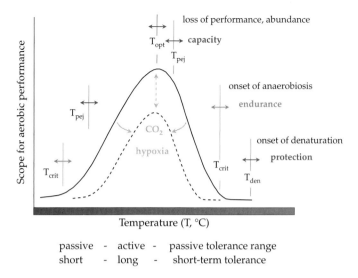

passive - active - passive tolerance range
short   - long - short-term tolerance

Figure 1.1 Relationship between organism performance and environmental temperature. From Pörtner and Farrell (2008), used with permission from AAAS.

environmental heterogeneity via processes such as reef building and bioturbation, i.e. 'ecosystem engineering' (Jones *et al.* 1994; Lawton 1994). Biogenic habitats such as coral reefs, mearl beds, and mussel banks are considered to be particularly important as 'hot spots' for biodiversity. However, many of these important habitats rely heavily on the health and survival of thermally sensitive, calcifying species which are considered to be at high risk from climate change (Raven *et al.* 2005). In addition, many keystone species and ecosystem engineers are predominantly large organisms and it might be considered that with larger organisms having generally lower metabolic rates than smaller organisms, increasing body size should enhance tolerance to increased temperature, hypercapnia, and anoxia. However, the current change in environmental conditions associated with climate change is more of a chronic pressure than an acute one with changes occurring over a timescale of tens to hundreds of years. Therefore, the long-term survival of any species is more likely to be determined by its ability to adapt rather than the ability of individuals to persist. Consequently, large animals with longer generation times will be at an adaptive disadvantage. There is already some evidence from the geological record that marine organisms do become smaller in

conjunction with many of the previous mass extinction events (Twitchett 2007); a phenomenon known as the 'Lilliput effect'. It should also be considered that a reduction in the top–down biological control of biodiversity could come about not just through the complete loss of important species, but also through changes in the rates of ecologically important activities, such as growth, feeding, bioturbation, and burrow irrigation.

1.9 Additional impacts of CO_2 on the marine environment

Whilst the measured rise in atmospheric CO_2 levels and global temperatures are dramatic, these increases would have been even greater were it not for the fact that almost half of all the anthropogenic CO_2 emitted over the last 250 years has been taken up by the world's oceans (Sabine *et al.* 2004). It is natural for atmospheric CO_2 levels to fluctuate through time creating CO_2 gradients between the atmosphere and the ocean as the air and seawater gradually equilibrate. These gradients will subsequently drive the oceans to absorb CO_2 during times of rising atmospheric concentrations and to release it during times of falling atmospheric concentrations. As we are currently in a period of rising atmospheric concentrations, CO_2 diffuses passively into ocean surface waters where it reacts with water molecules to form carbonic acid (H_2CO_3), a weak acid that rapidly dissociates to produce hydrogen (H^+) and bicarbonate (HCO_3^-) ions. This increase in H^+ ion concentration would cause seawater pH to fall—i.e. the acidity to increase—was it not for the saturation of surface waters with carbonate cations (CO_3^{2-}). These cations react with the H^+ ions to form more HCO_3^- and this entire reaction ($CO_2 + H_2O \leftrightarrow H_2CO_3 \leftrightarrow HCO_3^- + H^+ \leftrightarrow CO_3^{2-} + 2H^+$) is referred to as the carbonate buffering system; a process by which surface ocean pH has remained relatively stable (between 8.0 and 8.3 pH units) for the last 25 million years.

The current concentration of atmospheric CO_2 is not particularly unusual and is, in fact, still quite low compared to some periods in Earth's recent history, so it was assumed that the carbonate buffering system would continue to maintain seawater pH at 'normal' levels despite rising emissions. However, this assumption was challenged in 2003 when model predictions were used to show that, compared to pre-industrial times, ocean surface water pH had fallen by 0.1 pH unit, indicating a 30% increase in the concentration of H^+ ions (Caldeira and Wickett 2003); a phenomenon now widely recognized and referred to as 'the other CO_2 problem' or 'ocean acidification'. It is now clear that the major cause of ocean acidification is not the total amount of CO_2 entering the oceans, but the unprecedentedly short time over which this uptake is occurring. Atmospheric CO_2 concentrations are now increasing by more than 2 ppm per year. This rate exceeds that predicted by the Intergovernmental Panel on Climate Change (IPCC) under a 'worst case scenario' (IPCC 2007) and is nearly 100 times greater than any historical, naturally induced increase (Blackford and Gilbert 2007). As the carbonate buffering system relies on the availability of CO_3^{2-} cations, which are primarily derived from relatively slow geological process such as the weathering of carbonate rich rocks or the dissolution of carbonate sands, the Earth's oceans are rapidly becoming more acidic and experiencing increasing hypercapnia (elevated levels of dissolved CO_2). Recently published observational data have since confirmed

these model conclusions with declining values for pH being reported in a number of long running oceanographic time series, e.g. the Hawaiian Ocean Time-series (HOT) in the sub-tropical North Pacific (Brix et al. 2004), the European Station for Timeseries in the Ocean, Canary Islands (ESTOC) in the sub-tropical (East) North Atlantic (Santana-Casiano et al. 2007), the Bermuda Atlantic Time-Series (BATS) in the sub-tropical (West) North Atlantic (Bates and Peters 2007), and the Iceland Sea Polar (Nordic Seas) in the North Atlantic (Olafsson et al. 2009). All of these have reported a consistent trend for declining pH of between 0.0012 and 0.0024 pH units per year. If these trends continue, the pH of surface waters could be predicted to fall by on average up to 0.4 units before 2100 and by 0.7 units before 2250.

The potential physiological impacts of high levels of CO_2 on the health and function of marine species and communities have been detailed in a number of recent reviews (Seibel and Walsh 2001, 2003; Pörtner et al. 2004, 2005; Fabry et al. 2008; Widdicombe and Spicer 2008). In summary, when marine organisms are exposed to low pH seawater, the primary physiological effect is a decrease in the pH or an 'acidosis' of the extracellular body fluids such as blood, haemolymph, or coelomic fluid. In some species, this extracellular acidosis is fully compensated for as levels of extracellular bicarbonate are increased by either active ion transport processes in the gills or through passive dissolution of a calcium carbonate shell or carapace (see Widdicombe and Spicer 2008, and references therein). However, in other species from a variety of different taxa, such as the mussel *Mytilus edulis* (Michaelidis et al. 2005), the crabs *Callinectes sapidus* (Wood and Cameron 1985) and *Chionoecetes tanneri* (Pane and Barry 2007), and the sea urchin *Psammechinus miliaris* (Miles et al. 2007), studies have reported only partial, or no, compensation in the extracellular acid-base balance. Clearly if some species are physiologically better equipped to cope with elevated levels of CO_2 than others, the potential for species extinctions and biodiversity loss exists. At the phylum level, early evidence would suggest that echinoderms will be more vulnerable than molluscs, then crustaceans, with annelids showing the greatest tolerance to hypercapnia and acidification (Widdicombe *et al.* 2009). So in a future, more acidic ocean, marine communities could be made up of species from a limited number of tolerant taxonomic groups. Whilst this may or may not reduce the total number of species in any given area, as tolerant species could potentially move in to exploit vacated niches, it would certainly reduce taxonomic richness as well as species diversity both regionally and globally. However, even within taxonomic groups, variability in tolerance can exist between even closely related species with this variability seemingly linked to key elements of an organism's lifestyle. Organisms that already exist in habitats regularly exposed to highly variable levels of temperature, and CO_2 may be more likely to possess the physiological mechanisms necessary to cope with rapid changes in environmental conditions than organisms from areas with more stable conditions, such as the polar oceans, deep sea, or well oxygenated sands. For example, Pane and Barry (2007) reported marked differences in the acid-base responses to acute seawater hypercapnia between the deep sea Tanner crab *Chionoecetes tanneri* and the shallow water Dungeness crab *Cancer magister*. Also, as ocean warming and acidification will be most severe in shallow surface waters, pelagic organisms or benthic species that rely on a pelagic reproductive stage could be more vulnerable than species with direct benthic development. It seems clear that the likelihood that a species will be lost from an area as a result of increased environmental stress could be determined by both its phylogeny and its ecology. Consequently, the potential exists for climate change to reduce the total number of marine species, the taxonomic richness of the communities which remain and the diversity of functions that these communities can perform.

1.10   Hypoxia and 'dead zones'

Oxygen is a prerequisite for sustaining multicellular life in the oceans. However, dissolved oxygen concentrations are declining in a growing number of coastal areas. Over recent decades the number of coastal sites where hypoxia (DO concentration ≤ 2 mg O_2/litre) has been reported has increased

exponentially with Vaquer-Sunyer and Duarte (2008) estimating this increase to be around 5.5% every year. This hypoxia can not only cause large-scale mortality in metazoan animals, it can also exert sub-lethal stresses which inhibit growth and reproduction. Hypoxic stress can also force organisms from their normal habitats leaving them open to increased pressure from predation or competition. All of these processes can cause a loss of biodiversity and create the areas which are now termed 'dead zones' (Rabalais *et al.* 2001). The primary cause of this proliferation in dead zones is increasing eutrophication (Diaz and Rosenberg 2008), fuelled by an enormous growth in coastal cities and the increasing use of industrially produced nitrogen fertilizer by modern agriculture. As is likely to be the case for acidification, marine organisms exhibit a range of vulnerabilities to hypoxia and so the impacts on biodiversity are likely to be similar with a loss of species, taxonomic, and functional diversity.

There is strong evidence to suggest that climate change could exacerbate the problem of hypoxia in coastal marine ecosystems. Firstly, increased precipitation could increase riverine runoff and therefore transport more nutrients to the coastal zone. For example, Justić *et al.* (1996) suggested that under future climate scenarios, there would be a 20% increase in river discharge from the Mississippi and the nutrients associated with this would fuel a 50% increase in primary production and a decrease in subpycnocline dissolved oxygen of up to 60%. Secondly, over the range 0 °C to 15 °C the relationship between temperature and dissolved oxygen concentrations in seawater is approximately linear, with DO decreasing by about 6% for every 1° C rise in temperature. Increasing seawater temperatures could also promote thermal stratification and promote phytoplankton production. Finally, changes in wind strength and direction, or the frequency and intensity of storms could have contrasting effects. Off the western continental margins of South America, Africa, and India, offshore winds drive the upwelling of nutrient-rich bottom water and promote productivity. This process already creates large oxygen-minimum zones in these areas and any increase in wind strength could expand the extent of these zones. In other areas however, strong winds and intense storms could act to reduce hypoxia by removing stratification and enhance mixing of oxygen into deeper waters.

1.11   Summary

Biodiversity loss is happening and has been happening for hundreds of years. It is being driven by a multitude of anthropogenic stresses at a variety of spatial (local, regional, and global) and temporal (days to decades to centuries) scales. These stresses will act simultaneously to reduce taxonomic and functional diversity. The challenge for society now is to manage our exploitation of, and interaction with, the marine environment in a sustainable way so as to halt biodiversity loss and, where possible, promote recovery. However, the tools and policies necessary for such management must be based on a solid foundation of biodiversity-related science. As the following chapters will illustrate, there is substantial evidence that biodiversity is important for the delivery of ecosystem function; however, the mechanisms by which this occurs are still uncertain. Only by investing sufficient time and resources in studies which explore the fundamental mechanisms which underpin the biodiversity ecosystem function relationship will we generate the understanding necessary to predict the long-term impacts of man's activities on the marine environment.

Acknowledgements

This chapter is a contribution to the NERC funded Plymouth Marine Laboratory core project 'Oceans 2025'.

References

Alroy, J., Aberhan, M., Bottjer, et al. (2008) Phanerozoic trends in the global diversity of marine invertebrates. *Science* **321**, 97–100.

Angel, M.V. (1993) Biodiversity of the pelagic ocean. *Conservation Biology* **7**, 760–72.

Barrett, J.H., Locker, A.M. and Roberts, C.M. (2004) The origins of intensive marine fishing in medieval Europe: the English evidence. *Proceedings Royal Society London B* **271**, 2417–21.

Bates, N.R. and Peters, A.J. (2007) The contribution of atmospheric acid deposition to ocean acidification in the subtropical North Atlantic Ocean. *Marine Chemistry* **107**, 547–58.

Battistuzzi, U., Feijao, A. and Blair Hedges, S. (2004) A genomic timescale of prokaryote evolution: insights into the origin of methanogenesis, phototrophy, and the colonization of land. *BMC Evolutionary Biology* 2004, 4:44 doi:10.1186/1471-2148-4-44.

Berra, T.M. (1997) Some 20th Century fish discoveries. *Environmental Biology of Fishes*, **50**, 1–12.

Blackford, J.C. and Gilbert, F. (2007) pH variability and CO_2 induced acidification in the North Sea. *Journal of Marine Systems* **64**, 229–41.

Brierley, A.S. and Kingsford, M.J. (2009) Impacts of climate change on marine organisms and ecosystems. *Current Biology* **19**, R602–614.

Briggs, J.C. (1994) Species diversity: land and sea compared. *Systems Biology* **43**, 130–5.

Brix, H., Gruber, N. and Keeling, C.D. (2004) Interannual variability of the upper ocean carbon cycle at station ALOHA near Hawaii. *Global Biogeochemical Cycles* **18**, GB4019, doi: 10.1029/2004GB002245.

Bottjer, D.J. (2005) Geobiology and the fossil record: eukaryotes, microbes, and their interactions. *Palaeogeography, Palaeoclimatology, Palaeoecology* **219**, 5–21.

Bruggemann, J.H., Buffler, R.T., Guillaume, M.M.M. et al. (2004) Stratigraphy, palaeoenvironments and model for the deposition of the Abdur Reef Limestone: context for an important archaeological site from the last interglacial on the Red Sea coast of Eritrea. *Palaeogeography Palaeoclimatology Palaeoecology* **203**, 179–206.

Butterfield, N.J. (2010) Animals and the invention of the Phanerozoic Earth system. *Trends in Ecology and Evolution* **26**, 81–7.

Caldeira, K. and Wickett, M.E. (2003) Anthropogenic carbon and ocean pH. *Nature* **425**, 365.

Cavalier-Smith, T. (2006) Cell evolution and Earth history: stasis and revolution. *Philosophical Transactions of the Royal Society B* **361**, 969–1006.

Chisholm, S.W., Olson, R.J., Zettler, E.R. et al. (1988) A novel free-living prochlorophyte occurs at high cell concentrations in the oceanic euphotic zone. *Nature* **334**, 340–3.

Darwin, C. (1859) *On the origin of species by means of natural selection, or the preservation of favoured races in the struggle for life.* John Murray, London.

Diaz, R.J. and Rosenberg, R. (2008) Spreading dead zones and consequences for marine ecosystems. *Science* **321**, 926–9.

Fabry, V.J., Seibel, B.A., Feely, R.A. *et al.* (2008). Impacts of ocean acidification on marine fauna and ecosystem processes. *Journal of Marine Science* **65**, 414–32.

Feely, R.A., Sabine, C.L., Lee, K. *et al.* (2004) Impact of anthropogenic CO_2 on the $CaCO_3$ system in the oceans. *Science* **305**, 362–6.

Gaston, K.J. (1996) *Biodiversity: a biology of numbers and difference*. Blackwell Science, Oxford.

Gattuso, J.P., Frankignoulle, M. and Wollast, R. (1998) Carbon and carbonate metabolism in coastal aquatic ecosystems. *Annual Review of Ecological Systems* **29**, 405–34.

Grassle, J.F. (1991) Deep-sea benthic biodiversity. *BioScience* **41**, 464–9.

Gray, J.S. (1997) Marine biodiversity: patterns, threats and conservation needs. *Biodiversity Conservation* **6**, 153–75.

Haila, Y. and Kouki, J. (1994) The phenomenon of biodiversity in conservation biology. *Ann. Zool. Fennici* **31**, 5–18.

Harper, J.L. and Hawksworth, D.L. (1994) Biodiversity: measurement and estimation. *Philosophical Transactions of the Royal Society of London B* **345**, 5–12.

IPCC (2007) *Climate Change 2007: The Physical Science Basis. Contribution of Working Group I to the Fourth Assessment. Report of the Intergovernmental Panel on Climate Change.* Solomon, S., Qin, D., Manning, M., *et al.* (eds) Cambridge University Press, Cambridge, UK.

Jackson, J.B.C. (1991) Adaptation and diversity of coral reefs. *BioScience* **41**, 475–82.

Jackson, J.B.C. (2001) What was natural in the coastal oceans? *Proceedings of the National Academy of Sciences of the USA* **98**, 5411–18.

Jackson, J.B.C. (2008) Ecological extinction and evolution in the brave new ocean. *Proceedings of the National Academy of Sciences of the USA* **105**, suppl. 1, 11458–65.

Jackson, J.B.C., Kirby, M.X., Berger, W.H. *et al.* (2001) Historical overfishing and the recent collapse of coastal ecosystems. *Science* **293**, 629–38.

Jones, C.G., Lawton, J.H. and Shachak, M. (1994) Organisms as ecosystem engineers. *Oikos* **69**, 373–86.

Justić, D., Rabalais, N.N. *et al.* (1996) Effects of climate change on hypoxia in coastal waters: A doubled CO_2 scenario for the northern Gulf of Mexico. *Limnology and Oceanography* **41**, 992–1003.

Knauth, L.P. and Kennedy, M.J. (2009) The late Precambrian greening of the Earth. *Nature* **460**, 728–32.

Kump, L.R., Pavlov, A. and Arthur, M.A. (2005) Massive release of hydrogen sulfide to the surface ocean and atmosphere during intervals of oceanic anoxia. *Geology* **33**, 397–400.

Lawton, J.H. (1994) What do species do in ecosystems? *Oikos* **71**, 367–74.

Levin, L.A., Boesch, D.F., Covich, A. *et al.* (2001) The function of Marine Critical Transition Zones and the importance of sediment biodiversity. *Ecosystems* **4**, 430–51.

Liu, A.G., McIlroy, D. and Brasier, M.D. (2010) First evidence for locomotion in the Ediacara biota from the 565 Ma Mistaken Point formation, Newfoundland. *Geology* **38**, 123–6.

Logan, G.A., Hayes, J.M., Hieshima, G.B. *et al.* (1995) Terminal Proterozoic reorganization of biogeochemical cycles. *Nature* **376**, 53–6.

Lonsdale, P. (1977) Clustering of suspension-feeding macrobenthos near abyssal hydrothermal vents at oceanic spreading centers. *Deep Sea Research* **24**, 857–8.

Lotze, H.K., Lenihan, H.S., Bourque *et al.* (2006) Depletion, degradation and recovery potential of estuaries and coastal seas. *Science* **312**, 1806–9.

Lotze, H.K. and Worm, B. (2009) Historical baselines for large marine animals. *Trends in Ecology and Evolution* **24**, 254–62.

Love, G.D., Grosjean, E., Stalvies, C. *et al.* (2009) Fossil steroids record the appearance of Demospongiae during the Cryogenian period. *Nature* **457**, 718–21.

May, R.M. (1988) How many species are there on Earth? *Science* **241**, 1441–9.

May, R.M. (2010) Ecological science and tomorrow's world. *Philosophical Transactions of the Royal Society of London B* **365**, 41–7.

Meysman, F.J.R., Middleburn, J.J. and Heip, C.H.R. (2006) Bioturbation: a fresh look at Darwin's last idea. *Trends in Ecology and Evolution* **21**, 688–95.

Michaelidis, B., Ouzounis, C., Paleras, A. *et al.* (2005) Effects of long-term moderate hypercapnia on acid-base balance and growth rate in marine mussels *Mytilus galloprovincialis*. *Marine Ecology Progress Series* **293**, 109–18.

Miles, H., Widdicombe, S., Spicer, J.I. *et al.* (2007) Effects of anthropogenic seawater acidification on acid-based balance in the sea urchin *Psammechinus miliaris*. *Marine Pollution Bulletin* **54**, 89–96.

Millennium Ecosystem Assessment (2005) *Ecosystems and human well-being: Current state and trends, Volume 1.* Island Press, Washington DC.

Myers, R.A. and Worm, B. (2003) Rapid worldwide depletion of predatory fish communities. *Nature* **423**, 280–3.

Narbonne, G.M. and Gehlingl, J.G. (2003) Life after snowball: the oldest complex Ediacaran fossils. *Geology* **31**, 27–30.

National Research Council (1995) *Understanding marine biodiversity: a research agenda for the nation.* National Academy Press, Washington DC.

Olafsson, J., Olfsdottir, S.R., Benoit-Cattin, A. *et al.* (2009) Rate of Iceland Sea acidification from time series measurements. *Biogeosciences Discussions* **6**, 5251–70.

Paine, R.T. (1966). Food web complexity and species diversity. *American Naturalist* **100**, 65–75.

Pandolfi, J.M., Bradbury, R.H., Sala, E. *et al.* (2003) Global trajectories of the long-term decline of coral reef ecosystems. *Science* **301**, 955–8.

Pane, E.F. and Barry, J.P. (2007) Extracellular acid-base regulation during short-term hypercapnia is effective in a shallow-water crab, but ineffective in a deep-sea crab. *Marine Ecology Progress Series* **334**, 1–9.

Pauly, D. (1995) Anecdotes and the shifting baseline syndrome of fisheries. *Trends in Ecology and Evolution* **10**, 430.

Payne, J.L., Boyer, A.G., Brown, J.H. *et al.* (2009) Two-phase increase in the maximum size of life over 3.5 billion years reflects biological innovation and environmental opportunity. *Proceedings of the National Academy of Sciences of the USA* **106**, 24–7.

Pisani, D., Poling, L.L., Lyons-Weiler, M. *et al.* (2004) The colonization of land by animals: molecular phylogeny and divergence times among arthropods. *BMC Biology* 2:1 doi 10.1186/1741-7007-2-1.

Pörtner, H.O. and Farrell, A.P. (2008) Physiology and Climate Change. *Science* **322**, 690–2.

Pörtner, H.O., Langenbuch, M. and Michaelidis, B. (2005) Synergistic effects of temperature extremes, hypoxia, and increases in CO_2 on marine animals: From earth history to global change. *Journal of Geophysical Research—Oceans* **110**, 1–15.

Pörtner, H.O., Langenbuch, M. and Reipschläger, A. (2004) Biological impact of elevated ocean CO_2 concentrations: lessons from animal physiology and earth history. *Journal of Oceanography* **60**, 705–18.

Rabalais, N.N., Turner, R.E. and Wiseman, W.J. Jr (2001) Hypoxia in the Gulf of Mexico. *Journal of Environmental Quality* **30**, 320–9.

Raup D.M. and Sepkoski J.J. Jr (1982) Mass extinctions in the marine fossil record *Science* **215**, 1501–3.

Raven, J., Caldeira, K., Elderfield, H. *et al.* (2005) Ocean acidification due to increasing atmospheric carbon dioxide. The Royal Society policy document 12/05, Clyvedon Press, Cardiff.

Ray, G.C. (1991) Coastal-zone biodiversity patterns. *BioScience* **41**, 490–8.

Ray, G.C. and Grassle, J.F. (1991) Marine biological diversity. *BioScience* **41**, 453–61.

Richardson, A.J., Bakun, A., Hays, G.C. *et al.* (2009) The jellyfish joyride: causes, consequences and management responses to a more gelatinous future. *Trends in Ecology and Evolution* **24**, 312–22.

Sabine, C.L., Feely, R.A., Gruber, N. *et al.* (2004) The oceanic sink for anthropogenic CO_2. *Science* **305**, 367–71.

Santana-Casiano, J.M., González-Dávila, M., Rueda et al. (2007) The interannual variability of oceanic CO_2 parameters in the northeast Atlantic subtropical gyre at the ESTOC site. *Global Biogeochemical Cycles* **21**: GB1015, doi: 10.1029/2006GB00278.

Seibel, B.A. and Walsh, P.J. (2001) Potential impacts of CO_2 injection on deep sea biota. *Science* **294**, 319–20.

Seibel, B.A. and Walsh, P.J. (2003) Biological impacts of deep-sea carbon dioxide injection inferred from indices of physiological performance. *Journal of Experimental Biology* **206**, 641–50.

Sepkoski, J.J. Jr (1981) A factor analytic description of the Phanerozoic marine fossil record. *Paleobiology* **5**, 222–51.

Sepkoski, J.J. Jr (2002) A compendium of fossil marine animal genera. *Bulletin of American Paleontology* **364**, 1–560.

Snelgrove, P.V.R. (1999) Getting to the bottom of marine biodiversity: sedimentary habitats. *BioScience* **49**, 129–38.

Solbrig, O.T. (1991) *From genes to ecosystems: a research agenda for biodiversity.* IUBS, Cambridge, Mass.

Stanley, S.M. and Yang, X. (1994) A double mass extinction at the end of the paleozoic era. *Science* **266**, 1340–4.

Steele, J.H. (1985) A comparison of terrestrial and marine ecological systems. *Nature* **313**, 355–8.

Strathmann, R.R. (1990) Why life histories evolve differently in the sea. *American Zoologist* **30**, 197–207.

Suchanek, T.H. (1994) Temperate coastal marine communities: biodiversity and threats. *American Zoologist* **34**, 100–14.

Thomas, H., Bozec, Y., de Baar, H.J.W. *et al.* (2005) The carbon budget of the North Sea. *Biogeosciences* **2**, 87–96.

Thompson, R.C., Crowe, T.P. and Hawkins, S.J. (2002) Rocky intertidal communities: past environmental changes, present status and predictions for the next 25 years. *Environmental Conservation* **29**, 168–91.

Twitchett, R.J. (2007) The Lilliput effect in the aftermath of the end-Permian extinction event. *Palaeogeography, Palaeoclimatology, Palaeoecology* **252**, 132–44.

Vaquer-Sunyer, R. and Duarte, C. (2008) Thresholds of hypoxia for marine biodiversity. *Proceedings of the National Academy of Sciences of the USA* **105**, 15452–7.

Walter, R.C., Buffler, R.T., Bruggemann, J.H. *et al.* (2000) Early human occupation of the red Sea coast of Eritrea during the last interglacial. *Nature* **405**, 65–9.

Wang, D.Y., Kumar, S. and Hedges, S.B. (1999) Divergence time estimates for the early history of animal phyla and the origin of plants, animals and fungi. *Proceedings of the Royal Society of London Series B - Biological Sciences* **266**, 163–71.

Webb, T.J., Vanden Berghe, E. and O'Dor, R. (2010) Biodiversity's Big Wet Secret: The Global Distribution of Marine Biological Records Reveals Chronic Under-Exploration of the Deep Pelagic Ocean. *PLoS ONE* 5(8): e10223. doi:10.1371/journal.pone.0010223.

Weber, R.E. (1980) Functions of invertebrate hemoglobins with special reference to adaptations to environmental hypoxia. *American Zoologist* **20**, 79–101.

Widdicombe, S. and Austen, M.C. (2005) Setting diversity and community structure in subtidal sediments: The importance of biological disturbance, In: Kostka J., Haese R. and Kristensen E. (eds) *Interactions between macro- and microorganisms in marine sediments.* Coastal and Estuarine Studies: 60, American Geophysical Union, New York, 217–31.

Widdicombe, S., Dashfield, S.L., McNeill, C.L. *et al.* (2009) Effects of CO_2 induced seawater acidification on infaunal diversity and sediment nutrient fluxes. *Marine Ecology Progress Series* **379**, 59–75.

Widdicombe, S. and Spicer, J.I. (2008) Predicting the impact of Ocean acidification on benthic biodiversity: What can physiology tell us? *Journal of Experimental Marine Biology and Ecology* **366**, 187–97.

Wilson, E.O. and Peter, F.M. (1988) *Biodiversity.* National Academy Press, Washington DC.

Wood, C.M. and Cameron, J.N. (1985) Temperature and physiology of intracellular and extracellular acid-base regulation in the blue crab *Callinectes sapidus. Journal of Experimental Biology* **114**, 151–79.

Worm, B., Barbier, E.B., Beaumont, N. *et al.* (2006) Impacts of biodiversity loss on ocean ecosystem services. *Science* **314**, 787–90.

Xiao, S. and Laflamme, M. (2008) On the eve of animal radiation: phylogeny, ecology and evolution of the Ediacara biota. *Trends in Ecology and Evolution* **24**, 31–40.

CHAPTER 2

Biodiversity in the context of ecosystem function

Anne E. Magurran

2.1 Historical development of the concept

Biodiversity is a contraction of the phrase 'biological diversity', and many authors use the terms interchangeably, as I do here. The word biodiversity seems to have been coined by Walter G. Rosen during the planning of the 1986 National Forum on BioDiversity (1995). The publication of the proceedings of this meeting in a book entitled *Biodiversity*, edited by E.O. Wilson (1988), brought the word to a wider audience. Because of this relatively recent origin, some people mistakenly think that biodiversity is a new idea. However, the concept is probably as old as humankind and it is likely that our earliest ancestors judged hunting-grounds in relation to the abundance and diversity of the prey found there, and took biological diversity into account when deciding where to live and forage. The sense that some communities such as coral reefs and tropical rain forests are rich in species whereas others—e.g. boreal forests and salt marshes—are not, is one shared by many people and suggests that we have an intuitive ability to distinguish, and even to value, systems in terms of their diversity.

The nineteenth century marked a shift in thinking about biological diversity. This was a time when naturalists and explorers travelled the world and enthralled readers back home with tales of their journeys. Von Humboldt's description of the biological diversity of tropical South America led to latitudinal diversity gradients becoming 'ecology's oldest pattern' (Hawkins 2001). Darwin mentions diversity in many of his writings. The *Origin of*

Species (Darwin 1859), for example, discusses the 'diversity and proportion of kinds' of species found in entangled banks, and asserts that this pattern is not due to chance. He also touches on the relationship between diversity and productivity (Hector and Hooper 2002). Bates, who was a contemporary of Darwin, provides a detailed and readable account of his experiences in the Brazilian Amazon (Bates 1864); his estimates of insect species richness are, in some cases, still the best we have (Magurran and Queiroz 2010). Bates also raises a number of fundamental biodiversity questions. For instance, he ponders why a small area of the Amazon has a large number of butterfly species with no more individuals present than might be seen in southern Britain on a summer's day.

This expanding view of the world coincided with a movement towards experimental research. The motivation for this, as with many Victorian endeavours, was better productivity and efficiency, but the outcome was a new way of gathering data that led to the emergence of quantitative ecology as a vigorous side-shoot of natural history. Perhaps the most famous example of this is the Park Grass experiment at Rothamsted, established by Lawes and Gilbert in 1856 to examine the relationship between fertilizer use and yield. Within just two to three years it was apparent that the treatments were having a marked effect on species composition and abundance. These observations gave rise to a series of influential papers, and the Park Grass experiment, which continues to this day, has been used to answer a broad range of ecological questions (Crawley *et al.* 2005; Silvertown *et al.* 2006; Magurran *et al.* 2010).

Access to good data enabled researchers to develop models that could be used to describe and explain empirical patterns. An early goal was to model species abundance distributions. Motomura (1932), Fisher *et al.* (1943), and Preston (1948) introduced the geometric series, the log series model and the log normal model respectively. There is still debate about how best to assess species abundances and researchers continue to strive for improved descriptive and predictive models. Although these pioneering papers talked about diversity rather than biological diversity—Fisher, for instance, introduced what he called 'an index of diversity'—their species abundance models lie within the domain of biodiversity in its modern sense. Other early researchers also dealt with subject matter that would be immediately recognizable to ecologists today. Crozier (1923), for instance, was interested in seasonal variation in the composition and abundance of protozoa in sewage. The term 'biological diversity' was in use by the 1950s (Gerbilskii and Petrunkevitch 1955), and firmly established by the 1960s (e.g. Whiteside and Harmsworth 1967; Sanders 1968). Thus, while Harper and Hawksworth (1995) argue that the concept of biological diversity in its current form can be traced to 1980, specifically to papers by Lovejoy (1980a; 1980b) and Norse and McManus (1980), it is a concept with much deeper roots in ecology and natural history. The 1980s did, however, mark the point at which the concept of biological diversity broadened to include not just the diversity of species, but also the diversity of genes and communities. Norse *et al.* (1986) distinguished between genetic (within species), species (species richness), and ecological (community) diversity. This expanded view is encapsulated in the UNEP's definition of biological diversity:

'"Biological diversity" means the variability among living organisms from all sources including, inter alia, terrestrial, marine and other aquatic systems and the ecological complexes of which they are part; this includes diversity within species, between species and of ecosystems.'

2.2 Biological diversity—meaning and measurement

While the notion that biological diversity takes account of variety at different organizational levels is now well accepted, users can have strong and conflicting opinions about terminology. Harper and Hawksworth (1995) advocate the adjectives 'genetic', 'organismal', and 'ecological' diversity to refer to the three levels of diversity in the UNEP definition. Their 'organismal' diversity has traditionally been termed 'species' diversity (MacArthur and MacArthur 1961) or 'ecological' diversity (Pielou 1975). Harper and Hawksworth argue vigorously against the use of the term 'ecosystem diversity' to denote the diversity of communities, citing Tansley (1935), who introduced the word 'ecosystem' to embrace not just a community of organisms but also the physical environment in which they live. Since the physical environment does not have biodiversity, Harper and Hawksworth feel that the use of the term 'ecosystem diversity' devalues the concept of the ecosystem. By the same reasoning, the concept of ecosystem function could be said to be illogical, as it is the biological component of the ecosystem—i.e. the communities of organisms within an ecosystem—researchers focus on when investigating function. Nonetheless, the term 'biodiversity and ecosystem functioning' or BEF, has been universally adopted, not least because the alternatives of 'ecological function or 'community function' are problematic in other ways.

Harper and Hawksworth (1995) offer the very sensible and still relevant advice to users to be explicit about which level (or levels) of biodiversity they are dealing with. An alternative approach is to consider biodiversity on an integrated basis rather than examining each layer separately. Thus Noss (1990) advocates a nested hierarchical approach that begins with a coarse-scale assessment of landscape features and drills down through this at increasing levels of resolution. Chown and McGeoch (2011) use a similar approach to evaluate the diversity of modified landscapes.

The science of biodiversity is essentially a comparative one; we typically note that one woodland has more bird species than another, or that beetles or nematodes are particularly speciose, or that the variety of fish in a stream declined after it was polluted. A single measure of diversity, however it is obtained, is effectively meaningless unless there are benchmarks against which to judge it. Biodiversity measurement therefore usually means generating one or more numbers that can be compared with a null expectation, or a control treatment, or some other sample or plot (Magurran and McGill 2011).

Given the variety of ways in which it is possible to approach the investigation of biological diversity, it is not surprising that there are many competing measures. In essence, however, there are two main ways of quantifying this diversity: by assessing richness (or how many entities of interest there are), and by evaluating evenness (or the extent to which these entities vary in abundance). Traditionally, richness refers to number of species and this usage accords well with our intuitive sense that a coral reef with many species of fish is diverse, whereas a polluted inlet with only a handful of fish species is not. The concept of evenness also resonates with our innate sense of diversity. For example, a benthic assemblage with one highly dominant and many rare species appears to be less diverse than another assemblage in which there are the same numbers of species and individuals, but where the species are more equally represented. Species richness and evenness are often treated as independent or orthogonal measures of diversity, but in fact they represent two points in a continuum of measures, which are populated by many familiar statistics such as the Shannon, Simpson, and Margelef indexes of diversity. This continuum is formalized in diversity families such as those introduced by Rényi (1961) and Hill (1973). These families allow the user to build a profile of the diversity of an assemblage or sample using a series of measures that vary the weighting placed on rare species (Tóthmérész 1995). An important point to note is that different measures can rank assemblages in different ways (Hurlbert 1971; Tóthmérész 1995; Southwood and Henderson 2000; Nagendra 2002; Magurran 2004). Diversity statistics may have been used for decades but their performance continues to be hotly

debated (Beck and Schwanghart 2010; Magurran and McGill 2011).

Although species richness is the simplest and most intuitive measure of biological diversity, it is also one of the most difficult to quantify. This is because it is often impossible to provide an accurate count of the species that are present in an assemblage. Species-rich yet poorly documented habitats, such as tropical forests and deep-sea benthos, are the most obvious cases (e.g. Coddington *et al.* 2009) but tallying species can be difficult even in well-worked temperate and terrestrial communities. One issue here is what to do with species that are temporally absent; for example, as a result of migration or diapause, as well as those that, though present, occur at very low density and are not sampled. Since most species are rare, any decision to extend sampling duration, or increase sampling intensity will change the perception of species richness (Magurran 2011). There are circumstances where it is possible to identify and count every species. BEF experiments that use small numbers of pre-selected species are one of these exceptions. In cases where it is not possible to generate a robust measure of species richness, users have the option of either employing methods that estimate the minimum number of species present—e.g. Colwell and Coddington 1994; Gotelli and Colwell 2011—or of using rarefaction to make a fair comparison amongst sites or treatments based on sampling effort (Gotelli and Colwell 2001; Gotelli and Colwell 2011).

Rather than focusing on a single measure, or even a family of measures, researchers may choose to examine the overall species abundance distribution. A species abundance distribution shows the frequency of species in different abundance classes and has the advantage of giving a synoptic picture of the assemblage and its structure. There is a large volume of literature discussing the merits of various species abundance models, and it is now clear that the fact that a model predicts a distribution of relative species abundances is no guarantee that the mechanisms that it embodies are the correct ones (McGill *et al.* 2007; McGill 2011). However, approaches that capture the structure of the species abundance distribution provide a potentially powerful means of quantifying the patterns that do occur, and of enabling comparisons to be made between treatments in an experiment, or before and after an event such as a disturbance. Methods that can be used in this way include the log normal and log series models, and the Q statistic (Magurran 2004). Recently attention has turned the empirical cumulative distribution function (eCDF), an approach that appears to have great potential as a tool for assessing and comparing species abundance data (Dornelas *et al.* 2011; McGill 2011).

Although the concepts of richness and evenness and associated diversity statistics were developed in the context of species, there is no reason why other entities such as genes or traits cannot be evaluated in parallel ways. The past few years have seen the rapid development of methods that use other currencies of diversity. Phylogenetic diversity (Vellend *et al.* 2011) and trait diversity (Weiher 2011) are two areas that have benefited from this research. Moreover, the explosion of molecular tools means that it is now possible to quantify genetic diversity directly, and indeed this is the only feasible approach for investigating microbial diversity (Øvreas and Curtis 2011).

All of the above methods are used to measure what is usually called α diversity; that is, the diversity of a defined unit such as a quadrat or local habitat (Magurran 2004). However, as Whittaker (1960; 1972) recognized, it is also possible to think of diversity as the difference or dissimilarity in species composition amongst sites. Jost *et al.* (2011) provide an up-to-date overview of the methods of assessing β diversity. Both α and β diversity contribute to the diversity of a landscape (γ diversity); it is not just the diversity of local communities that are key here, but also how much they differ from one another in the species they support. While β diversity is typically assessed on the basis of species composition, other currencies, such as traits, can be used as well.

2.3 Biodiversity in the context of function

Concern about the rising tide of species extinctions has focused attention on the utilitarian rationale for biodiversity conservation. The need for a better understanding of the link between biodiversity and function has produced a flurry of modelling and experimental analyses. Yet function can mean any one of a wide range of services and products including carbon capture, nutrient cycling, genetic resources, improved resistance and resilience, natural harvests, and recreation. Investigators who examine biodiversity function thus need to make it clear which functions are targeted by their study. Less obviously, it is important to be explicit about which aspect of biodiversity is being assessed. While few would conclude that evidence of a link between a function and biodiversity means that every function behaves in the same way, there is a tendency to assume that one measure of biodiversity stands in for all measures of biodiversity.

As Srivastava and Vellend (2005) make clear, most BEF investigations are concerned with α diversity. The majority of studies equate biodiversity with the richness of species (Loreau *et al.* 2002), or to a lesser extent richness of genotypes (Srivastava and Vellend 2005). For example, Cardinale *et al.* (2009) provide information on 164 experimental manipulations of species richness. Biodiversity can also be assessed in terms of functional diversity; this is typically viewed as the number of functions that are represented by the species present and can be a stronger predictor of function than species richness (Hector *et al.* 1999; Tilman *et al.* 2001; Petchey and Gaston 2002). In contrast to community ecologists who devote considerable effort to explaining inequalities in species abundances, BEF studies generally pay little attention to evenness—though there are of course exceptions to this, such as Wilsey and Potvin 2000). There are grounds for believing that evenness is relevant (Chapin III *et al.* 2000), and that both common and rare species are important where function is concerned. For example, some services may be linked to common species (Gaston and Fuller 2008)—a natural harvest of fish would be one example—whereas in other cases, infrequent species can be important, for instance as pollinators. The uneven species abundances characterized by a typical species abundance distribution are universal and have the status of a law of ecology (McGill *et al.* 2007). Moreover, these species abundance distributions are temporally dynamic (Magurran and Henderson 2003). These are natural patterns that probably deserve more attention in BEF analyses and may help to make a stronger case for biodiversity conservation (Schwartz *et al.* 2000).

As we have already seen, different measures can lead to different conclusions about the diversity of a locality or assemblage, and ecologists are increasingly using a multi-pronged approach in their assessments (Magurran and McGill 2011). It is likely that new insights into the biodiversity–function relationship will emerge when a broader range of metrics are applied. This should certainly include the new range of trait (Weiher 2011), phylogenetic (Vellend *et al.* 2011), and molecular (Øvreas and Curtis 2011) approaches as well as the more familiar 'classical' species diversity measures (Magurran 2004).

A second important strand is species composition. As noted above, α diversity measures are blind to species identity such that two assemblages with equal numbers of species and equivalent species abundances would generate the same value of a Shannon index, even if one was composed entirely of invasive species. Ecologists have traditionally argued that there are advantages in having measures that ignore species composition since this facilitates comparisons amongst localities that share few or no species. However, environmental managers are increasingly interested in metrics that label species, and a new generation of diversity measures is being developed to meet this need (e.g. Mac Nally 2007; Collins *et al.* 2008). These measures bridge the traditional divide between α and β diversity and have considerable potential in BEF studies. A common BEF experimental approach is to tease apart diversity and composition by assembling communities in different ways; e.g. by random assembly, or by deciding which functional groups should be represented (Schmid *et al.* 2002). These analyses can suggest that biodiversity leads to improved function because the presence of more species increases that likelihood that taxa that perform important roles in a community will be present. Assemblages with higher complementarity—that is groups of species that perform complementary functions— appear to do especially well (Cardinale *et al.* 2007). Combining these approaches with methods that consider the relative abundance of species of inter-

est could be informative. For instance, does it matter if a taxon with a particular functional role is common or rare, or whether the shift in ranks through time is greater than expected by chance?

β diversity (*sensu* Whittaker) is also likely to be important. In BEF research, experiments are usually small in scale, short in duration, and contain a restricted trophic and taxonomic slice of the wider community (Cardinale *et al.* 2009). Drawing on these experiments to make inferences about the wider landscape can be challenging (Srivastava and Vellend 2005), not least because BEF relationships often flatten when species richness is well below the level seen in natural systems. However, if the goal is to unravel the relationship between biodiversity and function at the ecosystem level, the issue is not so much one of how local communities perform, but how they combine and interact at the metacommunity level. Local communities in nature can vary substantially in species richness, species composition, and in temporal biodiversity patterns (Matthews *et al.* 1994). Ecologists have long recognized that there is an interplay between local communities and the metacommunity, and that this is mediated by processes such as dispersal and local extinction (MacArthur and Wilson 1967; Hubbell 2001). Measures of β diversity enable researchers to quantify this variation over space and time. As Srivastava and Vellend (2005) point out, understanding how local communities jointly and individually deliver function across a landscape is as much a matter of addressing β diversity as it is of considering spatial scale. A challenge for BEF researchers then is to ask how the diversity of local communities (α diversity) is mediated by differences between local communities, and/or the extent to which they complement one another (β diversity). It is only by doing this that we can begin to understand the link between diversity and function at the landscape level, i.e. in terms of γ diversity. The term ecosystem usually implies a system with a reasonably large scale, such as the Greater Yellowstone National Park area (Hatala *et al.* 2011) or the Arctic Ocean. Landscape (γ) patterns of diversity and function, which can in turn be decomposed into α and β patterns of diversity and function, resonate with this usage.

2.4   Conclusions

Biodiversity or biological diversity is a concept with a long history. Ecologists have spent decades devising methods of assessing this diversity. Rapid advances in statistical and molecular techniques have led to the development of innovative methods, while greater computing power has allowed researchers to refine traditional methods; for example, by incorporating randomization tests and null models. Although there is still disagreement about which methods perform best, there is a growing realization that the assessment of biodiversity can be approached in different ways, and that by adopting a range of approaches it is possible gain new insights into the structure and diversity of ecological assemblages. There is also much greater appreciation that species richness—probably the most iconic measure of biodiversity—is also one of the most difficult to estimate accurately. In contrast, BEF research often focuses on species richness, or considers functional groups that emerge from the species that are present. This focus works well in small, controlled experimental set-ups, but it is worth remembering that it may well be impossible to quantify species richness at a metacommunity or landscape level. Instead, other approaches, including methods of assessing β diversity, and the conceptual links between α, β, and γ diversity, could be helpful here. There is encouraging evidence that BEF researchers are open to these ideas (Schmid *et al.* 2002). The biodiversity measurement toolkit provides rich opportunities for research, and for achieving a more nuanced view of biodiversity, and of its relationship with function.

References

Bates, H. W. (1864) *The naturalist on the river Amazon.* John Murray, London.

Beck, J. and Schwanghart, W. (2010) Comparing measures of species diversity from incomplete inventories: an update. *Methods in Ecology and Evolution* **1**, 38–44.

Cardinale, B. J., Srivastava, D. S., Duffy, J. E. *et al.* (2009) Effects of biodiversity on the functioning of ecosystems: a summary of 164 experimental manipulations of species richness. *Ecology* **90**, 854–4.

Cardinale, B. J., Wright, J. P., Cadotte, M. W. *et al.* (2007) Impacts of plant diversity on biomass production increase through time because of species complementarity. *Proceedings of the National Academy of Sciences* **104**, 18123.

Chapin III, F. S., Zavaleta, E. S., Eviner, V. T. *et al.* (2000) Consequences of changing biodiversity. *Nature* **405**, 234–42.

Chown, S. L. and McGeoch, M. A. (2011) Measuring biodiversity in managed landscapes. In *Biological diversity: frontiers in measurement and assessment.* In: *Biological diversity: frontiers in measurement and assessment* Magurran, A. E. and McGill, B. J. (eds), Oxford, Oxford University Press, 252–64.

Coddington, J. A., Agnarsson, I., Miller, J. A. *et al.* (2009) Undersampling bias: the null hypothesis for singleton species in tropical arthropod surveys. *Journal of Animal Ecology* **78**, 573–84.

Collins, S. L., Suding, K. N., Cleland, E. E. *et al.* (2008) Rank clocks and plant community dynamics. *Ecology* **89**, 3534–41.

Colwell, R. K. and Coddington, J. A. (1994) Estimating terrestrial biodiversity through extrapolation. *Philosophical Transactions of the Royal Society, London B.* **345**, 101–18.

Crawley, M. J., Johnston, A. E., Silvertown, J. *et al.* (2005) Determinants of species richness in the Park Grass Experiment. *American Naturalist* **165**, 179–92.

Crozier, W. J. (1923) On the abundance and diversity in the protozoan fauna of a sewage 'filter'. *Science* **58**, 424–5.

Darwin, C. (1859) *On the origin of species by means of natural selection, or the preservation of favoured races in the struggle for life.* London, John Murray.

Dornelas, M., Soykan, C. and Ugland, K. I. (2011) Biodiversity and disturbance. In *Biological diversity: frontiers in measurement and assessment.* In: *Biological diversity: frontiers in measurement and assessment,* Magurran, A. E. and McGill, B. J. (eds), Oxford, Oxford University Press, 237–51.

Fisher, R. A., Corbet, A. S. and Williams, C. B. (1943) The relation between the number of species and the number of individuals in a random sample of an animal population. *Journal of Animal Ecology.* **12**, 42–58.

Gaston, K. J. and Fuller, R. A. (2008) Commonness, population depletion and conservation biology. *Trends in Ecology and Evolution* **23**, 14–19.

Gerbilskii, N. L. and Petrunkevitch, A. (1955) Intraspecific biological groups of Acipenserines and their reproduction in the low regions of rivers with biological flow. *Systematic Zoology* **4**, 86–92.

Gotelli, N. J. and Colwell, R. K. (2001) Quantifying biodiversity: procedures and pitfalls in the measurement and comparison of species richness. *Ecology Letters* **4**, 379–91.

Gotelli, N. J. and Colwell, R. K. (2011) Estimating species richness. In: *Biological diversity: frontiers in measurement and assessment*, Magurran, A. E. and McGill, B. J. (eds), Oxford, Oxford University Press, 39–54. Oxford, Oxford University Press.

Harper, J. L. and Hawksworth, D. L. (1995) Preface. In *Biodiversity: measurement and estimation*, Hawksworth, D. L. (ed.), 5–12. London, Chapman and Hall.

Hatala, J.A., Dietze, M.C., Crabtree, R.L. *et al.* (2011) An ecosystem-scale model for the spread of a host-specific forest pathogen in the Greater Yellowstone Ecosystem. *Ecological Applications* 21, 1138–53.

Hawkins, B. A. (2001) Ecology's oldest pattern. *Trends in Ecology and Evolution* 16, 470.

Hector, A. and Hooper, R. (2002) Darwin and the first ecological experiment. *Science* 295, 639–40.

Hector, A., Schmid, B., Beierkuhnlein, C. *et al.* (1999) Plant diversity and productivity experiments in European grasslands. *Science* 286, 1123.

Hill, M. O. (1973) Diversity and evenness: a unifying notation and its consequences. *Ecology* 54, 427–31.

Hubbell, S. P. (2001) *The unified neutral theory of biodiversity and biogeography*. Princeton, Princeton University Press.

Hurlbert, S. H. (1971) The non-concept of species diversity: a critique and alternative parameters. *Ecology* 52, 577–86.

Jost, L., Chao, A. and Chazdon, R. L. (2011) Compositional similarity and beta diversity. In: *Biological diversity: frontiers in measurement and assessment*, Magurran, A. E. and McGill, B. J. (eds), Oxford, Oxford University Press, 66–84.

Loreau, M., Naeem, S. and Inchausti, P. (2002) *Biodiversity and ecosystem functioning: synthesis and perspectives*. Oxford, Oxford University Press.

Lovejoy, T. E. (1980a) Changes in biological diversity. In *The Global 2000 Report to the President, Vol. 2 (The technical report)*, G. O. Barney, G. O. (ed.), Harmondsworth, Penguin, 327–32.

Lovejoy, T. E. (1980b) Foreword. In *Conservation biology: an evolutionary-ecological perspective*, Soulé, M. E. and Wilcox, B. A.), Sunderland, Massachusetts, Sinauer Associates, v–ix.

Mac Nally, R. (2007) Use of the abundance spectrum and relative abundance distributions to analyze assemblage change in massively altered landscapes. *American Naturalist* 170, 319–30.

MacArthur, R. H. and MacArthur, J. W. (1961) On bird species diversity. *Ecology* 42, 594–8.

MacArthur, R. H. and Wilson, E. O. (1967) *The theory of island biogeography*. Princeton, Princeton University Press.

Magurran, A. E. (2004) *Measuring biological diversity*. Oxford, Blackwell Science.

Magurran, A. E. (2011) Measuring biological diversity in time (and space). In: *Biological diversity: frontiers in measurement and assessment*, Magurran, A. E. and McGill, B. J. (eds), Oxford, Oxford University Press, 85–93.

Magurran, A. E., Baillie, S. R., Buckland, S. T. *et al.* (2010) Long-term datasets in biodiversity research and monitoring: assessing change in ecological communities through time. *Trends in Ecology and Evolution* 25, 574–82.

Magurran, A. E. and Henderson, P. A. (2003) Explaining the excess of rare species in natural species abundance distributions. *Nature* 422, 714–16.

Magurran, A. E. and McGill, B. J. (eds.) (2011) *Biological diversity: frontiers in measurement and assessment*. Oxford, Oxford University Press.

Magurran, A. E. and Queiroz, H. (2010) Evaluating tropical biodiversity: do we need a more refined approach? *Biotropica* 42, 537–9.

Matthews, W. J., Harvey, B. C. and Power, M. E. (1994) Spatial and temporal patterns in the fish assemblages of individual pools in a midwestern stream (USA). *Environmental Biology of Fishes* 39, 381–97.

McGill, B. J. (2011) Species abundance distributions. In: *Biological diversity: frontiers in measurement and assessment*, Magurran, A. E. and McGill, B. J. (eds), Oxford, Oxford University Press, 105–22.

cGill, B. J., Etienne, R. S., Gray, J. S. *et al.* (2007) Species abundance distributions: moving beyond single prediction theories to integration within an ecological framework. *Ecology Letters* 10, 995–1015.

Motomura, I. (1932) On the statistical treatment of communities. *Zoological Magazine Tokyo (in Japanese)* 44, 379–83.

Nagendra, H. (2002) Opposite trends in response for Shannon and Simpson indices of landscape diversity. *Applied Geography* 22, 175–86.

Norse, E. A. and McManus, R. E. (1980) Ecology and living resources biological diversity. In *Environmental quality 1980: The eleventh annual report of the Council on Environmental Quality*. Washington, DC: Council on Environmental Quality.

Norse, E. A., Rosenbaum, K. L., Wilcove, D. S. *et al.* (1986) *Conserving biological diversity in our national forests*. Washington, DC: The Wilderness Society.

Noss, R. F. (1990) Indicators for monitoring biodiversity: a hierarchical approach. *Conservation Biology* 4, 355–64.

Øvreas, L. and Curtis, T. P. (2011) Microbial diversity and ecology. In: *Biological diversity: frontiers in measurement and assessment*, Magurran, A. E. and McGill, B. J. (eds), Oxford, Oxford University Press, 221–36.

Petchey, O. L. and Gaston, K. J. (2002) Functional diversity (FD), species richness and community composition. *Ecology Letters* **5**, 402–11.

Pielou, E. C. (1975) *Ecological diversity*. New York, Wiley Interscience.

Preston, F. W. (1948) The commonness, and rarity, of species. *Ecology* **29**, 254–83.

Rényi, A. (1961) On measures of entropy and information. In *Proceedings of the 4th Berkeley symposium on mathematical statistics and probability*, J. Neyman (ed.), Berkeley, University of California Press. 547–61.

Sanders, H. L. (1968) Marine benthic diversity: a comparative study. *Amer. Nat.* **102**, 243–82.

Schmid, B., Hector, A., Huston, M. A. *et al.* (2002) The design and analysis of biodiversity experiments. In *Biodiversity and ecosystem functioning*, Loreau, M., Naeem, S., Inchausti, P. (eds), Oxford, Oxford University Press, 61–75 .

Schwartz, M. W., Brigham, C. A., Hoeksema, J. D. *et al.* (2000) Linking biodiversity to ecosystem function: implications for conservation ecology. *Oecologia* **122**, 297–305.

Silvertown, J., Poulton, P., Johnston, E. *et al.* (2006) The Park Grass Experiment 1856–2006: its contribution to ecology. *Journal of Ecology* **94**, 801–14.

Southwood, R. and Henderson, P. A. (2000) *Ecological methods*. Oxford, Blackwell Science.

Srivastava, D. S. and Vellend, M. (2005) Biodiversity-ecosystem function research: is it relevant to conservation?

Tansley, A. G. (1935) The use and abuse of vegetational concepts and terms. *Ecology* **16**, 284–307.

Tilman, D., Reich, P. B., Knops, J. *et al.* (2001) Diversity and productivity in a long-term grassland experiment. *Science* **294**, 843.

Tóthmérész, B. (1995) Comparison of different methods for diversity ordering. *Journal of Vegetation Science* **6**, 283–90.

Vellend, M., Cornwell, W. K., Magnuson-Ford, K. *et al.* (2011) Measuring phylogenetic biodiversity. In: *Biological diversity: frontiers in measurement and assessment*, Magurran, A. E. and McGill, B. J. (eds), Oxford, Oxford University Press, 194–207.

Weiher, E. (2011) A primer of trait and functional diversity. In: *Biological diversity: frontiers in measurement and assessment*, Magurran, A. E. and McGill, B. J. (eds), Oxford, Oxford University Press, 175–93.

Whiteside, M. C. and Harmsworth, R. V. (1967) Species diversity in Chydorid (Cladocera) communities. *Ecology* **48**, 664–7.

Whittaker, R. H. (1960) Vegetation of the Siskiyou Mountains, Oregon and California. *Ecological Monographs* **30**, 279–338.

Whittaker, R. H. (1972) Evolution and measurement of species diversity. *Taxon* **21**, 213–51.

Wilsey, B. J. and Potvin, C. (2000) Biodiversity and ecosystem functioning: importance of species evenness in an old field. *Ecology* **81**, 887–92.

Wilson, E. O. (ed.) (1988) *Biodiversity*. Washington, DC, National Academy Press.

CHAPTER 3

Ecosystem function and co-evolution of terminology in marine science and management

David M. Paterson, Emma C. Defew, and Julia Jabour

3.1 Introduction

In 1992, Brian Walker addressed the issue of how a 'functional approach' to the analysis of biodiversity was required in order to maximize society's chances of managing the increasing anthropogenic pressures on natural ecosystems (Walker 1992). He advocated this approach from a clearly conservational point of view, to provide what he considered to be our best opportunity for maintaining ecosystem resilience and preserving ecosystem function. Since the 1990s, the literature on biodiversity and the implication of biodiversity science to the control of processes that occur within ecosystems has expanded rapidly (Solan *et al.* 2009; Loreau *et al.* 2002). More recently this has included the attempt to explicitly link ecological processes more closely with management approaches to enhance their stability and future ecosystem health (Hobbs *et al.* 2011; Paterson *et al.* 2011; Samhouri *et al.*, 2010), and also with the relative value that can be attributed to natural ecosystems (Naeem *et al.* 2009). This represents a complex pathway of co-evolution of science and policy from the development of questions in fundamental science, through the development of new approaches to those questions, and then the translation of that research into policy and management strategies. However, the language of policy is evolving rapidly and changing as fast, or even faster, than the science has developed and the mismatches that occur in usage have now been recognized (Holt *et al.* 2011). Reference is now made to 'biodiversity stocks', 'flow of ecosystem services', and of 'natural capital'

(Bhaskar and Adams, 2009), but as terms change and concepts are expanded, they often become confused and this has led to some debate in the literature (Flint and Kalke 2005; Raffaelli *et al.* 2005) where original meanings have become obscured or forgotten. As the drive toward a holistic approach to ecosystem management accelerates, it can therefore be hard to retain a clear view of the current state of knowledge and of what science can be expected to deliver realistically (Srivastava and Vellend 2005). A logical pathway from science to policy support might be illustrated as:

Phase 1: Formulating relevant hypotheses. For example, increasing biodiversity increases ecosystem function and ecosystem resilience.

Phase 2: Research: Targeted analysis (laboratory, field, and modelling studies) to examine these hypotheses.

Phase 3: Use the results of the science to support policy development. Biodiversity policy emerges. Marine protected areas, regulation of exploitation, protective habitat classification.

Phase 4: Management. Embed protection of biodiversity (as a proxy for ecosystem services) into policy and design suitable mechanisms for sustainable development and ongoing monitoring— e.g. water framework directive.

Of course, this is highly simplified, and follows a 'bottom up' approach, whereas policy development may also demand scientific advances in support of management approaches—e.g. fisheries legislation—and

Marine Biodiversity and Ecosystem Functioning. First Edition. Edited by Martin Solan, Rebecca J. Aspden, and David M. Paterson.
© Oxford University Press 2012. Published 2012 by Oxford University Press.

we recognize the much more complex feedback between policy requirements and scientific development. However, even this simplified pathway demonstrates the hierarchy of decision that have to be made which should be supported by fundamental science. Even in well-researched fields, the relative maturity or even veracity of the scientific knowledge at the base of this hierarchy is sometime heavily questioned—e.g. the global warming 'debate'—but at other times, terminology and concepts are accepted at a policy level before the science is fully resolved—e.g. the BEF debate. Scientists now agree that there is a link between biodiversity and ecosystem services, but also recognize the variability and context dependency of that link (Bulling *et al.* 2010). Many different hypothesis around the biodiversity–ecosystem function (BEF) question have been examined (Naeem *et al.* 2009), using varied measures of biodiversity and ecosystem function (Cardinale *et al.* 2006). There has been detailed consideration of the empirical process, including the confounding effects of sampling design, sampling effects (Houston, 1997), trophic influences, habitat structure (Godbold *et al.* 2011), and species elimination pathways (Solan *et al.* 2004). There has been much less consideration of the mechanistic side of the equation: the functional 'products' of individuals; the resultant combined response of the system; the precision and accuracy at which these measures might be determined; and the theoretical value of the measures in terms of the research question. This was understandable at the onset of BEF research where a simple measure of the functional capacity of a system was required against which to determine the effects of biodiversity manipulation. In the rush to establish the theory, the choice of response variable was often pragmatic and based on the ability to produce a relatively straightforward measure of system turnover rather than a measure selected for its particular applicability to the objective (see Chapter 10). In addition, while it is implicitly recognized that ecosystems are multifunctional with many processes occurring at once, most studies to date measure very few, and often only a single, functional response which can be used as an implicit and unstated proxy measure of the entire system response. As the BEF debate matures, it is timely for researchers to consider the relative value of different and multiple functional responses and consider their value for the theory experiments in question.

3.2 What's in a name? Ecosystem function

The debate over an acceptable definition for the science of ecology took many years (Begon *et al.* 1990) but has reached a fairly comfortable status where a number of definitions are acceptable, often with the implicit understanding that some are less than fully adequate in all circumstances but generally workable and probably 'as good as it gets'. There is therefore a sensible common tendency to redefine or refine quite well-known terms to insure that the context of the work is clearly communicated and fits well with the objectives of the study. This problem, like the original definitions of ecology, has already emerged in the BEF debate and the solution will probably be similar. For example, the following, often divergent definitions, of ecosystem function are commonly available.

3.2.1 Ecosystem function defined

- The collective intraspecific and interspecific interactions of the biota, such as primary and secondary production, and mutualistic relationships (W 1).
- The interactions between organisms and the physical environment, such as nutrient cycling, soil development, water budgeting, and flammability (W 1).
- The physical, chemical, and biological processes or attributes that contribute to the self-maintenance of the ecosystem; in other words, what the ecosystem does. Some examples of ecosystem functions are wildlife habitat, carbon cycling, or trapping nutrients (W 2).
- The characteristic exchanges within an ecosystem are called ecosystem functions and in addition to energy and nutrient exchanges, involve decomposition and production of biomass (W 3).
- The biophysical processes that take place within an ecosystem. These can be characterized apart from any human context—e.g. fish and waterfowl habitat, cycling carbon, trapping nutrients. The level of function depends on the capacity of the ecosystem—onsite features—and certain aspects of its landscape context—e.g. connectedness to other natural/human features, accessibility to birds, fish (W 4).

The different definitions are frequently related to the interests and experience of the author—e.g. biotic interactions, fire as a structuring force, etc.—and while it is possible to argue against any single definition, most have some merit, although we would not limit ecosystem function to biotic processes (W 1). Care must be taken then to understand what each worker in different disciplines means by their usage. This problem is exacerbated by adoption of similar, if not identical, terminology emerging between different disciplines and at the cross-over or co-evolution of environmental science and governing policy (Table 3.1).

For example, De Groot *et al.* (2002) define ecosystem functions (plural) as *'the capacity of natural processes and components to provide goods and services that satisfy human needs, directly or indirectly'*, whereas ecosystem goods and services are more commonly defined as a subset of ecosystem functions that provides benefit to mankind (Hooper *et al.* 2005). De Groot therefore implies that only 'useful' ecosystem processes are ecosystem functions, while most ecologists, and a good many economists, would argue not all ecosystem functions are recognized as beneficial to mankind and not all functions are therefore services—e.g. Millennium Ecosystem Assessment (2003); see White *et al.* (2010). There is also the consideration that definition may also be context-dependant (Langenheder *et al.* 2010); different functions may become services depending on environmental conditions. For example, water retention may be of greater value to humankind (an ecosystem service) in arid desert regions than in the relatively rain-soaked highlands of Scotland.

Table 3.1 Broadly ecological concepts adopted, developed and adapted into the policy arena.

Terms	Ecological usage	Reference
Ecosystem approach Ecosystem management Integrated coastal zone management (ICZM)	The first two are often largely synonymous terms, widely used to describe an 'integrative' approach to ecosystem management where the system is recognized to be complex and interactive so all components must be considered and the system managed at a high level (products and services rather than species). ICZM has the same holistic philosophy but is usually developed in response to local management problems such as coastal erosion, harbour siltation, eutrophication, and explicitly including geographical and sociopolitical issues	Alpert 1995 W5: European Union. Coastal Zone Policy: Integrated Coastal Zone Management (ICZM): <http://ec.europa.eu/environment/iczm/home.htm>
Ecosystem function	Ecosystem function can be defined, inclusively as: 1. Stocks of energy and materials in the system; 2. The fluxes of energy and materials, and; 3. Relative stability over time	Pacala and Kinzig 2002 Paterson *et al.* 2009
Ecosystem services	Usually defined as those ecosystem functions that are 'useful' to humankind. They include exploitable (fisheries) and indirect resources (nutrient turnover, pollutant amelioration), and sometimes aesthetic value.	Watson 2005
Biodiversity	Complex, wide-ranging concept that includes phylogenetic diversity (omega diversity), through the number of species in a region (alpha diversity), comparison of diversity between regions (beta diversity), and each can be described by numerous metrics ranging from simple species counts (species richness) to complex formulae.	Magurran 2004
Ecosystem resistance and resilience	These terms are often combined or interchanged in the literature but they describe slightly different properties. Ecosystem resistance describes the ability of the ecosystem to remain unchanged despite external forcing (resisting stressors) while resilience describes the ability of a system to recover and return to 'normal' after a disturbance.	Gibbs 2009 Srivastava and Vellend 2005
Ecosystem vulnerability	Ecosystem vulnerability is a similar concept as resistance and resilience and applied at an ecosystem level but focuses on risk assessment, often used in terms of exposure to xenobiotic compounds.	De Lange *et al.* 2010

Note: These definitions vary across the literature but represent the authors' views of appropriate usage.

3.3   Measuring ecosystem function

In most approaches to the BEF debate, a hierarchy of ecosystem activity is recognized that leads eventually to ecosystem properties, goods, and services based on a subjective value to mankind. At the most basic level, energy or materials are transformed within the ecosystem and translocated due to physical, chemical, or biological action. These 'ecosystem processes' vary widely and using the most inclusive definition would include biochemical activities such as the electron transport pathways of photosynthesis, chemical reactions such as oxidation or reduction events, and also physical transformations such as sediment erosion and related particle advection and nutrient release. Many of these processes occur at a very small (sub-millimeter) scale, but the combined effect of large numbers of small reactions produce a transfer of energy or material recognized as ecosystem functioning (Figure 3.1) and resulting in flows of ecosystem services.

In terms of the BEF debate, the importance of the biota is in providing these transformations (functions) in the ecosystem that underpin ecosystem services. The transformations that are studied and recognized as important in this context are obviously a small fraction of the overall system performance.

Photosynthesis is a complex process constructed from many elements that leads to several critical ecosystem services including the fixation of atmospheric carbon into organic molecules and the production of oxygen (Figure 3.1). Electron transport pathways might not normally be considered as an ecosystem function but relatively fine-scale activity can be used as a proxy for larger scale processes. PAM fluorescence measurements (Consalvey *et al.* 2005) essentially determine the transfer of energy though the photochemical pathways of the chloroplast, and this measure is commonly reported as a proxy for biomass and/or photosynthetic activity of an assemblage (Jesus *et al.* 2006, Hicks *et al.* 2011), and hence as a measure of ecosystem function. There are quite a few assumptions inherent in this progression which are not often discussed, but this demonstrates that there is not always an absolute separation of ecosystem processes from ecosystem functions, although some definitions have been attempted (W 6) and it is sensible to recognize that turnover with the system occurs at many scales, and many processes are difficult to measure directly. Ecosystem functioning is usually determined by an attempt to capture a dynamic process of transformation of materials within an ecosystem. At first this would seem to allow for many types of measurements, but for practical

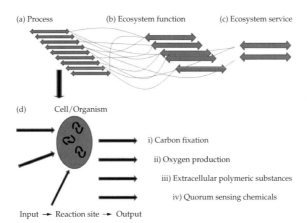

Figure 3.1   Ecosystem processes (a), ecosystem functions (b), and ecosystem services (c). Most processes within ecosystems are the result of chemical reactions driven by molecular interaction, both biotic and abiotic, which may include surface, solute, or cellular processes (d). Some outputs are readily recognized as ecosystem functions and services (i) and (ii), whereas other may only become important under certain circumstance depending on context. For example, extracellular polymeric substances (iii) have a number of important ecological roles (Underwood and Paterson 2003) but they are rarely measured as ecosystem functions. Other emerging ecosystem mechanism (iv) may rarely enter the BEF debate.

reasons the actual number of ecosystem functions generally measured in the literature is very limited (Table 3.2). This is because many ecosystem processes are hard to determine directly, and where repeated measures are required, techniques are further restricted by the need for non-destructive or disruptive methods. The majority of marine studies concentrate on primary production and nutrient turnover. Even so, the processes of transformation are rarely determined since the transfer of materials is hard to measure through the system. However, an indication of the material movement can be measured by variation in concentration of materials/products over time, such as nutrient flux between the bed and the water column, or increase of CO_2 from respiration. These measures require expansion to increase the understanding of overall ecosystem function.

This difficulty in the direct measurement of ecosystem processes also means that a number of proxies have been used to express ecosystem function. For example, primary production is often determined by measuring chlorophyll a (Chl a) which is taken to be a measure of the photosynthetic potential of the system. This is widely accepted, but care should be taken to recognize the difference between the proxy and the actual process. The proxy (in this case, amount of Chl a) responds relatively slowly to environmental change, whereas rates of photosynthesis are highly sensitive to light climate and environmental condition (Falkowski and Raven 2007).

The suggested link between Chl a and potential of the system to provide primary productivity is reasonable, but not precise.

The concept of functionality is therefore highly subjective and open to wide interpretation. There are two overall approaches that are often used in empirical studies of ecosystem function. At the extremes, these represent reductionist or holistic approaches. The reductionist views an organism as providing a set of capabilities, which may be internal or external, which contribute to ecosystem processes through the transformation of materials. This is a reflection of the organismal definition of niche as a set of capabilities for extracting of resources from a habitat (Hutchinsonian niche, Begon *et al.* 1990). The relevance of this idea to ecosystem function is that just as the resources required by an organism can be represented as an n-dimensional volume, with each required resource adding an axis to the theoretical multidimensional space taken to encapsulate the entire niche (Petchey *et al.* 2009), the same must be true of their resultant capacity to change these resources and contribute to the ecosystem processes driving system functionality. This highlights the complexity of the ecosystem processes that underlie functionality. At the other extreme, an ecosystem, including its biota, can be viewed as black box where material enters, is processed and transformed before being released from the system. Either of these extremes is unlikely to be of greatest benefit in terms

Table 3.2 Recent examples of empirical measures of ecosystem function.

Functional measure	via	Organism	Reference
NPP	Biomass	Microphytobenthos	Hicks *et al.* 2011
Ammonia flux	Nutrient analysis	Sediment infauna	Bulling *et al.* 2010
Phosphate flux	Nutrient analysis	Sediment infauna	Bulling *et al.* 2010
Biotubation	Luminophores	Sediment infauna	Bulling *et al.* 2010
Ammonia flux	Nutrient analysis	Sediment infauna	Bulling *et al.* 2008
Phosphate flux	Nutrient analysis	Sediment infauna	Bulling *et al.* 2008
Movement	Infaunal spp	Sediment infauna	Bulling *et al.* 2008
Surface adhesion	Magnetic particle induction (MagPI)	Biofilms	Larson *et al.* 2009
NPP	Biomass	Microphytobenthos	Dyson *et al.* 2007
NPP	Biomass	Algae	Zhang and Zhang, 2006
Decomposition	Biomass	Fungi/Bact	Constantini and Rossi 2010
Algal standing stock	Biomass	Algae	Walker 2010
2° Productivity	Biomass	Mollusc	Walker 2010

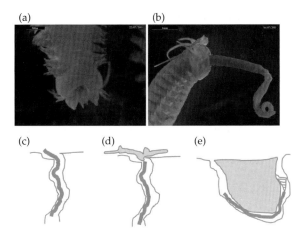

Figure 3.2   Functional plasticity in the feeding by the polycheate ragworm, *Hediste diversicolor*. *H. diversicolor* is an omnivore which may feed on vegetative material (a), become a carnivore (b), or use other methods of deposit feeding. It may forage over the local sediment surface from its tube (c), or if surface algal material is present, the worm pulls material into its burrows to consume it (d). However, when submerged and where the overlying water contains sufficient suspended organic material, the worm retreats into its tube and spins a mucilage net. The worm then undulates to create a water flow which is filtered through the mucilage net which is later consumed by the feeding worm (e). Images C. Wood and M. Chocholek, St Andrews (see Paterson 2005). See also Plate 1.

of the BEF debate. A reductionist approach quickly becomes impossibly complex, while the black box approach prevents the understanding of the importance of species identify and the value of inherent knowledge of species behaviour. An added complexity that is gaining increasing attention is the variability of species behaviour in term of environmental condition or context. (Riisgård *et al.* 2001). The functional plasticity inherent in the behaviour and lifestyle changes of many species (Boogert *et al.* 2006, Figure 3.2) also constrains the ability of researchers to recognize clear links between species identity and functional measures. The compromise to the reductionist/ black box quandary is the use of functional traits, where the functional abilities of the varied organisms are grouped together to provide a metric that reduces the complexity of the system but recognizes the functional abilities of the species (Hooper *et al.* 2005; Naeem *et al.* 2009). This approach is gaining prominence, but it also has its critics (Petchey *et al.* 2009), and does not deal easily with functional plasticity, although some methods of assigning traits allow organisms to be graded in terms of each trait providing some ability to classify individual species under multiple functional headings with different weightings (Solan *et al.* 2004).

3.4   Ecological terms and the co-evolutionary model

As the knowledge of BEF science developed—led by terrestrial work—the adoption of ecological terms and concepts in developing, explaining, and implementing environmental policy became more common (Table 3.1). However, the adoption of these terms may also form part of a science–policy communication problem (Holt *et al.* 2011) since they often have a precise meaning in ecology that is not well-translated in the policy realm. The 'ecosystem approach' is one of the concepts at fault here. It is widely promoted in marine as well as terrestrial policy (Garcia and Cochrane 2004; Thoms and Sheldon 2002), and is interpreted to encompass a paradigm of inclusivity. This requires that ecosystems be assessed at a higher-order level of organization, function, and scale rather than at the level of individual species' behaviour and response (Gibbs, 2009; Brussard *et al.* 1998). Yet the level of detail required in understanding an ecosystem, not to mention managing its uses and any consequent impacts, is almost overwhelming. The scientific basis for this concept includes theoretical approaches recognizing the complexity of ecosystems, and the multiple interdependencies between organisms and their environment that control the turnover of material and the production of biomass. Ecological analysis can also be represented as hierarchical, moving from genes, to individuals, to populations, to communities, and to ecosystems.

The move toward a higher-level approach (Ecosystem management, Table 3.1) also tends to encompass larger spatial scales and integration across natural heterogeneity (Moore *et al.* 2009). Ecosystem function (Table 3.1) is now used to describe ecosystem stocks, flows of services, processes, and resilience, but few, if any, ecologists would assert that that they can describe or predict the holistic functionality of even a simple ecosystem. Central to this debate is the contribution of species variety and abundance—the biodiversity—towards system functionality (Bracken *et al.* 2008). The BEF debate and its contribution to conservation (Srivastava and Vellend 2005) is still a vibrant but relatively immature area of marine research (Hendriks and Duarte 2008; Paterson *et al.* 2009), and the majority of studies are still empirical with few measures of system functionality (response variables). The choice of these determinants of 'ecosystem function' represents a balance between what can be measured relatively easily and the requirement to measure a variable relevant to the experimental hypothesis. Limitations like this are rarely mentioned at the science–policy interface yet the pressure for science to provide policy makers, and hence society, with better and clearer answers (Groffman *et al.* 2010) is growing. Ruijgrok *et al.* (1999) described the relationship between society and nature in these terms:

> *Both society and nature are allowed to change and to inflict change upon each other as long as neither of them suffers serious damage, threatening its existence; it is a matter of mutual benefit. Thus the term 'Coevolution' is used here to describe this interpretation of the functional view.*

The process of ecological concepts being transferred into the policy arena can be considered as a co-evolutionary process (Osterblom *et al.* 2010). In ecology, co-evolution describes where a development in one species results in a related development in another (Thompson 1994). Not all co-evolution is mutually beneficial—e.g. predator–prey relationships—and

whilst the strength of the interaction between the two species varies greatly, there is no implied dependency. Co-evolution as outlined by Ruijgrok is therefore a very constrained version of a natural ecological phenomenon. While co-evolution implies at least a two-way relationship, it is important to understand that there are numerous directional drivers or selective pressures of differing strengths that affect policy formulation.

3.5   Co-evolution, policy drivers, and opportunities

In an ideal world, scientific information alone would direct the development of policy. In the real world, however, multiple variables come into play in the development of policy, including not only scientific information but also economics, legal matters, resources, politics and political will, and timing, among others (May 2002). The integration of sometimes competing variables is required to form coherent policies when, or indeed, if, a policy window of opportunity (Kingdon 1984) becomes available. For example, there is little point having scientific evidence if there is no political will to follow through by making politically difficult decisions. Therefore, it is argued that true integration across many drivers will require a common purpose—something not readily embraced by those who place themselves on either side of the science–policy divide (May 2002). Once consensus is achieved then the objectives of a policy would be to eliminate or reduce the pressure or risk to an acceptable level.

Theoretically, the absolute impact of humans on the global environment is closely related to population and per capita resource utilization. This first was put into mathematical terms by Ehrlich and Holdren (1971) as:

$$I = PAT \qquad \text{(Equation 3.1)}$$

Where I = Human impact, P = human population, A = affluence (amount of material required by each individual), and T = Technology (related to the stage in technological development).

The carrying capacity of the system is strongly related to the effective functionality of the biota as these pressures increase. The IPAT equation has aroused considerable debate (Chertow 2001) and has been criticized for its simplicity, but it had been widely adapted and employed. Human population is increasing ($\approx$ 6.8 billion currently, reaching a predicted level of 9 billion by 2050), but also the average amount of resource that each human uses (A) is predicted to rise at a much greater rate (Watson 2005). This will be reflected in functional 'demands' (ecosystem services) from the system. This can be interpreted in terms of the co-evolutionary model where the drive to decrease impact stimulates both research for that purpose and concern over biodiversity threats. Offshore renewables present a current example. Clean power is needed and relevant guidance is rapidly being put in place to minimize the impact (W 7). The mitigation of environmental impact by management control can also be described mathematically:

$$I = PAT(1-E) \qquad \text{(Equation 3.2)}$$

Where E = effectiveness of a management approach, expressed as a value between 0 and 1 (0 = no effect, 1 = 100% effective).

E may increase with time as scientists respond to policy pressure and provide appropriate information—Ruijgrok's co-evolution (Ruijgrok *et al.* 1999). However, there can be a considerable lag before science is translated into policy, and this is sometimes in the order of decades rather than years. However, it is rarely necessary that impact be reduced to zero, but rather to an acceptable level which can be sustained by the system without deleterious effects. This emphasizes system resilience (Gibbs 2009), or as the more recently formulated 'ecosystem vulnerability' (De Lange *et al.* 2010). For marine systems, the OVI = Oil Vulnerability Index (King and Sanger 1979, cited in De Lange *et al.* 2010) and the VME = vulnerability of marine ecosystems (Halpern *et al.* 2007), are examples of the latter. The overall goal is to ensure that management control is in place and effective before systems become damaged and unsustainable. Like the original IPAT, this derivation is simplistic but emphasizes the role that policy can play in moderating impact.

3.6 Conclusions

The IPCC example may provide a warning in terms of developing marine science and policy. There is similarity in that the pressure to provide effective policy for the ocean grows with public awareness of global issues (sea-level rise, over-fishing, invasive species, ocean acidification). The move to a holistic (ecosystem management) approach, often based on BEF science, is to be supported but scientists must explain clearly the developmental stage of relevant theory in terms of what can be supplied by current BEF research. The ecosystem services which arise from the functioning of natural systems are the end of a complex conceptual chain of related processes both biogenic and abiotic that occur in any natural system. There is no easy or neat segregation that conveniently separates the transformations that take place in an ecosystem into chemical reactions, ecosystem processes and ecosystem services.

Acknowledgements

DMP received funding from the MASTS pooling initiative (the Marine Alliance for Science and Technology for Scotland), funded by the Scottish Funding Council (grant reference HR09011) and the work was supported by EU (FP7/2007-2013, No. 266445) Vectors of Change in Oceans and Seas (VECTORS).

References

Alpert, P. (1995) Incarnating Ecosystem Management. *Conservation Biol* **9**(4): 952–5.

Begon, M, Harper, J.L., Townsend, C.R. (1990) *Ecology: Individuals, populations and communities.* Blackwell Scientific Publications, London.

Bhaskar, V., Adams, W.A. (2009) Ecosystem services and conservation strategy: beware the silver bullet. *Conservation Letters* **2**(4): 158–62.

Boogert, N.J., Paterson, D.M., Laland, K.N. (2006) The implications of niche construction and ecosystem engineering for conservation biology. *Biosciences* **57**(7): 570–8.

Bracken M.E.S., Friberg, S.E., Gonzalez-Dorantes, C.A. et al. (2008) Functional consequences of realistic biodiversity changes in a marine ecosystem. *Proc Nat Acad Sci* **105**(3): 924–8.

Brussard P.F., Reed, J.M., Tracy, C.R. (1998) Ecosystem management: what is it really? *Landscape and Urban Planning.* **40**: 9–20.

Bulling, M.T., Solan, M., Dyson, K., *et al.* (2008) Species effects on ecosystem processes are modified by faunal responses to habitat composition. *Oecologia* **158**: 511–520.

Bulling, M.T., Hicks, N., Murray, L. *et al.* (2010) Marine biodiversity-ecosystem functions under uncertain environmental futures. *Philosophical Transactions of the Royal Society* **365**: 2107–16.

Cardinale, B.J., Srivastava, D.S., Duffy, J.E. *et al.* (2006) Effects of biodiversity on the functioning of trophic groups and ecosystems, *Nature* **443**, 989–92.

Chertow M.R. (2001) The IPAT Equation and Its Variants: Changing Views of Technology and Environmental Impact. *J Indust Ecol* **4**(4): 13.

Consalvey M., Perkins R.G., Underwood G.J.C. *et al.* (2005) PAM Fluorescence: A beginner's guide for benthic diatomists. *Diatom Research* **20**(1): 1–22.

Costantini, M.L., Rossi, L. (2010) Species diversity and decomposition in laboratory aquatic systems: the role of species interactions. *Freshwater Biology* **55**: 2281–2295.

De Groot R.S., Wilson, M.A., Boumans, R.M.J. (2002) A typology for the classification, description and valuation of ecosystem functions, goods and services. *Ecological Economics* **41**(3): 393–408.

De Lange, H.J., Sala, S., Vighi, M. *et al.* (2010) Ecological vulnerability in risk assessment—A review and perspectives. *Science of the Total Environment* **408**: 3871–9.

Dyson, K., Bulling, M.T., Solan, M., *et al.* (2007) Influence of macrofaunal assemblages and environmental heterogeneity on microphytobenthic production in experimental systems. *Proceedings of the Royal Society Series B-Biological Sciences* **274**: 2547–2554.

Ehrlich, P., Holdren, J. (1971) Impact of population growth. *Science* **171**: 1212–17.

Falkowski, P.G., and Raven, J.A. (2007) *Aquatic photosynthesis*. Princeton University Press, Princeton, 484 pp.

Flint, R.W., and Kalke, R.D. (2005) Re-inventing the wheel in Ecology. *Science* **307**: 1875.

Garcia, S.M., Cochrane, K.L. (2004) Ecosystem approach to fisheries: a review of implementation guidelines. *ICES J Marine Science* **62**(3): 311–18.

Gibbs, M.T. (2009) Resilience: What is it and what does it mean for marine policymakers? *Marine Policy* **33**(2): 322–31.

Godbold, J.A., Bulling, M., Solan, M. (2011) Habitat structure mediates biodiversity effects on ecosystem properties. *Proceedings of the Royal Society B-Biological Sciences* **278**(1717): 2510–18.

Groffman, P.M., Stylinski, C., Nisbet, M. *et al.* (2010) Restarting the conversation: challenges at the interface between ecology and society. *Frontiers in Ecol Environ* **8** (6): 284–91.

Halpern, B.S., Selkoe, K.A., Micheli, F. *et al.* (2007) Evaluating and ranking the vulnerability of global marine ecosystems to anthropogenic threats. *Conserv Biol* **21**: 1301–15.

Hendriks, I.E., Duarte, C.M. (2008) Allocation of effort and imbalances in biodiversity research. *J Exp Mar Biol Ecol* **360**: 15–20.

Hicks, N., Bulling, M.T., Solan, M. *et al.* (2011) Impact of biodiversity-climate futures on primary production and metabolism in a model benthic estuarine system. *BMC Ecology* **11**(7): <http://www.biomedcentral.com/1472-6785/11/7>

Hobbs, R.J., Hallett, L.M., Ehrlich, P.R. *et al.* (2011) Intervention Ecology: Applying Ecological Science in the Twenty-first Century. *BioScience* **61**(6):442–50.

Holt, A.R., Godbold, J.A., White, P.C.L. *et al.*. (2011) Mismatches between legislative frameworks and benefits restrict the implementation of the Ecosystem Approach in coastal environments. *Marine Ecology-Progress Series* **434**:213–28.

Hooper, D.U., Chapin III, F.S., Ewel, J.J. *et al.* (2005) Effects of Biodiversity on ecosystem functioning: a consensus of current knowledge. *Ecological Monographs* **75**(1) 3–35.

Huston, M.A. (1997) Hidden treatments in ecological experiments: re-evaluating the ecosystem function biodiversity. *Oecologia* **110**: 449–60.

Jesus, B., Perkins, R.G., Mendes, C.R. *et al.* (2006) Chlorophyll fluorescence as a proxy for microphytobenthic biomass: alternatives to the current methodology. *Marine Biology* **150**:17–28.

Kingdon, J.W. (1984) *Agendas, Alternatives and Public Policies*. Little, Brown & Co, Boston.

Langenheder, S., Bulling, M.T., Solan, M. *et al.* (2010) Bacterial biodiversity-ecosystem functioning relations are modified by environmental complexity. *PLoS ONE* **5**(5): e10834.

Larson, F., Lubarsky, H., Gerbersdorf, S.U., Paterson, D.M. (2009) Surface adhesion measurements in aquatic biofilms using magnetic particle induction: MagPI. *Limnology and Oceanography. Methods* **7**: 490–497.

Loreau, M., Naeem, S., Inchausti, P. (2002). *Biodiversity and ecosystem functioning: Synthesis and Perspectives*. Oxford University Press, London.

Magurran, A. (2004) *Measuring biological diversity*. Blackwell, Oxford.

May, A. (2002) Creating Common Purpose: The integration of science and policy in Canada's Public Service. Canadian Centre for Management Development, Ottawa.

Millennium Ecosystem Assessment. Ecosystems and Human Well-being. (2003). Washington, Island Press.

Moore, S., Wallington, T., Hobbs, R. *et al.* (2009) Diversity in Current Ecological Thinking: Implications for Environmental Management. *Environmental Management* **43** (1): 17–27.

Naeem, S., Bunker, D.E., Hector, A. *et al.* (2009) *Biodiversity, ecosystem functioning and human wellbeing.: an eco-*

logical and economic perspective. Oxford University Press, Oxford.

Osterblom, H., Gardmark, A., Bergstromc, L. *et al.* (2010) Making the ecosystem approach operational—Can regime shifts in ecological- and governance systems facilitate the transition? *Marine Policy* **34**: 1290–9.

Pacala S., Kinzig, A.P. (2002) Introduction to theory and the common ecosystem model. In: Kinzig, A.P., Pacala, S.W., Tilman, D. (eds), *Functional Consequences of Biodiversity: Empirical Progress and Theoretical Extensions.* Princeton, NJ, Princeton University Press, 169–74.

Paterson D.M. (2005) Biodiversity and Functionality of Aquatic Ecosystems. In: *Biodiversity: Structure and Function, Encyclopedia of Life Support Systems (EOLSS), UNESCO,* Eolss Publishers, Oxford, UK, <http://www.eolss.net>

Paterson D.M., Aspden R.J., Black K.S. (2009) Ecosystem functioning of soft sediment systems. In: Perillo G., Wolanski E., Cahoon D. *et al.* (eds), *Coastal wetlands: An Integrated Ecosystem approach: Intertidal Flats: Ecosystem functioning of soft sediment systems.* Elsevier Academic, Amsterdam, 317–43.

Paterson D. M., Hanley, N. D., Black, K. *et al.* (2011) Science and policy mismatch in coastal zone ecosystem management. *Marine Ecology-Progress Series* **43**: 201–2.

Petchey, O.L., O'Gorman, E.J., Flynn, F.B. (2009) A functional guide to functional diversity measures. In: Naeem, S., Bunker, D.E., Hector, A., *et al.* (eds), *Biodiversity, ecosystem Functioning and human wellbeing: an ecological and economic perspective.* Oxford University Press, Oxford, 49–59.

Raffaelli, D., Cardinale, B.J., Downing, A.L. *et al.* (2005) Reinventing the wheel in ecology research?—response. *Science* **307**, 1875–6.

Riisgård H. U., Kamermans, P. (2001) Switching Between Deposit and Suspension Feeding in Coastal Zoobenthos. In: Ecological Comparisons of Sedimentary Shores (Ed. Reise K.). *Ecological Studies* 2001, Volume **151**, Part I, 73–101.

Ruijgrok, E., Vellinga, P.M., Goosen, H. (1999) Dealing with nature. *Ecological Economics* **28**(3): 347–62.

Samhouri, J.F., Levin, P.S., Ainsworth, C.H. (2010) Identifying Thresholds for Ecosystem-Based Management. *PLoS ONE* **5**(1): e8907. doi:10.1371/journal.pone.0008907.

Srivastava, D.S., Vellend, M. (2005) Biodiversity-ecosystem function research: Is It Relevant to Conservation? *Annu Rev Ecol Evol Syst* **36**: 267–94.

Solan, M., Cardinale, B.J., Downing, A.L. *et al.* (2004) Extinction and ecosystem function in the marine benthos. *Science* **306** (5699): 1177–80.

Solan, M., Godbold, J.A, Symstad, A. *et al.* (2009) Biodiversity-ecosystem function research and biodiversity futures: early bird cathcs the worm or a day late and a dollar short? In: Naeem, S., Bunker, D.E., Hector, A. *et al.* (eds), *Biodiversity, Ecosystem functioning and human wellbeing: an ecological and economic perspective.* Oxford University Press, Oxford, 30–46.

Thompson, J.N. (1994) *The Coevolutionary Process.* University of Chicago Press, Chicago.

Thoms, M.C., Sheldon, F. (2002) An ecosystem approach for determining environmental water allocations in Australian dryland river systems: the role of geomorphology. *Geomorphology* **47**(2–4): 153–68.

Underwood, G.J.C., Paterson, D.M. (2003) The importance of extracellular carbohydrate production by marine epipelic diatoms. Advances in Botanical Research. **40**: 183–240.

Walker, B. (1992) Biodiversity and Ecological Redundancy. *Conservation Biology* **6** (1): 18–23.

Walker, A.B., Thompson, R.M. (2010) Consequences of realistic patterns of biodiversity loss: an experimental test from the intertidal zone. *Marine and Freshwater Research* **61**: 1015–1022.

Watson, T.R. (2005) Turning science into policy: challenges and experiences from the science–policy interface. *Phil Trans R Soc B* (2005) **360**: 471–77.

White, P.C.L., Godbold, J.A., Solan, M. *et al.* (2010) Ecosystem services and policy: a review of coastal wetland ecosystem services and an efficiency-based framework for implementing the ecosystem approach. In: *Ecosystem Services.* Hester R.E., Harrison R.M. (eds), Issues in Environmental Science and Technology, **30**, 29–51.

W 1: <http://www.biology-online.org/dictionary/Ecosystem_function>

W 2: <http://www.ecosystemvaluation.org/glossary.htm>

W 3: <http://www.sustainablescale.org/Conceptual-Frame-work/UnderstandingScale/BasicConcepts/EcosystemFunctionsServices.aspx>

W 4: <http://www.ecosystemvaluation.org/Indicators/economvalind.htm>

W 5: European Union. Coastal Zone Policy: Integrated Coastal Zone Management (ICZM): <http://ec.europa.eu/environment/iczm/home.htm>

W 6: <http://www.coastalwiki.org/coastalwiki/Ecosystem_function>

W 7: Crown Estate: <http://www.thecrownestate.co.uk/>

Zhang, Q.G., Zhang, D.Y. (2006) Species richness destabilizes ecosystem functioning in experimental aquatic microcosms. *Oikos* **112**: 218–226.

CHAPTER 4

Ecological consequences of declining biodiversity: a biodiversity–ecosystem function (BEF) framework for marine systems

Shahid Naeem

4.1 The significance of marine biological diversity

4.1.1 Significance

The word 'significance' is used in both social and scientific contexts, the former concerning the importance of something to human well-being, the latter the importance of something in governing a pattern or process in nature, so an evaluation of the significance of marine biological diversity should address both these concerns. Because biological diversity is linked to both human well-being and ecosystem functioning, I will address the question of the significance of marine biological diversity, or marine biodiversity—meaning the taxonomic and functional diversity of organisms in an ecosystem—as it relates to marine ecosystem functioning. Fortunately for me, I am familiar with the study of the relationship between biodiversity and ecosystem functioning (BEF), but I am less fortunate in two other ways. Firstly, I am not a marine biologist but a researcher whose focus concerns the ecosystem consequences of biotic impoverishment (local biodiversity loss). I have no particular prejudice towards any taxa (plant, animal, or microbial) or ecosystem (terrestrial, freshwater, or marine), but my experience has been largely with terrestrial systems. I do not have the grounding in the marine sciences that authors of other chapters have. Secondly, the field of BEF is dominated by studies of terrestrial and freshwater systems and, relatively speaking, only a few marine

studies have contributed to the growth and development of this field. These challenges, however, provide an incentive and opportunity to develop some basic, broadly applicable principles about the significance of marine biological diversity in terms of marine ecosystem functioning, which I hope can stimulate discussion and reflection, and encourage new research in the emerging field of marine BEF.

My approach will be to develop a marine BEF framework using the terrestrial BEF framework as a starting point. I will then review our current understanding of marine biodiversity and marine ecosystem functioning (here restricted to biogeochemical functioning) as it relates to this framework. This will entail developing a simple marine-based BEF model that will illustrate how contemporary changes in marine biodiversity are likely to influence marine ecosystem functioning, and perhaps how biodiversity may be included in marine models which often ignore biological diversity.

The over-arching idea of this BEF framework is that biodiversity is like a biogeochemical catalyst with Michaelis–Menten-like kinetics; the addition of biodiversity enhances ecosystem processes with diminishing returns (explained in detail in section 4.5.2, below). The converse, of course, is that the loss of biodiversity eventually has adverse impacts on ecosystem processes.

Anyone who knows the rather dire state of current marine biological diversity will guess that the logical conclusion of this exercise that links marine

Marine Biodiversity and Ecosystem Functioning. First Edition. Edited by Martin Solan, Rebecca J. Aspden, and David M. Paterson.
© Oxford University Press 2012. Published 2012 by Oxford University Press.

biodiversity to marine ecosystem functioning will be that biotic impoverishment of marine ecosystems is likely to dramatically alter their functioning. This is not a new message, but it will reinforce current efforts and support acceleration of marine BEF research so that we can predict more precisely what will happen. Such a conclusion can only suggest a conservative, precautionary approach to the conservation of marine biodiversity, but this hardly helps those on the front lines of managing and preserving marine ecosystems who need more specific advice—which species should be saved, how many, will protected areas work, how should fisheries policy be informed? For the time being, in the absence of further research, one can be armed with the message that nothing in marine science, conservation, or management should be done without the inclusion of marine biodiversity.

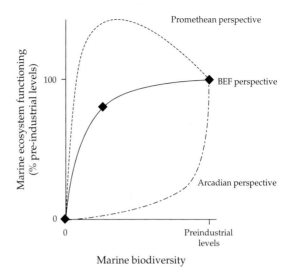

Figure 4.1 The relationship between marine ecosystem functioning and marine biodiversity as a function of pre-industrial levels.

4.1.2   A three-point framework for marine biodiversity

Consider three possibilities for marine biodiversity and its relationship to marine ecosystem functioning; (1) the complete extinction of every marine species, (2) the extinction of most marine species to a point where further loss leads to an acceleration in change in ecosystem functioning—i.e. an inflection point, and (3) restoration of marine biodiversity to its pre-industrial levels. These three possibilities describe a trajectory framed by three points in bivariate space whose end points are known, but whose mid-point (the inflection) is anyone's guess. These three points can describe many possible trajectories, the simplest being a saturating curve in which, after the inflection point, additional biodiversity has diminishing influence on ecosystem functioning. Rather than considering all possible relationships, of interest would be what the upper and lower boundaries for possible trajectories might be. An upper boundary might be one in which changes in biodiversity can actually lead to higher levels of ecosystem functioning than ever witnessed in nature—I refer to this as the Promethean trajectory for reasons explained below. A lower boundary might be where the loss of even a small amount of biodiversity causes a bottoming-out of ecosystem

functioning—I refer to this as the Arcadian trajectory, also explained below. This three-point framework for marine BEF, with these three trajectories (saturating, Promethean, and Arcadian) is illustrated in Figure 4.1.

It is important at this juncture to note that *biodiversity* in the context of ecosystem functioning refers more to the functional rather than the taxonomic component of biodiversity. Taxonomic diversity, however, is better known and functional diversity is, at least at crude levels, generally positively associated with taxonomic diversity. There are numerous issues surrounding the definition and quantification of functional diversity, but the field is advancing rapidly (for a review, see Petchey and Gaston 2006; Schleuter *et al.* 2010).

Given the diversity of studies in BEF literature, the compelling logic of its key mechanisms—i.e. niche complementarity and selection effects—scientific consensus (Hooper *et al.* 2005), its centrality to the Millennium Ecosystem Assessment Framework (MEA 2005), and its congruence with existing BRF marine studies—e.g. Solan *et al.* 2006; Worm *et al.* 2006—in the sense that it is well established that biodiversity influences marine function, I would argue that a rapid saturation of ecosystem functioning with biodiversity is the most conservative relationship to employ until further research can provide greater insights. By rapidly saturating, I mean the inflection occurs with just a fraction of extant biodiversity present, as illustrated in Figure 4.1.

This three-point framework needs two other features, as described above, to make it useful: an upper and lower boundary. Borrowing from M. W. Lewis' *Green Delusions* (1992), I will refer to the upper boundary as the Promethean trajectory, which is founded on the belief that human ingenuity can make marine ecosystems function at significantly higher levels than they currently do using engineering and management alone, much in the same the way that high-input, intensely managed monoculture crops in North American Midwestern croplands exceed the production of the prairies they replaced. Aquaculture, for example, is frequently founded on the idea that one can make marine ecosystems more productive than they are naturally by a series of interventions and management strategies. Farmed Atlantic salmon packed in Pacific aquaculture corrals, shrimp concentrated in aquaculture ponds, or tuna culled from the ocean and fed in cages (tuna ranching), can lead to enormous production per unit volume of sea water, thus one essentially shifts the inflection point to the upper left. Aquaculture, however, is heavily subsidized by a variety of inputs, especially feed that includes fishmeal (Naylor *et al.* 1998; Deutsch *et al.* 2007) and antibiotics, thus aquaculture ecosystem function is not likely to actually be above natural levels. Without subsidization, such high levels of fish production are not likely to be attained. I would argue that there is little evidence in support of the Promethean trajectory once one corrects for economic and ecological subsidies.

Of course production, which is frequently used as an ecosystem function in terrestrial BEF research that supports a Promethean perspective, is actually a complex function, the end point of many processes such as photosynthesis, respiration, growth rates of autotrophs, heterotrophic consumption, biotic interactions within and among autotrophs and heterotrophs—e.g. competition, facilitation, predation, or parasitism among species—and more. Terrestrial BEF research can be justifiably criticized for using primary production, or associated ecosys-

tem services such as food production, as an ecosystem function rather than more specific functions such as carbon sequestration, nitrogen mineralization, or community respiration. The Promethean perspective, however, would still argue that we could raise the magnitude of ecosystem functioning and the services derived from them, whether precisely or broadly defined, above what one sees in nature, by selecting certain combinations of species that collectively function better than natural combinations. For example, carbon sequestration might be lower in a natural community in which algal species compete for limiting mineral resources, such as iron, than in an artificial one created by selectively eliminating competitive species with low photosynthetic rates or increasing iron availability, acknowledging that there is some question about the likelihood of iron enrichment working, e.g. Hassler *et al.* (2011).

The lower boundary is what I call the Arcadian trajectory, which is based on the idea that marine ecosystems can provide everything humanity needs if it remains in its pristine state. While it is expected that species will vary in their ecological importance (Power *et al.* 1996), meaning that their contributions to ecosystem function will vary from large—e.g. a keystone species—to small—e.g. a rare and redundant species—loss of keystone species, almost by definition, would lead to dramatic declines in ecosystem functioning. Thus unless the majority of species have strong influences over ecosystem function, the Arcadian trajectory is unlikely, representing an extreme that is the roughly the inverse of the Promethean trajectory.

Together, the Promethean and Arcadian trajectory bound most of the bivariate space with the true marine BEF relationship somewhere in middle (Figure 4.1). What this framework is intended to convey is a large realm of possibilities, but also to draw attention to the most critical feature of marine BEF—the location of the inflection point. That is, the key question concerning the significance of marine biological diversity is:

Where's the marine BEF inflection point?
Arcadians see the point residing quite near the maximum level of biodiversity, while Prometheans see

it quite close to the minimum levels and above that of natural systems due to human ingenuity, ecological engineering, and aquaculture. In the absence of environmental ideologies such as Promethean and Arcadian environmentalism, ecologists imagine that marine systems are relatively robust to losses in biodiversity until it crosses the inflection point—the central trajectory illustrated in Figure 4.1. In all cases, however, knowing where and what influences the location of the inflection point is critical.

I selected 'Promethean' and 'Arcadian' as labels for the trajectories because they describe the beliefs underlying subscription to one or the other view, beliefs that are sometimes a mix of science and philosophy. The Promethean trajectory describes what one would expect if one believed that niche complementarity was prevalent in marine systems or believed in man's dominion over nature. In contrast, the Arcadian trajectory describes what one would expect if one believed species were unique or singular so that the loss of any species negatively impacts ecosystem function, or if one believes that man's tampering with nature always has negative impacts. We could call the Promethean trajectory the 'Complementarity' trajectory and the Arcadian trajectory the 'Singularity' trajectory, but in the absence of irrefutable support for the universality of species complementarity or singularity in marine systems, more than likely marine policy informed by this framework's trajectories will be the result of a mix of scientific and philosophical beliefs. In the United States, for example, climate change deniers consist of those who find climate-change science inconclusive and those who believe it is a liberal conspiracy. Both beliefs affect US policy, which has real consequences for sea-level rise, polar bear extinction, and ocean acidification. Biodiversity loss, like climate change, is similarly a mix of scientific—e.g. BEF—and environmental—e.g. business trumps conservation—beliefs which affect policy—e.g. US reticence to ratify the Convention on Biological Diversity.

With this framework in mind, which combines scientific and philosophical perspectives, I will provide an overview of marine BEF followed by a re-evaluation of the marine BEF framework in light of the overview.

4.2   Marine biodiversity and ecosystem function

The question of the trajectory relating ecosystem functioning to changes in biological diversity first arose in the 1990s, and its literature is extensively reviewed in four volumes (Schulze and Mooney 1993; Kinzig *et al.* 2001; Loreau et al. 2002; Naeem *et al.* 2009), thus I will not review it in detail here. Rather, I will provide a brief summary of its central findings as they relate to marine systems.

4.2.1   Daunting scales

By terrestrial standards, it seems impossible to get any kind of sensible take on marine biodiversity and ecosystem functioning (Naeem 2006). That the oceans represent $361 \times 10^6 \ km^2$ and contain 1.4 billion km^3 of water is beyond easy comprehension for the terrestrial ecologist who typically works on a scale of $1 \ m^2$ on a largely visible surface limited mostly to two dimensions. Even if we allow for the height of terrestrial ecosystems to be the biomass found above the substrate, and give it a generous approximation of 100 m in height, Denny estimates the terrestrial portion of the biosphere is one hundredth that of the marine portion (Deny 2008, p. 41).

Daunting though these scales may be, we can gain some broad insights into the relationship between marine biodiversity and ecosystem functioning by considering its microbial and macrobial (metazoan/metaphytan) diversity, basic marine ecosystem functioning, and what current marine BEF research tells us.

4.2.2   Marine biodiversity

4.2.2.1   Marine microbes
Microbes rule. They are not only likely to be the most diverse organisms in marine ecosystems, but they also control the bulk of marine biogeochemistry (Fenchel *et al.* 1998; Arrigo 2005; Falkowski *et al.* 2008). Their taxonomic biodiversity, based on molecular methods, is proving to be vastly more diverse than anyone ever imagined (Irigoien *et al.* 2004, Venter *et al.* 2004 Hong *et al.* 2006; Sogin *et al.*

2006; Huber *et al.* 2007). They are also structured strongly by trophic interactions, being fed on by protistan (Azam *et al.* 1983; Stone 1990), viral (Suttle 2005), and zooplankton consumers.

In addition to being extraordinarily diverse and the dominant biological component of marine biogeochemistry, they also dominate in numbers and mass. Whitman *et al.* (1998) estimated that there are 1.2×10^{29} prokaryotic cells in the open ocean and surface sediments, and 3.8×10^{30} cells in the oceanic subsurface, which totals 305 Gt (gigaton, or petagram, which is 10^{15} g) of carbon out of roughly 500 Gt for the total prokaryotic biosphere.

In spite of their extraordinary taxonomic diversity, mass, and dominance over marine ecosystem function, concern over microbial biodiversity loss is rarely voiced, in part because there is the perception that they are everywhere and capable of persisting after all environmental shocks. The familiar hypothesis among microbial ecologists is that *everything is everywhere, the environment selects*—i.e. the Bass-Becking hypothesis, see Martiny *et al.* (2006)—based on the ideas that microbes can disperse everywhere, and once there, remain quiescent until conditions favor germination. The Bass-Becking hypothesis has led to the notion that both α microbial diversity (diversity in place) and β diversity (turnover as one moves from one point in space to another) are low. The idea also contributes to the notion that microbial communities can exhibit a sort of extreme biorobustness in the sense that whatever environmental change occurs, from oil spills to changes in temperature to changes in host communities, if there is any microbial species in the world that could prosper under those conditions, they would either be there and wake from their slumber, or arrive soon as immigrants from elsewhere. And even if microbial species well adapted to these conditions did not exist in the world, they would evolve quickly, through rapid evolution because they can mutate and reproduce quickly, and because they can laterally exchange genetic material—pass genes among species.

Molecular microbial ecology over the last couple of decades, however, has radically altered our view of the Bass-Becking hypothesis (Martiny *et al.* 2006; Patterson 2009). Contemporary investigations of

microbial communities are revealing that they have ecologies and biogeographical patterns not unlike their macrobial counterparts—e.g. Horner-Devine *et al.* 2004; Irigoien *et al.* 2004; Furhman *et al.* 2006; Green and Bohannan 2006; Martiny *et al.* 2006). Microbes do have remarkable dispersal abilities. For example, Hubert *et al.* (2009) found a constant influx of thermophilic bacteria into the Arctic where they clearly cannot grow. Another example is the study by Horner-Devine *et al.* (2004) on the species–area relationship of salt-marsh bacteria which shows one of the lowest z values for organisms—using the well-known formula, $S = cA^z$, where S is the number of species, c is a constant, A is area sampled, and z is a constant that best describes the shape of the species–area relationship. Compared to most other taxa, bacterial diversity quickly saturates as one expands one's sample area—compare examples of marine fish, marine invertebrates, and marine bacteria in Figure 4.2. Yet there are extraordinary numbers of rare (Sogin *et al.* 2006) and endemic (Martiny *et al.* 2006) species, which is counter to what one would expect if the Bass-Becking hypothesis was fully correct (Patterson 2009).

Microbes also dominate primary production in marine systems. Terrestrial systems are dominated by macrobial plants numbering over 300 000 species, but in marine systems, with the exception of kelp and seagrass beds, primary production is dominated by a far less diverse microbial phytoplankton assemblage. Although accounting for nearly half of global annual net primary production, marine phytoplankton have fewer than 5000 described species (Simon *et al.* 2009). Like other microbial taxa, however, that number is probably a tiny fraction of true microbial phytoplankton diversity. For now, until greater taxonomic research is done, microbial primary producer diversity, even including macrobial primary producer species (see section 4.2.2.2, below) is low compared to their terrestrial counterparts (Simon *et al.* 2009). Also, like heterotrophic prokaryotes, they seem widely dispersed, even though they may show familiar macrobial biogeographic patterns, such as latitudinal gradients, over time (Cermeño and Falkowski 2009), and over space (Simon *et al.* 2009). Given that BEF research has focused heavily on terrestrial plants

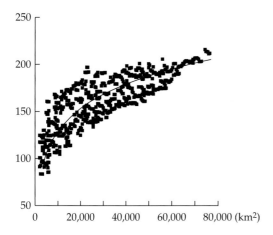

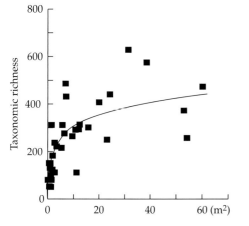

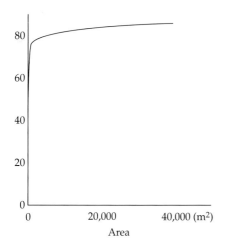

and primary production, how to relate this work to marine BEF remains unclear.

To summarize, marine microbial communities are different in many respects from macrobial communities, but they are similar in the most important ways, which means that basic BEF principles will apply. The small scale, evolutionary antiquity, mind-boggling magnitudes in numbers and mass, commonality of lateral gene transfer that is so rare in eukaryotes, extraordinary molecular diversity, dispersal rates unrivaled by the macrobial world, and lower taxonomic turnover—i.e. greater spatial homogeneity in diversity—when compared to macrobes, creates the sense that microbial communities are fundamentally different from macrobial communities. Marine microbial communities, however, are diverse, control biogeochemical or ecosystem processes, are structured by trophic and community interactions, and exhibit some ecological patterns that are reminiscent of their macrobial counterparts. It seems safe to say that the fundamental marine microbial BEF relationship is likely to be a saturating curve for primary producers and for heterotrophs, just as it is for macrobial taxa. As with macrobes, abiotic factors dominate microbial ecosystem function, but within a specific set of conditions, changes in microbial diversity will secondarily influence marine biogeochemistry.

4.2.2.2 Macrobial diversity
The lower macrobial diversity of marine ecosystems in comparison to terrestrial ecosystems means that changes in biodiversity may have stronger influences on ecosystem functioning. The World Registry of Marine Species (<http://www.marinespecies

Figure 4.2 Species area relations for marine species. Top: marine groundfish species from Cape Flattery, Washington to Point Conception, California, from trawl samples between 35–1200 m depth (for further detail, see Levin *et al.* 2009). Middle: marine soft sediment invertebrate species of the European continental shelf based on data from the MarBEF (EU Network of Excellence for Marine Biodiversity and Ecosystem Function) database (for further detail, see Renaud *et al.* 2009). Bottom: salt marsh bacterial taxonomic richness based on operational taxonomic units derived from analysis of 16s rDNA sequences, collected from Prudence Island, Rhode Island, USA (for further detail, see Horner-Devine *et al.* 2004). Note that unit for area varies among graphs, ranging from km² (top) to m² (middle and bottom).

.org>) currently lists over 250 000 animal species and over 11 000 plant species, which form the bulk of the nearly 286 000 described species. This pales in comparison to described terrestrial plant and animal species that number well over one million (MEA 2005). For example, macroalgal and plant species consist of ~ 50 species of seagrass (Duarte 2000), kelp and seaweeds comprise ~ 5000 species of Rhodophyta, ~ 1200 species of Phaeophyceae, and 1040 spp. Chlorophyta (Phillips 2001), yet add up to less than 3% of terrestrial plant diversity. Estimating actual species diversity—i.e. described and yet to be described—is, of course, notoriously difficult, but the prevailing view is that marine systems are less diverse than terrestrial ecosystems.

With respect to marine BEF, however, the magnitude of biodiversity is not as relevant as changes in biodiversity, especially functional diversity. Indeed, one could argue that lower taxonomic diversity in marine systems, if it translates into lower functional diversity, means that marine biotic impoverishment is of more concern than biotic impoverishment of terrestrial diversity. The less diversity an ecosystem has, the more valuable it is.

An extensive collaboration among marine biologists to share their taxonomic data (EU Network of Excellence for Marine Biodiversity and Ecosystem Function, or the MarBEF, database) has provided an invaluable picture of marine soft-sediment benthic macrofaunal biodiversity (Renaud *et al.* 2009). These fauna did not show the terrestrial latitudinal gradient of declining diversity as one moves north or south from the equator. There have been pole-ward decreases (though peaks may be mid latitude) detected in deep sea, estuarine tidal-flat, shallow subtidal hard substrate communities, and foraminifera, nematodes, gastropods, bivalves, and crustaceans, but Renaud *et al.* (2009) found little evidence for any such trends for the communities they analysed, which were based on nearly 2200 species.

Biomass of macrobes, on the other hand, may be higher than terrestrial systems. Both marine and terrestrial ecosystems exhibit about the same amount of global net primary productivity (~ 50 Gt C y^{-1}), but trophic transfer efficiencies are known to be greater in aquatic systems because primary producers are more readily consumed.

Focusing on vertebrates, for which there are better data and because they are large-bodied animals, the total biomass of teleost fish was estimated to be about 2 Gt C (see supplementary information in Wilson *et al.* 2009), all of which may be sustained by the consumption of less than 10% of marine primary production (Pauley and Christensen 1995). In contrast, terrestrial megafaunal mammalian biomass (which includes humans) is approximately only 1.45 Gt wet weight, which translates roughly to about 0.7 Gt, of which possibly two-thirds consists of livestock species (Barnosky 2008). Estimates of marine mammal biomass are on the order of 20–50 Mt (megaton, or teragram, which is 10^{12} g).

Trophic structure, as in terrestrial communities and marine microbial communities, is also extraordinarily important in marine macrobial communities. Paine (1966) set the stage with his experiments on marine intertidal invertebrates, demonstrating how single species could have strong influences on distribution and abundance as well as ecosystem function (Paine 2002). The change in North Atlantic ecosystems attributable to over-harvests of predatory fish—e.g. Frank *et al.* 2005; Frank *et al.* 2007—and large-scale changes in fish trophic structure (Pauly *et al.* 1998; Christensen *et al.* 2003; Pauly *et al.* 2003), is widely known.

In summary, although less diverse than terrestrial systems, macrobes play important roles in ecosystem functioning, and are structured by trophic interactions that influence ecosystem function, and species disperse widely, leading to weaker biogeographic patterns when compared to terrestrial systems.

4.2.3   Marine ecosystem functioning

As in the case of marine biodiversity, marine ecosystem functioning, restricted here to marine biogeochemical functioning, is a well-developed field that cannot be adequately reviewed in the limited space here, but drawing from general texts—e.g. Baskin 1997; Schlesinger 1997; Sarmiento and Gruber 2006—we can see that marine systems have smaller global fluxes than terrestrial systems, but have much larger pools and internal cycling. For example, over 90 Gt of carbon are annually cycled between the atmosphere and oceans, which is smaller than terrestrial fluxes (120 Gt y^{-1}), but the ocean carbon pool is an order of magnitude greater (38 000 Gt) than the terrestrial pool (2010 Gt), so potentially figures more prominently in the changing global carbon cycle than terrestrial systems. Similarly, annual biological fixation of atmospheric N by marine systems of 15 Mt is small compared to terrestrial systems (140 Mt), but within the oceans 8000 Mt of N are internally cycled every year, compared to only 1200 Mt in terrestrial systems. Thus changes in fluxes in marine systems can potentially lead to enormous change in atmospheric or hydrospheric geochemistry over long periods, even though current annual fluxes are smaller than those of terrestrial systems.

Global cycles miss the true complexity of marine biogeochemical cycles, of course. For example, carbon flow in marine phytoplankton includes the familiar oxygenic pathway, but anaerobic anoxygenic, aerobic anoxygenic, rhodopsin-based, and phytochrome-based pathways mean that both photo-autotrophy and photo-heterotrophy can occur, greatly complicating carbon cycling (Karl 2002). One might dismiss anoxygenic photo-heterotrophy as an exotic trophic group, but a recent study (Kolber *et al.* 2001) suggests that this group may represent 5–10% of chlorophyll found in the oceans. Another example of complexity concerns inverse food-webs, where heterotrophic microbial and zooplankton biomass is higher than producer biomass in coastal systems, but normal in open oceans (Gasol *et al.* 1997). As a final example of how complex things can get, in a recent study of calcium cycling by Wilson *et al.* (2009), calcium carbonate, a process regulated primarily by coccolithophores and foraminifera, which has been estimated to total to 0.7 to 1.4 Gt CaCO$_3$-C year^{-1}, is also affected by teleost fish that may contribute as much as 3–15% (0.04-0.11 Gt CaCO$_3$-C year^{-1}) of carbonate production.

Trophic structure has long been known to play a key role in marine biogeochemistry. Changes in North Atlantic ecosystems attributable to over-harvests of predatory fish are well known—e.g. Frank *et al.* 2005; Frank*et al.* 2007. Diversity and trophic structure interact with marine microbial

systems (Gamfeldt *et al.* 2005) as well as macrobial systems (Duffy *et al.* 2001; Duffy *et al.* 2005), and affect ecosystem function. Bottom-up regulation also occurs, as Ware and Thompson (2005) have shown for North American coastal fish, whose abundance appears to be regulated by phyto-plankton—zooplankton—fish linkages. And there is no question that changes in marine sediment macrofauna affect sediment biogeochemical processes (Waldbusser and Marinelli 2006). Marine biogeochemistry is clearly affected by vertebrate, invertebrate, and microbial diversity and trophic structure.

In combination, the fluxes, magnitudes of pools and internal cycling, complexity, and trophic structure, make predicting marine ecosystem function challenging. For example, Wohlers *et al.'s* (2009) mesocosm study, in which warming of marine phytoplankton to levels expected under current scenarios of global warming decreased the draw-down of surface inorganic carbon, and significantly increased accumulated dissolved organic carbon which could hypothetically reduce the ocean's biological carbon pump, and reduce carbon transfer to higher trophic levels. While missing biological diversity in their deliberations, this study nevertheless illustrates the importance of linking marine biogeochemical processes to trophic structure.

In summary, marine biogeochemical fluxes may be smaller than terrestrial systems, but given the large pools and the complexity and sensitivity of these fluxes to changes in climate and trophic structure, the influence of biodiversity over marine biogeochemical processes is likely to be as important as it is in terrestrial systems.

4.3   Marine biotic impoverishment

The primary motivation for BEF research at the beginning was the concern that terrestrial systems were changing at extraordinary rates, with the number of species per ecosystem, or unit area or volume, declining rapidly; i.e. the emphasis was on the loss of local diversity, not global diversity. The chief driver of local diversity loss in terrestrial systems continues to be changes in land use from natural or unmanaged systems, such as rainforests, grasslands, and deserts, to managed systems, such as farms, plantations, and pastures. Other drivers include pollution, over-harvesting, and habitat degradation, but with 40% of Earth currently made up of farms and pastures (Foley *et al.* 2005), land conversion has clearly been the dominant driver of biodiversity loss. Other drivers include elevated CO_2 and climate change, enhanced nitrogen deposition, and biotic exchange—i.e. the spread of exotics and invasive species, each of which has different effects on different ecosystems (Sala *et al.* 2000). Note that these drivers of terrestrial biodiversity loss are all anthropogenic.

The story for marine biodiversity loss is the same as it is for terrestrial systems, in that the drivers are also anthropogenic. Several high-profile papers have drawn attention to the extraordinary degradation of the marine biome—especially coral reef, fish, and mammalian diversity—at the hands of humans (Malakoff 1997; Pauly *et al.* 1998; Roberts and Hawkins 1999; Jackson *et al.* 2001; Christensen *et al.* 2003; Myers and Worm 2003; Pauly *et al.* 2003; Bellwood *et al.* 2004; Hutchings and Reynolds 2004; Worm *et al.* 2005; Sala and Knowlton 2006; Schipper *et al.* 2008). Coastal systems experience roughly 40% losses of species when polluted (Johnston and Roberts 2009). J. Jackson (2008) provides a detailed account of the extraordinary degradation of the marine biome at our hands, suggesting that it is akin to the threats tropical rainforests are experiencing, but without the press attention this attracts.

Given the staggering erosion of marine biodiversity, the inflection point in Figure 4.1 is likely to be near current biodiversity levels; that is, it seems likely that we are at the point where further biotic impoverishment will lead to accelerating declines in ecosystem functioning.

4.4 Marine BEF findings

BEF in marine research has been extensively reviewed in a series of eleven open-access papers in the journal *Marine Ecological Progress Series* (Solan *et al.* 2006). Marine BEF contributions—e.g. Duffy *et al.* 2001; Emmerson *et al.* 2001; Stachowicz *et al.* 2002; Solan *et al.* 2004) have played key roles in

the development of the field. In most cases, these studies have advanced basic BEF research as well as contributed more specifically to understanding marine BEF. More recent studies have examined decomposition and sediment biogeochemistry in intertidal mudflats—(Godbold *et al.* 2009),—the relationship between taxonomic and functional diversity in kelp-bed and fish communities (Micheli and Halpern 2005), and the importance of functional trait distributions (Hewitt*et al.* 2008; Bracken *et al.* 2008), performed what represents possibly a model for contemporary marine BEF research. They demonstrated that seaweed diversity influences N uptake, and built their case by using observational studies, experimental manipulations, and trait-based scenarios (Naeem 2008).

As the previous sections have all shown, trophic structure strongly influences marine BEF and the study by Duffy *et al.* (2005) provides some important insights into marine BEF. Using seagrass mesocosms, Duffy *et al.* found that food chain length and biodiversity interact; that is, the impacts on ecosystem functioning by one trophic level were affected by number of trophic levels as well as the diversity within each trophic level—in this study, there were three trophic levels; crabs as top predators, arthropod grazers, and producers consisting of eelgrass and algal species.

Though BEF has been dominated by terrestrial studies, enough marine studies have been conducted to allow for meta-analyses and syntheses done by Worm *et al.* (2006). These authors found that mesocosm studies, observational studies, and restoration studies, where recovery from biodiversity loss has occurred, support the general conclusion that biodiversity plays roles not unlike those documented for terrestrial systems (Worm *et al.* 2006); that is, on average, over space and time, the supply and stability of marine resources decline when biodiversity declines (Worm *et al.* 2006). Figure 4.3, based on data from Worm *et al.* (2006), shows two representative examples of production—in this case, secondary production in terms of fish harvested by fisheries in 64 large marine ecosystems—in relation to both the richness of fish taxa harvested—i.e. biodiversity—and the coefficient of variation, often considered a

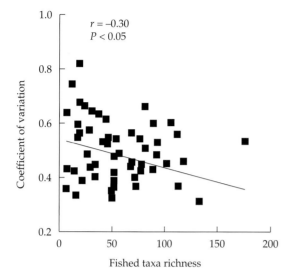

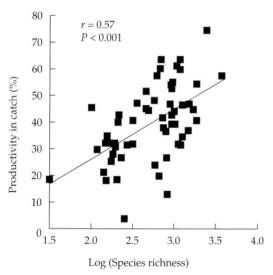

Figure 4.3   Biodiversity and ecosystem functioning in harvested marine fish species in large marine ecosystems. The top figure shows the relationship between the coefficient of variation in total catch for each ecosystem. Bottom figure shows average catch as percent of maximum catch in each ecosystem. Data from Worm *et al.* (2006).

measure of ecosystem stability. Note that the biodiversity x productivity plot is log-linear, which would yield a saturating curve if plotted as a linear-linear plot, the association most commonly found in terrestrial biodiversity and ecosystem functioning studies.

4.5   The fundamental marine BEF relationship in abstraction

4.5.1   Where's the inflection point?

I return now to the main point made by the framework for marine BEF (Figure 4.1)—where is the inflection point? From the preceding brief review, it is clear that marine organisms are diverse, trophically structured, disperse widely, and have strong controls over ecosystem processes—microbes directly as the engines of biogeochemistry and macrobes indirectly through biotic regulation. We also know that the scales are enormous, thus the answer to the question of where the inflection is located is very important, because even if biodiversity effects are locally small, the multipliers over the enormous scales add up to potentially large impacts on ecosystem functioning.

My goal here is to incorporate the basic trophic structure of marine systems into the basic BEF relationship of a saturating curve. There are two shortcomings to this approach which I want to highlight. The first shortcoming of applying the basic saturating BEF curve is, as mentioned at the outset, that the bulk of the work has been based on terrestrial and freshwater systems. Another shortcoming is that theory behind this curve is based on mechanisms that have never been truly tested. Experimental tests of BEF have been predominantly confirmatory, simply showing that loss of biodiversity impacts ecosystem functioning. Theory, however, is based on explicit mechanisms. For example, Tilman and Lehman's (Tilman *et al.* 1997) theory is based on competition for limiting resources, Loreau's theory (1998) is based on local resource depletion zones, neither of which has actually been tested. Other approaches include niche-packing models based on Lotka-Volterra approaches (2009), Michaelis-Menten kinetics (Naeem 2002), and metacommunity dynamics (Gonzalez *et al.* 2009), that similarly lack direct empirical confirmation. The assumption has always been that because theoretical predictions approximate the saturating curves observed in many studies, they are likely candidates for explaining the cause. Of course, this situation is generally true of theory and empiricism in ecology in general; mechanistic theory is often supported by experimental and observational studies whose findings match the predictions. These shortcomings aside, the prevailing theoretical BEF relationship that is consistent with empirical findings is a simple saturating function for ecosystem function as a dependent variable, but since the mechanism behind this curve remains open, we can use any saturating function for the purposes of this exercise.

4.5.2   The BEF curve for marine systems

BEF relationships, based on 15 years of research and some 900 studies, have been either linear, decelerating—i.e. asymptotic, saturating, log-linear, meaning that each species contributes less to overall functioning of the ecosystem—or something else. In the most recent meta-analysis, Schmid et al. (2009) found that that the average R^2 for studies that fitted a log-linear curve to the relationship between biodiversity and ecosystem functioning (25 studies) was 0.68, while the average R^2 for studies that fitted some form of a saturating curve was 0.69 (49 studies). Linear studies were rare, though idiosyncratic relationships—i.e. not linear or decelerating—were not uncommon. Theory (e.g. Tilman et al. 1997; Loreau 1998; Naeem 2002; Cardinale et al. 2009; Gonzalez et al. 2009) similarly supports the saturating function, though theory has been influenced by empirical work since it came first. There is always the danger that theory, which is infinitely malleable to our machinations, can take its shape based on what we observe, which then questions if one confirms the other or if both are part of some tautological exercise. Again, however, this is a general issue that all of ecology faces.

Given that theory and empiricism provide a robust argument for the BEF relationship to be some form of a simple saturating function, we can borrow directly from the simplest of current models proposed, a phenomenological Michaelis-Menten formula for enzyme kinetics, where the rate of an ecosystem function (F), in relation to biodiversity (D) is:

$$F(D) = \frac{F_{max}D^c}{D_{min} + D^c} \quad \text{(Equation 4.1),}$$

where c is a constant, and $c \geq 1$ and governs the shape of the saturating curve, D_{min} is the minimum

diversity needed to sustain ecosystem functioning, and F_{max} is the saturation or maximum value for ecosystem functioning. When $c = 1$, species contributions to ecosystem functioning are additive, and when species interact with one another, their impacts are multiplicative, leading to a more rapid increase in functioning with each additional species until saturation. I will not develop this approach further here, but any theory, as those cited above, can be used in this exercise. Compared to other formulae, however, the Michaaleis-Menten formula is appealing because it sees marine biodiversity as a sort of biogeochemical catalyst where some biodiversity immediately increases ecosystem function, but this positive influence diminishes as more and more species are added.

To complete the exercise, we need a model of marine ecosystem function. Since fisheries production is one of the ecosystem functions that is related to a major ecosystem service of much current interest, and one that I have argued here fits the BEF pattern—e.g. Figure 4.2—I borrow directly from Duffy and Stachowicz's (2006) graphic model of fish production—i.e. their Figure 4.1. Their model sees the biomass of higher trophic-level fish (F_F), the target species of marine fisheries, as that portion of eukaryotic consumer's biomass (heterotrophic protists, zooplankton, and small fish, or F_E) that was consumed by the fisheries species. F_E, in turn, is that fraction of primary producer mass (F_P), and that fraction of heterotrophic prokaryotic biomass (F_H) consumed and converted by F_E. Finally, both F_P and F_H are a function of the dissolved organic matter they consume, but we can leave this abiotic component out of our deliberations for now. This static version of Duffy and Stachowicz's dynamic model can be written as:

$$F = m_V F_F m_F F_E m_E (F_H m_H + F_P) \quad \text{(Equation 4.2).}$$

That is, the total biomass of fish that could be taken by all fisheries (F) is equal to the proportion (m_V) of biomass of all fish species the fisheries (F_F) consume, which is a proportion (m_F) of the total biomass of eukaryotic species (F_E) consumed, which is a proportion (m_E) of the total biomass of biomass ($F_H + F_P$) derived from DOM. Because heterotrophic prokaryotes (i.e. F_H) consume DOM, they are multiplied by a conversion factor as well, but primary producers (i.e. F_p), which do not consume DOM, have no conversion factors. Note that the conversion factors (m) are proportions, so they can range in value from 0 to 1.

Substituting the simple Michaelis-Menten formula (Equation 4.1) for each instance where diversity of harvested fish, eukaryotic prey, primary producers, and heterotrophic prokaryotes are controlled by biodiversity of each respective group, we would get,

$$F(D) = m_V \left(\frac{F_{F\max} D_F^C}{D_{F\min} + D_F^C} \right) m_F \left(\frac{F_{E\max} D_E^C}{D_{E\min} + D_E^C} \right) m_E$$
$$\left[m_H \left(\frac{F_{H\max} D_H^C}{D_{H\min} + D_H^C} \right) + \left(\frac{F_{p\max} D_P^C}{D_{P\min} + D_P^C} \right) \right]$$

(Equation 4.3).

What this formula does for us is decompose the biomass-associated ecosystem function of interest, in this case the standing stock of marine fish that can be harvested by fisheries, into sub biomass-associated ecosystem functions (each term in parentheses) that each represent a key trophic component (F, E, H, and P). By simply making each trophic component a saturating function of biodiversity, where D is biodiversity as in Equation 4.1, this formula now links all marine biodiversity to an important ecosystem function.

A more generic form provides a simpler view, though it requires several significant simplifying assumptions. First, sub ecosystem functions or trophic levels that are added are combined to form a single sub function. For example, primary producers and heterotrophic prokaryotes which are consumed by heterotrophic eukaryotes are treated as one level that is the sum of both. Second, all biomass consumption factors are considered equivalent. Third, biomass and biodiversity within each trophic level is scaled as the proportion of the total within each trophic level. This means that all maxima for biomass in each trophic level are equal to 1 and D ranges from 0 to 1 within each trophic level. Fourth, the minimum biodiversity needed in each trophic level is a small proportion of the total in each trophic level and roughly equivalent among trophic levels. With this heavy dose of assumptions, we would get,

$$F(D_T) = \left(\frac{mD^c}{D_{min} + D^c} \right)^T \qquad \text{(Equation 4.4)},$$

where T is the number of trophic levels, m is the universal conversion factor—note that we have considered the proportion of fish human fisheries take as equivalent to all other inter-trophic conversions.

What is gained by this exercise is a simple way to show how biodiversity can inform us about the quantity of fish we are likely to be able to harvest from marine ecosystems in the face of declining biodiversity. The formula acknowledges that the biomass we harvest in fisheries is a small fraction of the total biomass of the ecosystem. What controls where the inflection occurs in reference to ecosystem functioning is how interactive species are (determined by the value of c, where values much less than one mean species hardly impact each other's influence on ecosystem services, and values much greater than one mean species influence each other strongly). What controls where the inflection occurs in reference to biodiversity is the number of trophic levels.

Leaving out the abiotic influences does not diminish their importance. Indeed, in marine as in terrestrial systems, abiotic factors, such as climate and geochemistry, play the dominant role in determining the magnitude and rates of ecosystem processes. The BEF approach, however, concerns changes within a defined set of abiotic conditions, so we can treat them as controlled factors for a within-ecosystem model. For example, the location and potential productivity of a coral reef is primarily determined by abiotic factors, including climate, geography, geology, and water chemistry, but changes in the composition of coral, other invertebrate, fish, and alagal species, will secondarily affect the production of coral reefs. Ocean acidification will dominate processes that affect coral reef function, but the crown of thorn starfish (*Acanthaster planci*) can have strong impacts too, and the loss of corals would have a cascading effect on the ecosystem, which would change production as well. Marine or terrestrial, BEF largely concerns within-ecosystem biotic impacts that are secondary to abiotic factors.

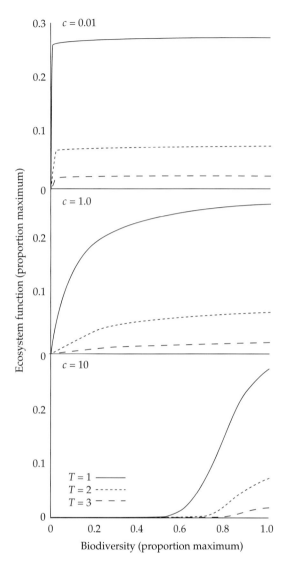

Figure 4.4   The shape of the marine BEF curve in relation to trophic complexity (*T*) and species interactions (*c*).

4.6   Synthesis

4.6.1   A simple but telling marine BEF framework

The significance of marine biological diversity can be measured in many ways, but I have focused on marine biodiversity's influence over biogeochemical or ecosystem functioning, inspired by the recent and increasingly robust science of BEF that links

biodiversity to human well-being. Borrowing from terrestrial BEF, which is consistent with current marine BEF research, I suggest a simple three-point marine BEF framework. The middle point is the inflection point of a saturating function that relates marine biodiversity and ecosystem functioning. A major effort in marine biodiversity research should be launched to locate the inflection point, given that the science is incomplete and opinions range between Promethean and Arcadian views. If the inflection is near current levels of biodiversity, given the dramatic impoverishment climate change, pollution, over-harvesting, disease, and other anthropogenic drivers have created, then the Promethean perspective is less persuasive. If the inflection is far to the left of current levels of marine biodiversity, then the Arcadian perspective is less persuasive. The future of marine diversity, ecosystem functioning, and global biogeochemical processes that affect land as much as the sea, will rest heavily on the inflection point.

When one examines marine microbial and macrobial biodiversity, marine biogeochemistry, and how the three are linked, several observations emerge:

1. The marine microbial world is extraordinarily important in terms of diversity, mass, and influence over biogeochemistry.
2. The diversity of the marine macrobial world is low in comparison to terrestrial systems.
3. Strong trophic structuring is omnipresent.
4. The vast scales of the ocean mean that the densities of life are low, thus biodiversity's influence locally might be small, but scaled up to the appropriate level, their impact could be substantial.
5. Marine biotic impoverishment is at an all-time high, possibly worse so than for terrestrial systems.

These observations suggest a modified marine BEF framework. Starting with the saturating function at the heart of the basic framework shown in Figure 4.1, one has to first select an appropriate function. Several candidates exist, but I selected the simplest, phenomenological Michaelis-Menten function in which biodiversity catalyzes biogeochemical function. Then, using a simple multi-trophic structure provided by Duffy and Stachowicz (2006), the saturating function was applied to biodiversity in every trophic compartment. With a hefty dose of simplifying assumptions, one arrives at Equation 4.4, which shows how trophic structure, energy transfer, and biotic interactions determine the location of the inflection point.

The modified framework suggests an obvious solution of including biodiversity in marine ecosystem modeling even when information about marine biodiversity and marine ecosystem functioning remains poor. The important terms are the biodiversity minimum for an ecosystem to function at some level, the biodiversity maximum (I suggest pre-industrial levels), how many trophic levels are involved (don't forget the microbes, including viruses), how interactive species are with one another, and what the mass conversion ratios are from one trophic level to the next (often presumed to be 10% at high levels and 30% at lower levels). These terms in the framework point to areas where marine BEF research should focus.

The main point that emerges from the simple heuristic exercise is that the inflection point is pulled in multiple directions by changes in biodiversity.

4.6.2   Remember the humongous multipliers

Given the magnitude of marine systems and the lower levels of macrobial biodiversity (point 4 above), small changes in biodiversity will have impacts that are so dilute as to be difficult to detect at a global level. A farmer might immediately recognize the benefits of diversity, a nation might pick up the benefits in carbon credits, but the potentially tiny influences phytoplankton diversity will have in the upper few meters of the water column will not be missed by any investigation of conventional size. The multipliers, however, are enormous.

Do not be fooled by 'small' marine BEF effects. Too often, the terrestrial BEF literature fails to point to the significance of its multipliers, leading to mistaken impressions that biodiversity's influence on ecosystem functioning is not significant for conservation biology (Schwartz *et al.* 2000; Srivastava and Vellend 2005), in spite of the prevailing paradigm of conservation biology being 'act locally, but think globally.' Multipliers matter. Small changes

observed in model native grasslands of the US—e.g. Tilman *et al.* 1996; Knops *et al.* 1999; Reich *et al.* 2001; Tilman *et al.* 2002—when multiplied by the extent to which land has been transformed, with less than 1% of the American prairie left, translates into an extraordinarily massive change. Extant biodiversity in the former American prairie is now represented by a handful of domestic plant species, like corn, wheat, soybean, cattle, hogs, and chickens. It would be a pity to make the same mistake for the marine biome where the overwhelming evidence is that marine biodiversity is collapsing everywhere, and that even if the influence of biodiversity per unit area or unit volume seems small, the multipliers are huge.

4.6.3   Future directions

The analyses presented here indicate a number of research directions for marine BEF research (which is guided by the framework developed in Equation 4.4), some of which have independently been anticipated and are indeed currently being promoted to marine research and policy agencies, and which are brought together here as a synthesized and prioritized list:

1. Use saturating functions to include biodiversity in marine biogeochemical models, including coupled biosphere–geosphere climate models. Base scenarios on where the inflection point is likely to be given current marine BEF findings.
2. Build a universally accessible database of marine biodiversity that includes taxonomic identity, phylogenetic position, population densities, geographic location, and basic traits—e.g. body size, metabolic rate, trophic position, growth rates. These data provide the means for estimating the basic BEF curves for marine ecosystems—examples, see Solan *et al.* 2004; Bunker *et al.* 2005; McIntyre *et al.* 2007; Bracken *et al.* 2008. Without this resource, marine BEF will continue to be severely hindered in providing insight and precise forecasts demanded by policy makers and managers.
3. Integrate microbial and macrobial communities in BEF models. The dominant biomass and diversity is in the microbes while the dominant

trophic regulatory elements are among the macrobes—it does not make sense to conduct research on macrobes in the absence of microbial studies and vice versa.

4. Design marine protected areas to preserve marine BEF, not diversity or function separately. Protected areas should not just be storage houses for diversity, but back-up and safety systems for marine biogeochemical functioning, otherwise marine environmental management will fail to serve its full purpose in securing human well-being.

5. Determine how anthropogenic drivers of biodiversity change are impacting microbial biodiversity.

Each of the five suggested future directions for research should serve not just the scientific community, but provide effective scientific foundations for ecosystem-based management. Ecosystem-based management attempts to strike a balance between opposing environmental and socioeconomic objectives—e.g. maximize harvest for fisheries to sustain local community livelihoods while preserving biological diversity and ecosystem function—or seek optima in situations where multiple tradeoffs among objectives exist (Levin *et al.* 2009). A marine BEF framework is the best way to achieve ecosystem-based management. For example, marine protected areas can maintain biodiversity and ecosystem functioning within them, but can only partially restore fish yields after fishery collapse and are unlikely to boost yields. These are what local communities, whose populations are growing and income is fishery-based, expect of such areas (Levin *et al.* 2009). The only way to insure that a marine protected area is doing its job is to be certain that it is large enough and species-rich enough to provide the necessary ecosystem functioning local communities expect. To succeed requires information on local microbial and macrobial taxonomic and functional diversity, measures of ecosystem services, and an understanding of how anthropogenic drivers may impact the system.

4.7 Conclusions

Marine BEF research represents one of the biggest gaps in contemporary ecology, though some important and high-profile studies have begun and show the way forward. The significance of marine biological diversity certainly goes beyond its influence on marine biogeochemistry. Not only are terrestrial systems intimately linked through global biogeochemical processes and climate, but people depend strongly on the seas for food, recreation, and many cultural values that are more the domain of social science, arts, and the humanities. The challenge is that there is a long way to go before we have the necessary taxonomic and functional data, biodiversity-explicit models, and empirical validation of an abundance of BEF theory to provide policy makers and environmental managers with what they need to do their important work. The good news is that the drivers are all anthropogenic, and with worldwide calls and commitments to shifting from traditional, exploitative development economics to sustainable development, we can solve all the problems before us. Truth lies somewhere between the Promethean and Arcadian perspectives, and an all-out expansion of marine BEF research in the coming decades will find that truth.

References

Arrigo, K. R. (2005) Marine microorganisms and global nutrient cycles. *Nature* **437**:349–55.

Azam, F., Fenchel, T., Field, J. G. *et al.* (1983) The ecological role of water-column microbes in the sea. *Marine Ecology Progress Series* **10**:257–63.

Barnosky, A. D. (2008) Megafauna biomass tradeoff as a driver of Quarternary and future extinctions. In Avise, J. C., Hubbell, S. P. and Ayala, F. J. (eds), *In Light of Evolution*. National Academies Press Washington, DC, 227–41.

Baskin, Y. (1997) *The Work of Nature: How the Diversity of Life Sustains Us*. Island Press, Washington DC.

Bellwood, D. R., Hughes, T. P., Folke, C. *et al.* (2004) Confronting the coral reef crisis. *Nature* **429**:827–33.

Bracken, M. E., Friberg, S. E., Gonzales-Dorantes, C. A. *et al.* (2008) Functional consequences of realistic biodiversity changes in a marine ecosystem. *Proceedings of the National Academy of Sciences* **105**:924–8.

Bunker, D. E., DeClerck, F. Bradford, J. C. *et al.* (2005) Species Loss and Aboveground Carbon Storage in a Tropical Forest. *Science* **310**:1029–31.

Cardinale, B. J., Duffy, J. E., Srivastava, D. S. *et al.* (2009) Towards a food web perspective on biodiversity and ecosystem functioning. In: Naeem, S., Bunker, D. E., Hector, A. *et al.* (eds), *Biodiversity, Ecosystem Functioning, and Human Well-being: An Ecological and Economic Perspective.* Oxford University Press, Oxford, 105–20.

Cermeño, P. and Falkowski, P. G. (2009) Controls on Diatom Biogeography in the Ocean. *Science* **325**:1539–41.

Christensen, V., Guenette, S., Heymans, J. J. *et al.* (2003) Hundred-year decline of North Atlantic predatory fishes. *Fish and Fisheries* **4**:1–24.

Deny, M. (2008) How the Oceans Work. Page 320. Priceton University Press, Princeton.

Deutsch, L., Gräslund, S., Folke, C. et al. (2007) Feeding aquaculture growth through globalization: Exploitation of marine ecosystems for fishmeal. Global Environmental Change 17:238–49.

Duarte, C. M. (2000) Marine biodiversity and ecosystem services: an elusive link. *Journal of Experimental Marine Biology and Ecology* **250**:117–31.

Duffy, J. E., McDonald, S. K., Rhode, J. M. *et al.* (2001) Grazer diversity, functional redundancy, and productivity in seagrass beds: An experimental test. *Ecology* **82**:2417–34.

Duffy, J. E., Richardson, J. P. and France, K. E. (2005) Ecosystem consequences of diversity depend on food chain length in estuarine vegetation. *Ecology Letters* **8**:301–9.

Duffy, J. E. and Stachowicz, J. J. (2006) Why biodiversity is important to oceanography: potential roles of genetic, species, and trophic diversity in pelagic ecosystem processes. *Marine Ecology Progress Series* **311**:179–89.

Emmerson, M. C., Solan, M., Emes, C. *et al.* (2001) Consistent patterns and the idiosyncratic effects of biodiversity in marine ecosystems. *Nature* **411**:73–7.

Falkowski, P. G., Fenchel, T. and Delong, E. F. (2008) The Microbial Engines That Drive Earth's Biogeochemical Cycles. *Science* **320**:1034–9.

Fenchel, T., G. King, M. and Blackburn, T. H. (1998) Bacterial biogeochemistry: the ecophysiology of mineral cycling. Academic Press, San Diego.

Foley, J. A., DeFries, R., Asner, G. P. *et al.* (2005) Global Consequences of Land Use. *Science* **309**:570–4.

Frank, K. T., Petrie, B., Choi, J. S. *et al.* (2005) Trophic Cascades in a Formerly Cod-Dominated Ecosystem. *Science* **308**:1621–3.

Frank, K. T., Petrie, B. and Shackell, N. L. (2007) The ups and downs of trophic control in continental shelf ecosystems. *Trends in Ecology and Evolution* **22**:236–42.

Furhman, J. A., Hewson, I., Schwalbach, M. S. *et al.* (2006) Annually reoccurring bacterial communities are predictable from ocean conditions. *Proceedings of the National Academy of Sciences* **103**:13104–9.

Gamfeldt, L., Hillebrand, H. and Jonsson, P. R. (2005) Species richness changes across two trophic levels simultaneously affect prey and consumer biomass. *Ecology Letters* **8**:696–703.

Gasol, J. M., d. Giorgio, P. A. and Duarte, C. M. (1997) Biomass Distribution in Marine Planktonic Communities. *Limnology and Oceanography* **42**:1353–63.

Godbold, J. A., Solan, M. and Killham, K. (2009) Consumer and resource diversity effects on marine macroalgal decomposition. *Oikos* **118**:77–86.

Gonzalez, A., N. Mouquet, and Loreau, M. (2009) Biodiversity as spatial insurance: the effects of habitat fragmentation and dispersal on ecosystem functioning. In: *Biodiversity, Ecosystem Functioning, and Human Well-being: An Ecological and Economic Perspective.* Naeem, S. Bunker, D. E., Hector, A. *et al.* (eds), Oxford University Press, Oxford, 134–46.

Green, J. and Bohannan, B. J. M. (2006) Spatial scaling of microbial biodiversity. *Trends in Ecology and Evolution* **21**:501–7.

Hassler, C. S., Schoemann, V., Nichols, C. M. et al. (2011) Saccharides enhance iron bioavailability to Southern Ocean phytoplankton. *Proceedings of the National Academy of Sciences* **108**:1076–81.

Hewitt, J. E., Thrush, S. F. and Dayton, P. D. (2008) Habitat variation, species diversity and ecological functioning in a marine system. *Journal of Experimental Marine Biology and Ecology* **366**:116–22.

Hong, S.-H., Bunge, J., Jeon, S.-O. *et al.* (2006) Predicting microbial species richness. Proceedings of the National Academy of Sciences **103**:117–22.

Hooper, D. U., J. J. Ewel, J. J., A. Hector, A. *et al.* (2005) Effects of biodiversity on ecosystem functioning: a consensus of current knowledge and needs for future research. Ecological Monographs **75**:3–35.

Horner-Devine, M. C., Lage, M., Hughes, J. B. *et al.* (2004) A taxa-area relationship for bacteria. *Nature* **432**:750–3.

Huber, J. A., Mark Welch, D. B., Morrison, H. G. *et al.* (2007) Microbial Population Structures in the Deep Marine Biosphere. *Science* **318**:97–100.

Hubert, C., Loy, A. Nickel, M. *et al.* (2009) A Constant Flux of Diverse Thermophilic Bacteria into the Cold Arctic Seabed. *Science* **325**:1541–4.

Hutchings, J. A. and Reynolds, J. D. (2004) Marine Fish Population Collapses: Consequences for Recovery and Extinction Risk. *BioScience* **54**:297–309.

Irigoien, X., J. Huisman, and R. P. Harris. (2004) Global biodiversity patterns of marine phytoplankton and zooplankton. *Nature* **429**:863–7.

Jackson, J. B. C. (2008) Ecological extinction and evolution in the brave new ocean. In: *In the Light of*

Evolution. Avise, J. C., Hubbell, S. P. and Ayala, F. J. (eds), National Academy of Sciences Press, Washington DC, 414.

Jackson, J. B. C., Kirby, M. X., Bergre, W. H. *et al.* (2001) Historical overfishing and the recent collapse of coastal ecosystems. *Science* **293**:629–38.

Johnston, E. L. and Roberts, D. A. (2009) Contaminants reduce the richness and evenness of marine communities: A review and meta-analysis. *Environmental Pollution* **157**:1745–52.

Karl, D. M. (2002) Microbiological oceanography: Hidden in a sea of microbes. *Nature* **415**:590–1.

Kinzig, A., Pacala, S. W. and Tilman, D. (eds) (2001) *The functional consequences of biodiversity.* Princeton University Press, Princeton, New Jersey.

Knops, J. M. H., Tilman, D. Haddad, N. M. *et al.* (1999) Effects of plant species richness on invasion dynamics, disease outbreaks, insects abundances and diversity. *Ecology Letters* **2**:286–93.

Kolber, Z. S., Plumley, F. G., Lang, A. S. *et al.* (2001) Contribution of aerobic photoheterotrophic bacteria to the carbon cycle in the ocean. *Science* **292**:2492–5.

Levin, P. S., Kaplan, I., Grober-Dunsmore, R. *et al.* (2009) A framework for assessing the biodiversity and fishery aspects of marine reserves. *Journal of Applied Ecology* **46**:735–42.

Lewis, M. W. (1992) *Green delusions: an environmentalist critique of radical environmentalism.* Duke University Press, Durham.

Loreau, M. (1998) Biodiversity and ecosystem functioning: a mechanistic model. *Proceedings of the National Academy of Sciences* **95**:5632–6.

Loreau, M., Naeem, S. and P. Inchausti, P. (eds) (2002) *Biodiversity and Ecosystem Functioning: Synthesis and Perspectives.* Oxford University Press, Oxford.

Malakoff, D. 1997. Extinction on the high seas. Science **277**:486–8.

Martiny, J. B. H., Bohannan, B. J. M. Brown, J. H. *et al.* (2006) Microbial biogeography: putting microorganisms on the map. *Nature Review Microbiology* **4**:102–12.

McIntyre, P. B., Jones, L. E., Flecker, A. S. *et al.* (2007) Fish extinctions alter nutrient recycling in tropical freshwaters. *PNAS* **104**:4461–6.

MEA, M. E. A. (2005) Ecosystems and Human Well-Being: A Framework for Assessment. Island Press, Washington DC.

MEA, M. E. A. (2005) Ecosystems and Human Well-Being: Current State and Trends: Findings of the Condition and Trends Working Group (Millennium Ecosystem Assessment Series). Island Press, Washington DC.

Micheli, F. and Halpern, B. S. (2005) Low functional redundancy in coastal marine assemblages. *Ecology Letters* **8**:391–400.

Myers, R. A. and Worm, B. (2003) Rapid worldwide depletion of predatory fish communities. Nature **423**:280–3.

Naeem, S. (2002) Ecosystem consequences of biodiversity loss: The evolution of a paradigm. Ecology **83**:1537–52.

Naeem, S. (2006) Expanding scales in biodiversity-based research: challenges and solutions for marine systems. *Marine Ecology Progress Series* **311**:273–83.

Naeem, S. (2008) Advancing realism in biodiversity research. *Trends in Ecology and Evolution* **23**:414–16.

Naeem, S., Bunker, D. E., Hector, A. *et al.* (eds), (2009) *Biodiversity, ecosystem functioning, and human wellbeing: an ecological and economic perspective.* Oxford University Press, Oxford.

Naylor, R. L., Goldburg, R. J. Mooney, H. *et al.* (1998) Nature's subsidies to shrimp and salmon farming. *Science* **282**:883–4.

Paine, R. T. (1966) Food web complexity and species diversity. *American Naturalist* **100**:65–75.

Paine, R. T. (2002) Trophic control of production in a rocky intertidal community. *Science* **296**:736–9.

Patterson, D. J. (2009) Seeing the Big Picture on Microbe Distribution. *Science* **325**:1506–7.

Pauley, D. and Christensen, V. (1995) Primary production required to sustain global fisheries. *Nature* **374**:255–7.

Pauly, D., Alder, J., Bennett, E. (2003) The Future for Fisheries. *Science* **302**:1359–61.

Pauly, D., Chrinstensen, V. Dalsgaard, J. *et al.* (1998) Fishing down marine food webs. *Science* **279**:860–3.

Petchey, O. L. and K. J. Gaston (2006) Functional diversity: back to basics and looking forward. Ecology Letters **9**:741–58.

Phillips, J. (2001) Marine macroalgal biodiversity hotspots: why is there high species richness and endemism in southern Australian marine benthic flora? *Biodiversity and Conservation* **10**:1555–77.

Power, M. E., Tilman, D, Estes, J. A. *et al.* (1996) Challenges in the quest for keystones. *BioScience* **46**:609–20.

Reich, P. B., Knops, J., Tilman, D. *et al.* (2001) Plant diversity influences ecosystem responses to elevated CO_2 and nitrogen enrichment. *Nature* **410**:809–12.

Renaud, P., Webb, T., Bjorgesaeter, A. *et al.* (2009) Continental-scale patterns in benthic invertebrate diversity: insights from the MacroBen database. *Marine Ecology Progress Series* **382**:239–52.

Roberts, C. M. and Hawkins, J. P. (1999) Extinction risk in the sea. *Trends in Ecology and Evolution* **14**:241–6.

Sala, E. and Knowlton, N. (2006) Global Marine Biodiversity Trends. *Annual Review of Environment and Resources* **31**:93–122.

Sala, O. E., Stuart-III, S. F., Armesto, J. J. *et al.* (2000) Biodiversity: Global biodiversity scenarios for the year 2100. *Science* **287**:1770–4.

Sarmiento, J. L. and Gruber, N. (2006) *Ocean Biogeochemical Dynamics*. Princeton University Press, Princeton.

Schipper, J., Chanson, J. S., Chiozza F. *et al.* (2008) The Status of the World's Land and Marine Mammals: Diversity, Threat, and Knowledge. *Science* **322**:225–30.

Schlesinger, W. H. (1997) *Biogeochemistry*, 2nd edition. Academic Press, San Diego.

Schleuter, D., Daufresne, M., Massol, F. *et al.* (2010) A user's guide to functional diversity indices. *Ecological Monographs* **80**:469–84.

Schmid, B., Balvanera, P., Cardinale, B. J. *et al.* (2009) Consequences of species loss for ecosystem functioning: meta-analyses of data from biodiversity experiments. In: *Biodiversity, ecosystem functioning, and human wellbeing: an ecological and economic perspective.* Naeem, S., Bunker, D. E., Hector, A. *et al.* (eds), Oxford University Press, Oxford, 14–29.

Schulze, E. D. and Mooney, H. A. (eds) (1993) *Biodiversity and Ecosystem Function.* Springer Verlag, New York.

Schwartz, M. W., Brigham, C. A., Hoeksema, J. D. *et al.* (2000) Linking biodiversity to ecosystem function: implications for conservation ecology. *Oecologia* **122**:297–305.

Simon, N., Cras, A.-L., Foulon, E. *et al.* (2009) Diversity and evolution of marine phytoplankton. *Comptes Rendus Biologies* **332**:159–70.

Sogin, M. L., Morrison, H. G., Huber, J. A. *et al.* (2006) Microbial diversity in the deep sea and the underexplored rare biosphere. *Proceedings of the National Academy of Sciences* **103**:12115–20.

Solan, M., Cardinale, B. J.. Downing, A. L. *et al.* (2004) Extinction and ecosystem function in the marine benthos. *Science* **306**:1177–80.

Solan, M., Raffaelli, D., Paterson, D. M. *et al.* (2006) Introduction: Marine biodiversity and ecosystem function: empirical approaches and future research needs. *Marine Ecology Progress Series* **311**:175–8.

Srivastava, D. S. and Vellend, M. (2005) BIODIVERSITY-ECOSYSTEM FUNCTION RESEARCH: Is It Relevant to Conservation? *Annual Review of Ecology, Evolution, and Systematics* **36**:267–94.

Stachowicz, J. J., Fried, H., Osman, R. W. *et al.* (2002) Biodiversity, invasion resistance, and marine ecosystem function. Reconciling pattern and process. *Ecology* **83**:2575–90.

Stone, L. (1990) Phytoplankton-bacteria-protozoa interactions: a qualitative model portraying indirect effects. *Marine Ecology Progress Series* **64**:137–45.

Suttle, C. A. (2005) Viruses in the sea. *Nature* **437**:356–61.

Tilman, D., Knops, J., Wedin, D. *et al.* (2002) Plant diversity and composition: effects on productivity and nutrient dynamics of experimental grasslands. In:. *Biodiversity and Ecosystem Functioning: Synthesis and perspectives.* Loreau, M., Naeem, S. and Inchausti, P. (eds), Oxford University Press, Oxford, 21–35.

Tilman, D., Lehman, C. L. and Thomson, K. T. (1997) Plant diversity and ecosystem productivity: theoretical considerations. *Proceedings of the National Academy of Science* **94**:1857–61.

Tilman, D., Wedin, D. and Knops, J. (1996) Productivity and sustainability influenced by biodiversity in grassland ecosystems. *Nature* **379**:718–20.

Venter, J. C., Remington, K., Heidelberg, J. F. *et al.* (2004) Environmenal genome shotgun sequencing of the Sargasso Sea. *Science* **304**:66–74.

Waldbusser, G. G. and Marinelli, R. L. (2006) Macrofaunal modification of porewater advection: role of species function, species interaction, and kinetics. *Marine Ecology Progress Series* **311**:217–31.

Ware, D. M. and Thomson, R. E.. (2005) Bottom-Up Ecosystem Trophic Dynamics Determine Fish Production in the Northeast Pacific. *Science* **308**: 1280–4.

Whitman, W. B., Coleman, D. C. and Wiebe, W. J. (1998) Prokaryotes: The unseen majority. *PNAS* **95**:6578–83.

Wilson, R. W., Millero, F. J., Taylor, J. R. *et al.* (2009) Contribution of Fish to the Marine Inorganic Carbon Cycle. *Science* **323**:359–62.

Wohlers, J., Engel, A., Zöllner, E. *et al.* (2009) Changes in biogenic carbon flow in response to sea surface warming. *Proceedings of the National Academy of Sciences* **106**:7067–72.

Worm, B., Barbier, E. B., Beaumont, N. *et al.* (2006) Impacts of Biodiversity Loss on Ocean Ecosystem Services. Science **314**:787–90.

Worm, B., Sandow, M., Oschlies, A. *et al.* (2005) Global Patterns of Predator Diversity in the Open Oceans. Science **309**:1365–9.

CHAPTER 5

Lessons from the fossil record: the Ediacaran radiation, the Cambrian radiation, and the end-Permian mass extinction

Stephen Q. Dornbos, Matthew E. Clapham, Margaret L. Fraiser, and Marc Laflamme

5.1 Introduction

Ecologists studying modern communities and ecosystems are well aware of the relationship between biodiversity and aspects of ecosystem functioning such as productivity (e.g. Tilman 1982; Rosenzweig and Abramsky 1993; Mittelbach *et al.* 2001; Chase and Leibold 2002; Worm *et al.* 2002), but the predominant directionality of that relationship, whether biodiversity is a consequence of productivity or vice versa, is a matter of debate (Aarssen 1997; Tilman *et al.* 1997; Worm and Duffy 2003; van Ruijven and Berendse 2005). Increased species richness could result in increased productivity through 1) interspecies facilitation and complementary resource use, 2) sampling effects such as a greater chance of including a highly productive species in a diverse assemblage, or 3) a combination of biological and stochastic factors (Tilman *et al.* 1997; Loreau *et al.* 2001; van Ruijven and Berendse 2005).

The importance of burrowing organisms as ecosystem engineers that influence ecosystem processes in the near-seafloor environment has gained increasing recognition (e.g. Thayer 1979; Seilacher and Pflüger 1994; Marinelli and Williams 2003; Solan *et al.* 2004; Mermillod-Blondin *et al.* 2005; Ieno *et al.* 2006; Norling *et al.* 2007; Solan *et al.* 2008). Increased bioturbation influences the benthic ecosystem by increasing the water content of the uppermost sediment layers, creating a diffuse mixed layer and altering substrate consistency (Thayer 1979; Seilacher and Pflüger 1994). In addition, burrowing organisms play a crucial role in modifying decomposition and enhancing nutrient cycling (Solan *et al.* 2008).

Increased species richness often enhances ecosystem functioning, but a simple increase in diversity may not be the actual underlying driving mechanism; instead an increase in functional diversity, the number of ecological roles present in a community, may be the proximal cause of enhanced functioning (Tilman *et al.* 1997; Naeem 2002; Petchey and Gaston, 2002). The relationship between species richness (diversity) and functional diversity has important implications for ecosystem functioning during times of diversity loss, such as mass extinctions, because species with overlapping ecological roles can provide functional redundancy to maintain productivity (or other aspects of functioning) as species become extinct (Solan *et al.* 2004; Petchey and Gaston, 2005).

5.2 Strengths and limitations of the geological record

The fossil record provides an unparalleled, nearly 600 million year-long history of changing diversity in marine animal ecosystems. This extensive deep-time history provides a series of natural experiments in changing biodiversity, ranging from rapid

Marine Biodiversity and Ecosystem Functioning. First Edition. Edited by Martin Solan, Rebecca J. Aspden, and David M. Paterson.
© Oxford University Press 2012. Published 2012 by Oxford University Press.

diversification during the radiation of metazoans in the Ediacaran and the Cambrian explosion to extensive species losses during mass extinctions such as the Permian-Triassic crisis. The fossil record is also the only source of potential analogues to challenges facing the biosphere in the near future. Situations such as ice-free climates, rapid greenhouse gas emission and global warming, or the extinction of nearly 80% of marine invertebrate genera are not available for study on the modern Earth, yet are all represented in the fossil record.

Despite incomplete preservation, most notably of species that lack hard parts like shells or bone, the fossil record retains a recognizable signal of changes in biodiversity. Measures of diversity are particularly reliable in exceptionally preserved localities (Lagerstätten) containing soft tissue preservation, such as the Cambrian Burgess Shale or Chengjiang Fauna (Allison and Briggs 1993; Dornbos et al. 2005) or the latest Precambrian Ediacara biota (Clapham et al. 2003; Droser et al. 2006), but localities with typical hard part-only preservation can still provide useful data, although with more caveats (Kidwell, 2002; Tomašových and Kidwell, 2010). The loss of soft tissues, along with preservation biases against certain types of shell mineralogy (Cherns and Wright 2009), precludes comparison of shelly fossil collections with extant communities but does not prevent comparison among different fossil assemblages. The typical shelly fossil assemblage contains a mixture of individuals that originally lived over a span of decades to millennia ('time averaging'), not preserved in life position but mostly derived from within a particular habitat ('spatial mixing'). Time averaging inflates species richness relative to any single life assemblage (Kidwell 2002), a problem that becomes increasingly pronounced with increasing temporal scales of time averaging (Tomašových and Kidwell 2010). These effects introduce random noise to the biodiversity signal, reducing the precision of potential analyses and conclusions. Biodiversity changes can nevertheless be resolved and other ecological attributes, such as the relative-abundance structure of the shell-bearing taxa within a community, are preserved with greater fidelity (Kidwell 2002).

Proxies for ecosystem functioning may be more difficult to obtain, however, and there are several factors that complicate reconstruction of the productivity-diversity relationship in nearly all fossil assemblages. Most experimental analyses of the productivity-diversity relationship are conducted in assemblages of primary producers (terrestrial plant assemblages); thus, productivity can be measured directly as biomass or the rate of biomass accumulation (e.g. Tilman et al. 1997; Worm and Duffy 2003). We will use this definition of productivity throughout the chapter. In contrast, the marine fossil record, at least in the Phanerozoic (the time of conspicuous animals, since the Cambrian Explosion), is dominated by heterotrophic, primarily suspension- and deposit-feeding, invertebrates. Because they are not primary producers, their diversity is unlikely to have direct effects on primary production as postulated for plant assemblages (Tilman et al. 1997; Worm and Duffy 2003; van Ruijven and Berendse 2005). The combined biomass of heterotrophic invertebrates is presumably influenced to some degree by food supply, however, suggesting that biotic or stochastic factors (e.g. facilitation, niche complementarity, sampling effects; Tilman et al. 1997; Loreau et al. 2001) could similarly influence the biomass-diversity relationship.

Although food supply is influenced by primary productivity, it is difficult to relate primary productivity to nutrient levels, which are more commonly inferred through geological proxies, because of the complicated controls on primary production (Braiser 1995). Even nutrient levels themselves are difficult to assess in the ancient record. First, quantitative geochemical proxies for paleoproductivity, such as accumulation rates of the mineral barite (Dymond et al. 1992; Paytan and Griffith 2007) or trace metal (Cd, Cu, Ni, Zn) concentrations (Piper and Perkins 2004; Piper and Calvert 2009), rely on accurate age models of sedimentation rates that are typically only available in deep-sea sediment cores or other fine-grained mudrocks (Paytan and Griffith 2007). However, nearly all sedimentary deposition on the continental shelf, the depositional environment for almost all invertebrate fossil localities, is episodic and punctuated by numerous hiatuses of unknown but typically geologically short duration.

Second, total organic carbon (TOC) concentrations reflect a complex signal of not only primary productivity but also dilution by accumulating inorganic sediment particles such as clay minerals, preservation versus degradation in surface sediments (Calvert *et al.* 1992; Lee 1992; Ganeshram *et al.* 1999), and subsequent alteration during deep burial (e.g. Pell *et al.* 1993). Although quantitative geochemical proxies are not available, time-averaged estimates of productivity can be obtained from calculations of biomass (e.g. Clapham *et al.* 2003) or trends in shell abundance or body size of marine invertebrates (e.g. Fraiser and Bottjer 2004). Shell abundance is affected by the same factors that complicate geochemical proxies (sediment dilution, shell destruction), and body size may reflect other influences, such as greater juvenile mortality or shortened lifespans due to harsh environmental conditions. Nevertheless, the density and/or size of fossil taxa provide a reasonable proxy for biomass, which itself is related to productivity (Waide *et al.* 1999).

Evidence for burrowing activity in marine sediments, or bioturbation, is readily observable in the fossil record through preserved trace fossils (also called ichnofossils). Although the uppermost levels of the mixed layer are rarely preserved, except when marine substrates were firmly consolidated (Droser *et al.* 2002), maximum burrow depth and burrow diameters can readily be measured from preserved trace fossils (Droser and Bottjer 1988; Droser *et al.* 2002; Marenco and Bottjer 2011). The intensity of sediment mixing can be measured independently using the semi-quantitative ichnofabric index method, which scores sedimentary beds on a scale of 1–5 depending on the amount of disruption to original depositional features (Droser and Bottjer 1986). An ichnofabric index of 1 corresponds to absent—or extremely minimal—bioturbation, such that original layering and features are preserved, whereas ichnofabric index 5 corresponds to complete homogenization of the layer. Thus, ichnofabric index measures and quantification of burrow size and depth provide a good proxy for sediment mixing rates and the ecosystem functions—such as nutrient recycling—associated with bioturbation.

Functional diversity is difficult to quantify even in modern settings (Petchey *et al.* 2004), and it is unlikely that the biological requirements and ecological contributions of extinct taxa can be known with similar precision. Paleontologists have a long history, however, of grouping extinct taxa into broader functional guilds using attributes such as life position—e.g. benthic versus planktonic, epifaunal versus infaunal—motility, and trophic mode (e.g. Bambach, 1983; Bambach *et al.* 2007; Bush and Bambach 2011; Bush *et al.* 2011). These characteristics can be reliably determined for the vast majority of fossil taxa, even for extinct groups such as trilobites (Fortey and Owens 1999), and provide a coarse proxy for functional diversity. Additional aspects of functional diversity have been described in terms of tiering—sometimes called stratification in the biological literature—the vertical subdivision of epifaunal or infaunal habitat space, presumably indicative of niche partitioning (—e.g. Ausich and Bottjer 1982; Clapham and Narbonne 2002).

In this chapter, we explore three of these critical biodiversity changes: 1) the Ediacaran radiation of large multicellular eukaryotes, 2) the Cambrian radiation of complex bilaterian-grade animals, and 3) the end-Permian mass extinction, the largest known extinction event in Earth history. For each of these events, we examine how changes in biodiversity were related to various factors involved in ecosystem functioning such as productivity, bioturbation (biogenic mixing depth), and functional diversity.

5.3   Ediacaran ecosystems

For the first three billion years of its existence, life on Earth was dominated by microorganisms. The appearance in the latest Ediacaran (578–542 Mya) of large, soft-bodied, and structurally complex multicellular eukaryotes colloquially called the Ediacara biota marked a fundamental change in the way ecosystems are built and sustained (Narbonne 2005; Xiao and Laflamme 2009). Despite sharing a similar age and mode of preservation, most Ediacara biota represent an assemblage of unrelated organisms. Some Ediacaran fossils represent the oldest animals and perhaps root stock to modern faunas (Fedonkin and Waggoner 1999), while others represent extinct clades lying outside of crown Metazoa (Narbonne 2004; Xiao and Laflamme 2009). Despite this taxonomic uncertainty, temporal, biogeographic, and species-diversity patterns in Ediacaran localities highlight the existence of three assemblages characterized by distinct evolutionary innovations (Waggoner 2003):

1) The Avalon assemblage of Mistaken Point in Newfoundland and Bradgate Park in England (578–560 Mya) is an entirely deep-water assemblage, having formed several hundred meters below storm wave base and the photic zone (Wood et al. 2003; Wilby et al. 2011). As a result, the Ediacara biota from these localities could not utilize sunlight as their primary source of energy. The Avalon assemblage is dominated by rangeomorphs, an extinct clade of organisms constructed almost entirely from repeatedly branched (fractal) units and temporally restricted to the Ediacaran (Figure 5.1; Narbonne 2004; Narbonne et al. 2009; Brasier and Antcliffe 2009). This fractal branching allowed the rangeomorphs to attain high surface-area to volume ratios necessary for diffusion-based feeding, termed osmotrophy (Sperling et al. 2007; Laflamme et al. 2009). In the absence of metazoan zooplankton and macropredators in the Ediacaran, and as a result a lack of sloppy feeding and egestion, a greater portion of labile dissolved organic carbon was able to reach the deep ocean to be consumed by osmotrophes (Sperling et al. 2011). Trace-fossils from the Avalon assemblage are rare and restricted to the surface (Liu et al. 2010), while carbonate skeletons are unknown.

2) The White Sea assemblage (560–550 Mya), best known from the Flinders Ranges in South Australia and the White Sea coast of Russia, marks a significant step in metazoan evolution, with the first appearance of likely bilaterians (Fedonkin and Waggoner 1997; Fedonkin 2007) and a diverse array of novel trace fossils. The highest diversity of fossils from the White Sea assemblage occurs in rocks that were deposited in well-lit and energetically active shallow water (Gehling 2000). The White Sea assemblage postdates a major change in the stable carbon isotope composition (ratio of ^{13}C to ^{12}C) of lime-

stones that has been (contentiously) interpreted as an oxidation of a large deep ocean dissolved organic carbon pool (Rothman et al. 2003; Fike et al. 2006).

3) The Nama assemblage (550–541 Mya) of Namibia marks a further step in the progression towards modern ecosystems with the advent of biologically mediated secretion of calcium carbonate skeletons (Figure 5.1; Grotzinger et al. 1995).

The underlying cause for the segregation of the Ediacara biota into three assemblages is difficult to isolate (Waggoner 2003; Grazhdankin 2004; Narbonne 2005) and could have been influenced, at least in part, by important differences in the way fossils were preserved in each locality (Narbonne 2005). Biological responses to limiting factors associated with changes in bathymetry could also explain the diversity pattern, as the Avalon assemblage represents a deep-water community, while the most diverse and fossiliferous White Sea and Nama assemblages occupied significantly shallower water settings. Most intriguingly, however, the increased diversity and ecological complexity could instead reflect evolutionary advances from the oldest Avalon assemblage, through the younger White Sea assemblage containing the first mobile bilaterians, to the youngest Nama assemblage with its biomineralizing taxa (Xiao and Laflamme 2009; Brasier et al. 2010a).

The unusual style of fossil preservation in Ediacaran communities allows potential relationships between diversity and ecosystem functioning, including productivity, bioturbation, and functional diversity, to be examined more rigorously. In contrast to virtually all other fossil assemblages, which are time-averaged and have undergone some degree of spatial mixing, most Ediacaran assemblages record a snapshot of the original in situ community of soft-bodied organisms as it appeared at the moment of death (Seilacher 1992; Gehling 1999; Clapham et al. 2003). Any dead specimens (Liu et al. 2010) would decay within days to weeks (Allison 1988; Brasier et al. 2010b; Darroch et al. submitted), orders of magnitude smaller than typical time averaging in shelly communities (several hundred years; Kowalewski et al. 1998), and on the same timeframe as modern ecological census population studies.

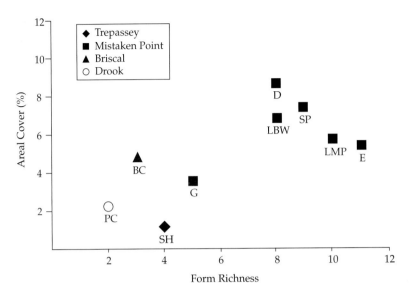

Figure 5.1   Richness-areal coverage relationship for Ediacaran communities from the Avalon Assemblage, Newfoundland. Genus richness is approximated by form richness, although informal groups such as 'dusters' may actually include several genera. The probable microbial structures *Ivesheadia* and the informally named 'lobate discs' are excluded. Areal coverage is a proxy for three-dimensional biomass and is measured as the percentage of the rock outcrop covered by fossil impressions, also excluding the area accounted for by *Ivesheadia* and 'lobate discs.' Although areal coverage in the most diverse surfaces is lower than the peak value, maximum likelihood estimation strongly supports a linear model fit (Akaike weight 0.91, compared to 0.055 for Gaussian and 0.035 for quadratic fits).

This nearly unique style of preservation allows investigation of ecological questions that may be obscured in the younger fossil record. For example, did more productive Ediacaran communities have higher species richness? What effects did the onset of bioturbation have on diversity in younger Ediacaran communities? Was there positive feedback between increasing functional diversity and higher species richness, in the form of facilitation or niche partitioning (Laflamme and Narbonne 2008b)?

5.3.1   Productivity–biodiversity relationship

Using organic carbon (TOC) as a proxy for productivity in Ediacara biota assemblages is not possible because of complications from dilution by sediment input and degradation during burial of the sedimentary strata, as discussed above. In addition, TOC levels in fossiliferous Ediacaran units are uniformly low: < 0.2 weight % in Newfoundland (Canfield *et al.* 2007), < 0.2 wt % in South Australia

(Calver 2000), and < 0.1 wt % in Namibia (Kaufman *et al.* 1991). Although paleoproductivity cannot be quantified in any Ediacaran fossil assemblage, the unusual preservation of the autotrophic (likely osmotrophic) fossils as 'snapshots' of the living community does allow estimation of standing biomass, analogous to ecological approaches (e.g. Tilman *et al.* 1997), and its relationship to diversity (Clapham and Narbonne, 2002; Clapham *et al.* 2003; Droser *et al.* 2006; Wilby *et al.* 2011). As productivity is the rate of biomass creation, biomass measures provide an approximation of time-averaged productivity over the lifespan of the organisms (Waide *et al.* 1999), although biomass and productivity at any given instant may themselves be interrelated (Guo 2007). Because the three-dimensional shape and material properties are unknown for most Ediacaran fossils, biomass can only be approximated by calculating the percentage of the rock surface covered by fossils—percent areal cover (Clapham *et al.* 2003). Areal cover is compared with 'form diversity,' an approximation of species diversity that takes the presence of undescribed taxa—such as 'dusters', which may represent several genera (Clapham *et al.* 2003)—into account. Taxa such as *Ivesheadia* and the undescribed 'lobate disc' were included in the areal coverage calculations in Clapham *et al.* (2003) but are excluded here—except for Long Beach West (LBW) and Spaniard's Bay (SP) surfaces as taxon-specific areal coverage data are not available for those localities—because those taxa likely represent microbial mat structures rather than multicellular eukaryotes (Laflamme *et al.* in press; although see Liu *et al.* 2010). The exclusion of ivesheadiomorphs from the analysis also eliminates some of the time-averaging issues raised by Lui *et al.* 2011.

Sample size is small and the data are noisy, but there is a weakly significant positive relationship ($r = 0.68$, $p = 0.04$) between form richness—'diversity'—and areal cover—'biomass'—in Avalon assemblages at Mistaken Point (Figure 5.1). Although there is a significant linear correlation, peak areal coverage does not occur at maximum form richness (Figure 5.1), potentially suggestive of a unimodal—'hump-shaped'—relationship commonly, but not always, observed in modern communities (e.g. Waide *et al.* 1999; Worm and Duffy 2003; van Ruijven and Berendse 2005; Guo 2007). Maximum likelihood estimation was used to fit and compare linear, Gaussian, and quadratic models to discriminate between linea-r and unimodal-richness areal cover relationships at Mistaken Point. The linear model receives the substantial proportion of model support (Akaike weight 0.91), whereas Gaussian (Akaike weight 0.055) and quadratic (Akaike weight 0.035) models are poor fits to the data. However, Ediacaran communities span a limited range of richness values—less than 10 form taxa per site—and could potentially occupy one limb of a larger unimodal trend. Ongoing paleoecological research in the Flinders Ranges, South Australia (White Sea Assemblage; Droser *et al.* 2006) may also help extend and clarify trends interpreted from Avalonian assemblage communities at Mistaken Point. Nevertheless, the correlation based on existing data tentatively suggests that richer Ediacaran communities had greater biomass,

therefore also were more productive, although the existence or directionality of causation (whether productivity led to greater diversity or vice versa) cannot be determined.

5.3.2 Influence of bioturbation on ecosystem functioning

The Ediacaran interval marks the onset of animal bioturbation and its ecological effects: infaunal bioturbation is absent in the oldest Ediacaran communities (Avalon assemblage) and simple horizontal traces are first found in the younger White Sea assemblage (Seilacher *et al.* 2005; Jensen *et al.* 2006). The first appearance of trace fossils in the White Sea assemblage corresponds with an increase in ecological complexity and the appearance of a variety of new functional groups, including mobile grazing metazoans, but does not correspond with an increase in within-community richness (Narbonne 2005; Bottjer and Clapham 2006; Droser *et al.* 2006; Shen *et al.* 2008). A causal relationship between the two events is unlikely, however, because Ediacaran bioturbation was extremely sparse and entirely represented by surficial traces that did not rework the sediment (Seilacher *et al.* 2005; Jensen *et al.* 2006). It is more likely that the appearance of new functional groups, including mobile animals, reflects some combination of permissive environmental conditions—such as an increase in oxygen levels—ecological feedbacks, or simply continuing biological evolution. Frondose taxa, which had been highly dominant in the oldest Avalon assemblage (Laflamme *et al.* 2004; 2007; Narbonne *et al.* 2009) and were highly adapted to firm microbial substrates (Seilacher 1999; Laflamme *et al.* 2010), were also less important constituents of the White Sea assemblage (Bottjer and Clapham 2006; Droser *et al.* 2006). Frondose taxa and mobile grazers appear to have been largely segregated in White Sea assemblage communities (Droser *et al.* 2006), suggesting that the decline of erect frondose forms may have resulted from sediment disturbance by mobile matgrazers and/or burrowers. However, bioturbation did not have a significant vertical component that would lead to appreciable sediment reworking until

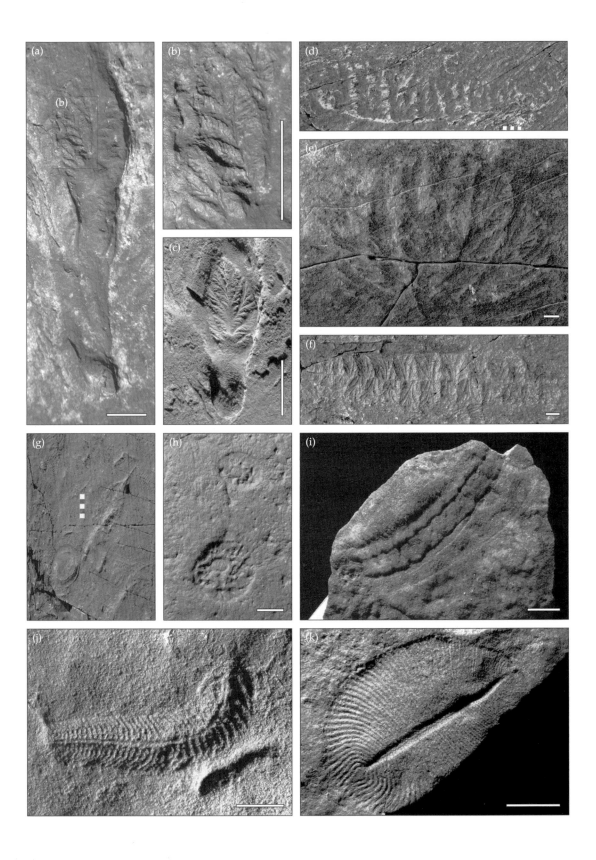

the Cambrian (Droser and Bottjer 1988; Droser *et al.* 2002), at which point it did exert an important control on marine ecosystems (Bottjer *et al.* 2000).

5.3.3   Species richness–functional diversity relationship

Ediacaran communities display a trend of increasing species richness from the earliest Avalonian examples to the White Sea assemblage, accompanied by a corresponding trend towards greater functional diversity. Ediacaran functional diversity cannot be quantified due to the uncertain life habits of many of the taxa (Xiao and Laflamme 2009), but it is possible to make some qualitative generalizations regarding changes in the types of functional groups represented. Early Avalon assemblage communities, in the Drook and Briscal Formations, had low species richness (3–5 form species present) and also had low functional diversity and poorly developed epifaunal stratification (referred to as tiering in the paleontological literature). The two dominant taxa in the Pigeon Cove community—*Charnia* and *Thectardis*—were both relatively unspecialized erect mid-to-upper-tier feeders (Clapham and Narbonne, 2002; Laflamme *et al.* 2007; Sperling *et al.* 2011). The third, *Ivesheadia*, is unlikely to be a metazoan (Liu *et al.* 2010; Laflamme *et al.* in press). The younger Bristy Cove community was nearly exclusively populated by low-tier rangeomorphs (e.g. *Fractofusus*, *Beothukis*) that fed from the basal few centimeters of the water column. In contrast, richness and functional diversity were both higher in later Avalonian communities, especially in the Mistaken Point Formation. Although those communities were still dominated by rangeomorphs, taxa began to display morphological adaptations for subdividing ecospace such as a differentiation of

their body into a separate food-gathering apparatus and elevating stem (Laflamme *et al.* 2004; Laflamme and Narbonne 2008a, 2008b). Well-developed epifaunal tiering also resulted in a much more precise subdivision of the vertical habitat: the D surface community at Mistaken Point contained *Fractofusus* in the lowest tier, *Bradgatia* at an intermediate level, *Pectinifrons* at greater heights, and *Beothukis* in the upper level at heights of nearly 30 cm (Figure 5.2; Clapham and Narbonne 2002; Laflamme and Narbonne, 2008b). Comparison of the ecological structure of low- and high-diversity Avalon communities indicates that high-diversity communities had a broader range of organism heights subdivided into a more complex and regular tiering structure. Thus, the positive relationship between species richness and functional diversity was plausibly driven by niche partitioning and development of a more finely subdivided tiering structure in Avalon assemblage communities.

In the White Sea assemblage, fronds still dominated the upper tiers while the lowermost tiers were colonized by attached epibenthic recliners such as *Tribrachidium, Arkarua, Parvancorina* (Figure 5.2), and *Phyllozoon* (Droser *et al.* 2006). Mobile *Kimberella* (Figure 5.2; Fedonkin and Waggoner 1997; Fedonkin 2007; Ivantsov 2009) and *Dickinsonia* (Figure 5.2; Gehling *et al.* 2005; Sperling and Vinther 2010), both of which grazed on abundant microbial mats, also make their appearance in the White Sea assemblage; as the first mobile, non-suspension feeding (or absorbing) taxa, they indicate a major increase in functional diversity. However, this increase in functional diversity—perhaps the most significant ecological change within the Ediacaran—was not accompanied by an increase in within-community species richness, or morphospace expansion (Shen *et al.* 2008), although the regional species pool was

Figure 5.2   Diversity of Ediacaran organisms. (a–f) represent various members of the rangeomorpha from Newfoundland. (a) *Avalofractus abaculus* from the Trepassey Fm. at Spaniard's Bay, with long stem and close-up (b) of a fractal, rangeomorph frondlet. (c) *Beothukis mistakensis* (NFM F-758) from the Trepassey Fm. at Spaniard's Bay. (d) Fence-shaped *Pectinifrons abyssalis* from the Mistaken Point Fm. at Mistaken Point North. (e) Cabbage-shaped *Bradgatia linfordensis* (ROM36500) from the Mistaken Point Fm. 'd' surface at Mistaken Point. (f) Spindle-shaped *Fractofusus misrai* from the Mistaken Point Fm. 'e' surface at Mistaken Point. (g) Frondose *Charniodiscus spinosus* from the Mistaken Point Fm. 'e' surface at Mistaken Point. (h–k) Assortment of classic Ediacaran organisms from the Flinders Ranges in South Australia. (h) Two specimens of the probable sponge *Palaeophragmodictya reticulata* (P32352a–f). (i) Probable stem-group mollusk *Kimberella quadrata* (P23532). (j) Bilaterally-symmetrical *Spriggina floundersi* (P12771) with anterior-posterior differentiation. (k) Probable placozoan *Dickinsonia costata* (P18888). Scale bars 1 cm or 1 cm increments.

substantially larger in the White Sea assemblage. Richness and evenness values in White Sea communities from Australia—mean richness of 5.25 species per bedding plane—are not significantly different from those in Avalon communities from Newfoundland (Droser *et al.* 2006).

Increased functional richness in younger White Sea assemblage Ediacaran communities, even though assemblage-level species richness was similar to the older Avalon assemblage, implies that Australian communities had lower functional redundancy. Australian assemblages also had more variable species composition and relative abundances (Droser *et al.* 2006), whereas Mistaken Point communities contain many of the same species, albeit at different abundances, over more than 10 million years of their history (Clapham *et al.* 2003). Community stability in the Avalon assemblage may reflect the stabilizing role of functional redundancy, allowing maintenance of community structure during disturbances (Naeem 1998). However, it is difficult to definitively reconstruct the true relationship between stability and functional redundancy because of the overprint of environmental parameters such as the frequency of disturbance. The significantly different habitats of the two assemblages—deep marine continental slope for the Avalon assemblage, coastal shallow marine for the White Sea assemblage—may instead be primarily responsible for increased volatility in community composition in the Australian examples.

5.4 Cambrian ecosystems

Ecosystems continued to increase in complexity during the Cambrian radiation (starting at 542 Mya). This event involved the geologically rapid diversification of marine animal life into relatives of nearly all modern phyla. This increase in diversity was coupled with increasing ecological complexity as active animals began interacting with each other on a large scale. Although their precursors might lie in the Ediacaran, it is in the Cambrian that we find the first definitive evidence for predation, the first widespread biomineralization, the first macroscopic sensory organs, and the first development of the seafloor mixed layer through bioturbation (Bottjer

et al. 2000; Bengtson 2002, 2004; Plotnick *et al.* 2010; Callow and Brasier 2009). It was indeed a time of rapid restructuring of marine ecosystems.

Lagerstätten deposits, which provide the highest definition view of animal life during the Cambrian radiation due to their preservation of soft tissues, are particularly abundant during the early and middle Cambrian, and then fade into comparative rarity through the remainder of the Phanerozoic Eon (Allison and Briggs 1993). Two of the most well studied Lagerstätten assemblages are the early Cambrian Chengjiang biota of Yunnan Province, China, and the middle Cambrian Burgess Shale biota of British Columbia, Canada. Priapulid worms and arthropods are dominant in the Chengjiang biota, and arthropods dominate the Burgess Shale biota (Dornbos *et al.* 2005; Caron and Jackson 2008; Zhao *et al.* 2009).

5.4.1 Productivity–biodiversity relationship

The rapid early Cambrian diversification and radiation of phytoplankton fossils provides evidence for a dramatic increase in primary productivity during the Cambrian radiation (e.g. Butterfield 1997). These fossils consist of small acanthomorphic arcritarchs, thought to be fossil remnants of planktonic eukaryotic algae (Butterfield 1997). These fossils are accompanied by the first evidence for mesozooplankton, consisting of minute feeding structures preserved alongside the acritarchs (Butterfield 1997).

The development of multiple trophic levels in the planktonic realm was an important characteristic of the Cambrian radiation (Butterfield 1997), and most likely would have facilitated the further revolution of suspension-feeding benthic organisms, as consolidated organic particles became more common in marine waters. It is also likely that the evolution of zooplankton led to the restriction of a large percentage of primary productivity, and as such labile dissolved organic carbon (DOC), to the surface ocean where it is rapidly consumed and recycled through the microbial loop (Azam *et al.* 1983; Pomeroy *et al.* 2007). Furthermore, the evolution of fecal pellets clumped organic waste into ballasted packages that could travel through the water column to accumulate as detritus on the seafloor

(Logan *et al.* 1995). As such, the deep oceans likely became impoverished in labile DOC, which would have had a dramatic effect on deep ocean osmotrophic Ediacarans which relied on DOC and could not capture and process larger particulates favored by filter-feeding cnidarians and sponges (Sperling *et al.* 2011). The increasing complexity that developed near the base of the marine trophic web during the Cambrian radiation clearly played a role in the dramatic increase in biodiversity that is one of the signatures of this critical event in the evolutionary history of animals (Butterfield 2009).

5.4.2 Influence of bioturbation on ecosystem functioning

Trace fossil taxa diversified rapidly during the Cambrian radiation, reflecting the increasing complexity of animal behavior (e.g. McIlroy and Logan 1999; Droser *et al.* 2002). Bioturbation depth and intensity increased accordingly within a broad spectrum of Cambrian depositional environments (Droser 1987). Evidence for a well-developed mixed layer in the upper few centimeters of the seafloor is found for the first time in the Cambrian, and paleoecological studies of benthic suspension feeders indicates that it had a profound effect on ecosystem functioning (Seilacher 1999; Bottjer *et al.* 2000; Dornbos 2006).

Echinoderms are among the earliest skeletonized sessile benthic suspension feeders in the Phanerozoic, and have therefore been the focus of many studies aimed at understanding the evolutionary response of benthic organisms to mixed layer development during the Cambrian. Early and middle Cambrian echinoderms include many genera adapted to living directly attached to, or inserted in, firm unlithified sediments (Bottjer *et al.* 2000; Dornbos 2006). By the late Cambrian, however, all sessile benthic echinoderms lived attached to hard substrates, either rare carbonate hardgrounds or loose shell material. This trend away from direct interaction with unlithified sediments and toward hard substrate attachment through the Cambrian is consistent with an evolutionary response to the development of the mixed layer in marine sediments.

Examination of the substrate adaptations of sessile benthic suspension feeders in the early Cambrian Chengjiang biota supports the pattern seen in early echinoderms. From the perspective of both generic diversity and relative abundance, forms adapted to firm unlithified sediments dominate the Chengjiang biota (Dornbos *et al.* 2005; Dornbos and Chen 2008). This would be expected given the early Cambrian age of the fossil assemblage, as mixed layer development was only beginning. Direct examination of the sediments in which these fossils are preserved indeed reveals that bioturbation levels were minimal (Dornbos *et al.* 2005). Similar data from the Burgess Shale biota is limited, but suggests that fewer sessile suspension-feeding genera were adapted for direct substrate interaction (Dornbos *et al.* 2005).

The increasing bioturbation depth and intensity through the Cambrian clearly was one critical factor in shaping ecosystem functioning. Sessile benthic taxa were placed under strong selective pressure to adapt to increased sediment mixing, resulting in the evolution of new attachment structures and strategies.

5.4.3 Species richness–functional diversity relationship

There was a remarkable increase in animal species richness from the Ediacaran into the Cambrian. This increase in richness was also associated with a marked increase in functional diversity, as measured by ecospace occupation (e.g. Bambach *et al.* 2007). When comparing ecospace occupation during the Ediacaran and Cambrian Periods, it becomes clear that a major shift took place from relatively large, mostly immobile animals in the Ediacaran, to somewhat smaller but more mobile animals in the Cambrian (Bambach *et al.* 2007; Xiao and Laflamme 2009). Most of the immobile Ediacaran animals were most likely absorbing dissolved organic carbon from the seawater, while the smaller immobile Cambrian animals are interpreted as suspension feeders on organic particles. Mobile Cambrian animals occupied a wide range of ecospace compared their mobile Ediacaran counterparts (Bambach *et al.* 2007), which were also much more rare.

Indeed, it is during the Cambrian in which we find the first definitive evidence for macrophagous predation, widespread skeletonization, macroscopic sensory organs, and intense bioturbation of the seafloor (Bottjer *et al.* 2000; Bengtson 2002, 2004; Plotnick *et al.* 2010). Evidence for these ecological changes is found throughout the early and middle Cambrian fossil record, but is even more apparent in deposits of exceptional preservation, such as the Chengjiang and Burgess Shale biotas. The older Chengjiang biota is particularly informative because of its early Cambrian age, which places it during the heart of Cambrian radiation. In this assemblage, we find the oldest evidence for the largest known Cambrian predators, the well-known stem group arthropods known as the anomalocaridids (e.g. Whittington and Briggs 1985). Well-developed eyes and mechano- and chemosensory organs are also found preserved in many Chengjiang taxa, indicating the importance of navigating the Cambrian marine landscape in search of prey, mates, or protection from predators (Plotnick *et al.* 2010).

5.5 The end-Permian mass extinction and its aftermath

The Great Ordovician Biodiversification Event (GOBE) 485–460 Mya exhibited a greater increase in biodiversity at the family, genus, and species levels than the Cambrian Explosion, and was associated with increases in tiering, epifaunal modes of life, and bioturbation (Servais *et al.* 2010). The GOBE established the Paleozoic Evolutionary Fauna that was dominant approximately 488–251 Mya (Ogg *et al.* 2008), and characterized as having intermediate-diversity communities, common epifaunal suspension feeders, developing tiering, and intermediate complexity food webs (Sheehan 1996). The end of the Paleozoic era and the dominance of the Paleozoic Evolutionary Fauna were marked by the end-Permian mass extinction ~252 Mya. This extinction was the largest mass extinction since the Cambrian radiation, whereby 78% of marine genera, and 49% and 63% of marine and terrestrial families disappeared (Raup and Sepkoski 1982; Clapham *et al.* 2009).

5.5.1 Environmental changes during the late Paleozoic to early Mesozoic

By the Late Carboniferous to Early Permian, Earth's major continents were configured as the supercontinent Pangea (Scotese 1997). Atmospheric carbon dioxide levels were low during the Permo-Carboniferous, while atmospheric oxygen levels were at their zenith in the Early Permian (Berner 2004). The precise cause of the end-Permian mass extinction remains controversial, but it is generally well-accepted that the ultimate cause of the extinction was massive volcanism leading to a cascade of environmental effects (Riccardi *et al.* 2007). Some combination of hypercapnia (carbon dioxide toxicity), hydrogen sulfide toxicity, and climate change, triggered or catalyzed by the eruption of the Siberian Traps, appears to be the most likely kill mechanism (Kump *et al.* 2005; Knoll *et al.* 2007). Deep-water anoxic and euxinic conditions developed in the Middle (Isozaki 1997) and Late Permian (Nielsen and Shen 2004), indicating that environmental stress that led to the Paleozoic-Mesozoic biotic crisis was likely to have been initiated several million years before the end-Permian mass extinction. Geochemical and sedimentological data also indicate that anoxia, euxinia, and high CO_2 concentrations in the oceans persisted at least intermittently through the Early Triassic (Pruss *et al.* 2006; Takahashi *et al.* 2009).

5.5.2 Permian–Triassic marine nutrient levels and primary productivity

Modeling supports the hypothesis that global warming during the Permian through Triassic would have decreased the equator to pole temperature gradient, thereby lowering wind velocities, thermohaline circulation, upwelling intensities, transport of nutrients to the deep ocean and to surface waters, and thus primary productivity (Kidder and Worsely 2004). However, some sedimentological and geochemical proxies indicate that nutrient levels and primary productivity may have fluctuated spatially and temporally during the late Permian through Early Triassic.

Hypotheses have been proposed for both enhanced and decreased input of continental nutrients into the marine realm during the Late Permian to Early Triassic. Modeling suggests that because mountain building had ceased, silicate weathering decreased, and thus nutrient input into the oceans was minimal and led to a crash in primary productivity in the Late Permian (Kidder and Worsley 2004). On the contrary, the extinction of land plants, the occurrence of pedoliths, and a rise in strontium isotopic values through the Early Triassic indicate that terrestrial primary productivity decreased rapidly, soil erosion was drastic, and continental weathering increased (Korte *et al.* 2003; Sephton *et al.* 2005; Sheldon 2006; Hu *et al.* 2008; Huang *et al.* 2008; Sedlacek *et al.* 2008; Ward *et al.* 2005; Retallack *et al.* 2005). The increased transport of terrestrial sediments and nutrients into the marine realm is proposed to have facilitated eutrophication, possibly increasing primary productivity in the ocean (Sephton *et al.* 2005).

The Early Triassic has a unique sedimentological record for the Phanerozoic that can also serve as a proxy for nutrient levels (e.g. Pruss *et al.* 2006). There is little documented organic matter in Lower Triassic strata, despite conditions that would have contributed to organic matter production and preservation—including low oxygen conditions, the phosphate cycle, and low bioturbation (e.g., Berner and Canfield 1989; Wignall and Twitchett 1996). Organic-rich black shales have been found in southwest Japan, Australia, and the Sverdrup basin in northern Canada, and are hypothesized to have resulted from increased nutrients via rivers or upwelling, or enhanced preservation under anoxic conditions (Suzuki *et al.* 1998; Thomas *et al.* 2004; Grasby and Beauchamp 2009). The possibility exists that high TOC was recorded in some Lower Triassic strata that has not yet been discovered (Suzuki *et al.* 1998; Golonka *et al.* 1994; Thomas *et al.* 2004).

Prior to the end-Permian mass extinction, there was a long episode of global chert accumulation termed the Permian Chert Event (PCE) (Isozaki 1994; Beauchamp and Baud 2002). It is proposed that the PCE was facilitated by upwelling of nutrient- and silica-rich water along north-western Pangea as seasonal melting of northern sea ice led to circulation of these waters (Beauchamp and Baud 2002; Kidder and Worsley 2004). The Early Triassic

was a time of very little biogenic chert production and/or preservation around the world, termed the Early Triassic Chert Gap (ETCG) (Racki 1999; Beauchamp and Baud 2002; Kidder and Worsely, 2004; Takemura *et al.* 2003). Sluggish thermohaline circulation caused by global warming, and low bioavailable nutrients that must be present in sufficient amounts for biogenic siliceous sediments to precipitate during the Early Triassic may have caused a breakdown in the conditions favorable for chert production, accumulation, and preservation (Laschet 1984; Racki 1999; Gammon *et al.* 2000). Alternate hypotheses for the ETCG include a decrease in biogenic silica due to the extinction of some sponge groups (Kato *et al.* 2002; Sperling and Ingle 2006) and a poor record of radiolarians (De Wever *et al.* 2006).

Extensive pyrite deposition is another characteristic feature of rocks deposited during the end-Permian mass extinction and post-extinction aftermath (Grasby and Beauchamp 2009). Prominent pyrite deposition suggests euxinic conditions that would have stripped essential bioavailable trace elements from the seawater (Grasby and Beauchamp 2009) that would have stressed marine primary producers (Anbar and Knoll 2002).

Phosphorites are rare during the Early Triassic, possibly indicating decreased nutrient input and upwelling intensity (Trappe 1994; Kidder and Worsely 2004). The distribution of phosphorites in mid-high latitudes suggests that there may have been a shift in ocean productivity to high latitudes because of changes in chemical weathering patterns (and the resultant P-flux to the ocean) due to the hot and arid climate (Trappe 1994) and wind-driven upwelling. Modeling by Hotinski *et al.* (2001) show that nutrients in high latitudes must have been high in order for H_2S and CO_2 to build up in the ocean.

Permian-Triassic sections around the world record several negative and positive excursions in $d^{13}C$ (Corsetti *et al.* 2005 and references therein; Kaiho *et al.* 2005; Galfetti *et al.* 2007; Haas *et al.* 2007; Hays *et al.* 2007; Grasby and Beauchamp; 2009), which reflect, in part, fluctuations in primary productivity (Suzuki *et al.* 1998; Berner 2002; Payne and Kump 2007). A drop in marine primary productivity has been proposed as one major cause of the negative shifts (e.g. Grard *et al.* 2005; Rampino and Caldeira 2005; Haas *et al.* 2006, 2007). In euxinic settings, trace metals become scarce and lead to severe nitrogen limitation, which could reduce primary productivity (Anbar and Knoll 2002). That euxinia is recorded in strata immediately before the negative shift in C-isotopes strongly suggests that a global euxinic event may have had a significant negative impact on bioavailable nutrients required to support ocean primary productivity during the Late Permian (Grasby and Beauchamp 2009). $\Delta^{34}S$ and $\Delta^{13}C$ values are interpreted in some strata to record a crash in primary productivity in the Late Permian perhaps of up to 50% (Algeo *et al.* 2010).

The anoxic conditions that would have facilitated decreasing primary productivity would have also increased it through phosphorus recycling, since phosphorus is a key nutrient driving productivity (Payne and Kump 2007). A productivity bloom during the Permian-Triassic interval would have occurred when autotrophs recovered, creating a positive excursion in the C-isotope record (Grard *et al.* 2005). Thus productivity crashes were likely to have been followed by productivity blooms that could have been major factors in causing the delayed biotic recovery by consuming oxygen in the photic zone and reducing Earth's capability of CO_2 drawdown (Winguth and Maier-Reimer 2005; Suzuki *et al.* 1998; Kidder and Worsely 2004).

Direct evidence of productivity is indicated by fossils of planktonic and benthic producers. Early-Middle Triassic phytoplankton assemblages were characterized by low biodiversity and widespread blooms of opportunistic acritarchs and prasinophytes that were likely to have formed the base of Early Triassic marine ecosystem (Eshet *et al.* 1995; Krassilov *et al.* 1999; Afonin *et al.* 2001; Grice *et al.* 2005; Payne and van de Schootbrugge 2007). Intermittent euxinic surface water and nitrogen limitation could have facilitated the blooms of cyanobacteria recorded in Lower Triassic strata, as they are able to fix nitrogen (Anbar and Knoll 2002; Xie *et al.* 2005; Grice *et al.* 2005).

Calcareous algae are rare in the benthic realm during the Early Triassic (see Payne and Van de Schootbrugge 2007). However, microbialites (microbially mediated calcareous structures, *sensu*

Riding 2000) and other microbially mediated structures (such as wrinkle structures) are widespread geographically and occur throughout the Early Triassic (Pruss *et al.* 2006; Baud *et al.* 2005). The fabrics and frameworks of most Early Triassic microbialites were most likely to have been built by calcified coccoid and filamentous cyanobacteria (Lehrmann 1999; Thomas *et al.* 2004; Hips and Haas 2006; Jiang *et al.* 2008). The proliferation of benthic primary producers is hypothesized to have been facilitated by decreased bioturbation, and upwelling of supersaturated, nutrient-rich anoxic marine water (e.g. Pruss *et al.* 2006; Kershaw *et al.* 2007).

5.5.3 Productivity–biodiversity–biomass relationship

Approximately 78% of marine invertebrate genera went extinct at the end of the Permian period (Clapham *et al.* 2009). Molluscs, brachiopods, echinoderms, bryozoans, and Porifera are the only higher taxa with known macroscopic benthic fossil representatives in the Early Triassic (Griffin *et al.* 2010); 143 genera of bivalves, gastropods, and brachiopods have been identified (Fraiser *et al.* 2010). The mean alpha diversity of Early Triassic collections of skeletonized benthic organisms ranges from 3.67 to 4.19 depending on the preservation style of the fossils (Fraiser *et al.* 2010). The diversity of trace fossils, the tracks and trails of organisms, was also low, ranging from 0–11 ichnogenera (Fraiser and Bottjer 2007). The end-Permian extinction has been correlated with a decline in primary productivity. Some nektonic organisms, like cephalopods, diversified quickly after the extinction, and may indicate an increase in diversity and abundance of primary producers (Brayard *et al.* 2009).

Decreases in bioturbation, animal size, and the abundance of fossil grains in thin- sections suggest that a decrease in animal biomass occurred across the Permian-Triassic boundary and persisted nearly to the end of the Early Triassic (e.g. Payne *et al.* 2006; Fraiser and Bottjer 2007). Tallies from bulk samples and analysis of fossil accumulations indicate that bivalves and microgastropods—gastropods with greatest dimension less than 1 cm—comprised the majority of animal biomass during the Early Triassic (Fraiser and Bottjer 2004, 2007; Fraiser *et al.* 2005). It has been proposed that reduced Early Triassic animal biomass can be attributed to increased global temperatures, small body sizes, and decreased growth efficiency of bacterial heterotrophs that could not sustain large numbers of molluscs with high metabolic demands (Payne and Finnegan 2006). Opportunistic blooms of cyanobacteria did not necessarily facilitate animal diversification or an increase in biomass because they are not as effective of a food source as eukaryotic algae (see references in Payne and Finnegan 2006).

Yet a decline in abundance and quality of primary productivity cannot explain why organisms with low growth efficiency (i.e. molluscs), instead of those with high growth efficiency (i.e. brachiopods), proliferated around the globe during the aftermath of the end-Permian mass extinction. Fraiser and Bottjer (2007) proposed that microgastropods and bivalves were able to survive the chemically and/or physiologically harsh environmental conditions during the Early Triassic better than most skeletonized benthic marine invertebrates because of their distribution in shallow marine environments and their physiological attributes. Thus, though diversity was very low, some taxa were present in very high numbers likely because they were able to cope with the environmental conditions (e.g. Fraiser and Bottjer 2004).

5.5.4 Discussion

Though fluctuations in nutrient levels and primary productivity contributed to the biotic crisis during the Late Permian through Early Triassic, it is more likely that other factors were more important influences on the extinction and its recovery. Environmental conditions, such as euxinia, most likely directly caused extinctions among both microscopic primary producers and macroscopic organisms (Bottjer et al. 2008). Conversely, nutrient levels and primary productivity could have been affected by the decline in biodiversity and biomass as well. Low bioturbation levels might have reduced nutrient recycling in the water column (e.g. Thayer 1983).

Some changes in community structure persisted only during the Early Triassic, while other aspects of benthic level-bottom shallow marine communities were permanently altered. A decrease in biodiversity, the numerical dominance of a few taxa, and the proliferation of microbially-mediated structures represent perhaps the most significant short-term changes in ecosystem functioning that lasted up to 5 million years (Pruss *et al.* 2006; Fraiser and Bottjer 2008; Alroy *et al.* 2008). Some short-term changes, such as a decrease in bioturbation and the proliferation of microbial structures, resemble Cambrian ecosystems (Bottjer *et al.* 1996; Fraiser and Bottjer 2007). Examples of lasting changes in ecosystem function related to the end-Permian mass extinction and its aftermath include the switch in taxonomic and ecologic dominants (Fraiser and Bottjer 2007) and a shift to more complex ecosystems (Bambach *et al.* 2002; Wagner *et al.* 2006). Some of the short-term changes may have facilitated the long-term changes: bivalves may have been aided by firm, microbially mediated substrates onto which they could byssally attach (Fraiser and Bottjer 2007). A complex interplay of environmental mechanisms directly and indirectly caused the extinction and delayed biotic recovery.

5.6 Conclusions

The biodiversity changes associated with the Ediacaran radiation, the Cambrian radiation, and the end-Permian mass extinction provide natural deep-time experiments on the impact of such biodiversity shifts on ecosystem functioning, informing our study of the modern biodiversity crisis. The biodiversity increases associated with the Ediacaran and Cambrian radiations show that ecosystem functioning is enhanced by such changes, making ecosystems more complex and stable in a series of positive feedback loops that increase productivity, bioturbation levels, and functional diversity. The opposite is true of the dramatic loss of marine biodiversity associated with the end-Permian mass extinction. Ecosystem instability lasted for up to 5 million years following this extinction event, with large crashes and blooms in productivity, decreased bioturbation of marine sediments, and sharply decreased functional diversity. These patterns validate the modern emphasis on protecting marine biodiversity, as its decline can foster devastating ecosystem instability that can last far beyond human timescales.

References

Aarssen, L. W. (1997) High productivity in grassland ecosystems: Effected by species diversity or productive species? *Oikos* **80**:183–4.

Afonin, S. A., Barinova, S. S., and Krassilov, V. A. (2001) A bloom of zygnematalean green algae *Tympanicysta* at the Permian-Triassic boundary. *Geodiversitas* **23**: 481–487.

Algeo, T. J., Hinnov, L., Moser, J., Maynard, J. B., Elswick, E., Kuwahara, K., and Sano, H. (2010) Changes in productivity and redox conditions in the Panthalassic Ocean during the latest Permian. *Geology* **38**: 187–190.

Allison, P. A. (1988) The decay and mineralization of proteinaceous macrofossils. *Paleobiology* **14**:139–54.

Allison, P.A. and Briggs, D.E.G. (1993) Exceptional fossil record: distribution of soft-tissue preservation through the Phanerozoic. *Geology* **21**: 605–8.

Anbar, A.D., and Knoll, A.H. (2002) Proterozoic ocean chemistry and evolution: a bioorganic bridge? *Science* **297**: 1137–1142.

Ausich, W. I. and Bottjer, D. J. (1982) Tiering in suspension-feeding communities on soft substrata throughout the Phanerozoic. *Science* **216**:173–4.

Azam, F., Fenchel, T., Field, J.G. *et al.* (1983) The ecological role of water-column microbes in the sea. *Marine Ecology Progress Series* **10**: 257–63.

Bambach, R. K. (1983) Ecospace utilization and guilds in marine communities through the Phanerozoic. In: *Biotic Interactions in Recent and Fossil Benthic Communities*, Tevesz, M. J. S. and McCall, P. L. (eds), Plenum Press, New York, 719–46.

Bambach, R. K., Knoll, A. H., and Sepkoski, J. J. Jr. (2002) Anatomical and ecological constraints on Phanerozoic animal diversity in the marine realm. *Proceedings of the National Academy of Sciences, USA* **99**: 6954–6959.

Bambach, R. K., Bush, A. M. and Erwin, D. (2007) Autecology and the filling of ecospace: key metazoan radiations. *Palaeontology* **50**:1–22.

Beauchamp, B., and Baud, A. (2002) Growth and demise of Permian biogenic chert along northwest Pangea: evidence for end-Permian collapse of thermohaline circulation. *Palaeogeography, Palaeoclimatology, Palaeoecology* **187**: 37–63.

Bengtson, S. (2002) Origins and early evolution of predation. *The Paleontological Society Special Papers* **8**: 289–317.

Bengtson, S. (2004) Early skeletal fossils. In: Neoproterozoic-Cambrian Biological Revolutions. Lipps, J.H. and Waggoner, B.M. (eds), *The Paleontological Society Special Papers* **10**: 67–77.

Berner, R. A. and Canfield, D. E. (1989) A new model for atmospheric oxygen over Phanerozoic time. *American Journal of Science* **289**: 333–381.

Berner, R.A. (2002) Examination of hypotheses for the Permo-Triassic boundary extinction by carbon cycle modeling. *Proceedings of the National Academy of Science USA* **99**: 4172–4177.

Berner, R.A. (2004) *The Phanerozoic Carbon cycle*: CO_2 and O_2. Oxford University Press, New York.

Bottjer, D.J., Schubert, J.K., and Droser, M.L. (1996) Comparative evolutionary palaeoecology: assessing the changing ecology of the past. *In*: Hart, M. B. (Ed.), *Biotic Recovery from Mass Extinction Events*. Geological Society Special Publication **102**: 1–13.

Bottjer, D. J., Hagadorn, J. W. and Dornbos, S. Q. (2000) The Cambrian substrate revolution. *GSA Today* **10**(9):1–7.

Bottjer, D. J. and Clapham, M. E. (2006) Evolutionary paleoecology of Ediacaran benthic marine animals. In: Xiao, S. and Kaufman, A. J. (eds), *Neoproterozoic Geobiology and Paleobiology*. Spr.ger, The Netherlands, 91–114.

Bottjer, D.J., Clapham, M.E., Fraiser, M.L., and Powers, C.M. (2008) Understanding mechanisms for the end-Permian mass extinction and the protracted Early Triassic aftermath/recovery. *GSA Today* **18**: 4–10.

Braiser, M.D. (1995) Fossil indicators of nutrient levels. 1: Eutrophication and climate change. In: Bosence, D.W.J. and Allison, P.A. (eds), *Marine Palaeoenvironmental Analysis from Fossils*. Geological Society Special Publication, London, **83**: 113–32.

Brasier, M.D. and Antcliffe, J. B. (2009) Evolutionary relationships within the Avalonian Ediacara biota: new insights from Laser Analysis. *Journal of the Geological Society*, **166**: 363–84.

Brasier, M.D., Antcliffe, J. B. and Callow, R. H. T. (2010a) Evolutionary trends in remarkable preservation across the Ediacaran-Cambrian transition and the impact of Metazoan Mixing. In: Allison, P. and Bottjer, D.J. (eds), *Taphonomy: Process and Bias Through Time*. Springer, Dordrecht, 519–67.

Brasier, M.D., Callow, R., Menon, L. *et al.* (2010b) Osmotrophic biofilms: from modern to ancient. In: Seckbach, J. and Oren, A. (eds), *Microbial Mats. Modern and Ancient Microorganisms in Stratifed Systems*, Springer, Dordrecht, 131–48.

Brayard, A., Escarguel, G., Bucher, H., Monnet, C., Bruhwiler, T., Goudemand, N., Galfetti, T., and Guex, J. (2009) Good genes and good luck: Ammonoid diversity and the end-Permian mass extinction. *Science* **325**: 1118–1121.

Bush, A.M. and Bambach, R. K. (2011) Paleoecologic Megatrends in Marine Metazoa. *Annual Review of Earth and Planetary Sciences* 39:241–69.

Bush, A.M., Bambach, R. K. and Erwin, D. (2011) Ecospace Utilization During the Ediacaran Radiation and the Cambrian Eco-explosion In: Laflamme, M., Schiffbauer, J.D. and Dornbos, S.Q. (eds), *Quantifying the Evolution of Early Life*: *Numerical Approaches to the Evaluation of Fossils and Ancient Ecosystems, Topics in Geobiology Series* **36**:111–33, Springer, Dordrecht.

Butterfield, N.J. (1997) Plankton ecology and the Proetrozoic-Phanerozoic transition. *Paleobiology* **23**: 247–62.

Butterfield, N.J. (2009) Macroevolutionary turnover through the Ediacaran transition: ecological and biogeochemical implications. In: Craig, J., Thurow, J., Thusu, B., Whitham, A., Abutarruma, Y. (eds), *Global Neoproterozoic Petroleum Systems*: *The Emerging Potential in North Africa*. Geological Society of London, London, 55–66.

Callow, R.H.T. and Brasier, M. D. (2009) Remarkable preservation of microbial mats in Neoproterozoic siliciclastic settings: Implications for Ediacaran taphonomic models: *Earth-Science Reviews* **96**: 207–19.

Calver, C. R. (2000) Isotope stratigraphy of the Ediacarian (Neoproterozoic III) of the Adelaide Rift Complex, Australia, and the overprint of water column stratification. *Precambrian Research* **100**:121–50.

Calvert, S. E., Bustin, R. M. and Pedersen, T. F. (1992) Lack of evidence for enhanced preservation of sedimentary organic matter in the oxygen minimum of the Gulf of California. *Geology* 20:757–60.

Canfield, D. E., Poulton, S. W. and Narbonne, G. M. (2007) Neoproterozoic deep-ocean oxygenation and the rise of animal life. *Science* 315:92–5.

Caron, J.B. and Jackson, D. A. (2008) Paleoecology of the Greater Phyllopod Bed community, Burgess Shale. *Palaeogeography, Palaeoclimatology, Palaeoecology* **258**: 222–56.

Chase, J. M. and Leibold, M. A. (2002) Spatial scale dictates the productivity-biodiversity relationship. *Nature* 416:427–30.

Cherns, L. and Wright, V. P. (2009) Quantifying the impacts of early diagenetic aragonite dissolution on the fossil record. *Palaios* 24:756–71.

Clapham, M. E. and Narbonne, G. M. (2002) Ediacaran epifaunal tiering. *Geology*, 30:627–30.

Clapham, M. E., Narbonne, G. M. and Gehling, J. G. (2003) Paleoecology of the oldest-known animal communities: Ediacaran assemblages at Mistaken Point, Newfoundland. *Paleobiology* 29:527–44.

Clapham, M.E., Shen, S.Z., and Bottjer, D.J. (2009) The double mass extinction revisited: re-assessing the severity, selectivity, and causes of the end-Guadalupian biotic crisis (Late Permian). *Paleobiology* **35**: 33–51.

Corsetti, F.A., Baud, A., Marenco, P.J., and Richoz, S. (2005) Summary of Early Triassic carbon isotope records. *Comptes Rendus Palevol* 4: 473–486.

Darroch, S., Laflamme, M., Schiffbauer, J. D. *et al.* Submitted. Experimental Formation of a Microbial Death Mask. *Palaios*.

De Wever, P., O'Dougherty, L., Gorican, S. (2006) The plankton turnover at the Permo-Triassic boundary, emphasis on radiolarians. *Eclogae geologicae Helvetiae* **99**: S49–S62.

Domke, K.L. and Dornbos, S.Q. (2010) Paleoecology of the middle Cambrian edrioasteroid echinoderm *Totiglobus*: Implications for unusual Cambrian morphologies. *Palaios* 25: 209–14.

Dornbos, S.Q. (2006) Evolutionary palaeoecology of early epifaunal echinoderms: Response to increasing bioturbation levels during the Cambrian radiation. *Palaeogeography, Palaeoclimatology, Palaeoecology* **237**: 225–39.

Dornbos, S.Q. and Bottjer, D.J. (2000) Evolutionary paleoecology of the earliest echinoderms: Helicoplacoids and the Cambrian substrate revolution. *Geology* **28**: 839–42.

Dornbos, S.Q. and Bottjer, D.J. (2001) Taphonomy and environmental distribution of helicoplacoid echinoderms. *Palaios* **16**: 197–204.

Dornbos, S.Q., Bottjer, D.J. and Chen, J.Y. (2005) Paleoecology of benthic metazoans in the Early Cambrian Maotianshan Shale biota and Middle Cambrian Burgess Shale biota: Evidence for the Cambrian substrate revolution. *Palaeogeography, Palaeoclimatology, Palaeoecology* **220**: 47–67.

Dornbos, S.Q. and Chen, J.Y. (2008) Community palaeoecology of the Early Cambrian Maotianshan Shale biota: Ecological dominance of priapulid worms. *Palaeogeography, Palaeoclimatology, Palaeoecology* **258**: 200–12.

Droser, M. L. and D. J. Bottjer. (1986) A semiquantitative field classification of ichnofabric. *Journal of Sedimentary Research* **56**:558–9.

Droser, M.L. (1987) *Trends in depth and extent of bioturbation in Great Basin Precambrian-Ordovician strata, California, Nevada, and Utah*. Ph.D. Dissertation, University of Southern California, 365 p.

Droser, M. L. and D. J. Bottjer. (1988) Trends in depth and extent of bioturbation in Cambrian carbonate marine environments, western United States. *Geology* **16**:233–6.

Droser, M. L., Gehling, J. G. and Jensen, S. R. (2006) Assemblage palaeoecology of the Ediacara biota: the unabridged edition? *Palaeogeography, Palaeoclimatology, Palaeoecology* **232**:131–47.

Droser, M. L., Jensen, S. and Gehling, J. G. (2002) Trace fossils and substrates of the terminal Proterozoic-Cambrian transition: implications for the record of early bilaterians and sediment mixing. *Proceedings of the National Academy of Sciences, USA* **99**:12572–6.

Dymond, J., Suess, E. and Lyle, M. (1992) Barium in deep-sea sediment: a geochemical proxy for paleoproductivity. *Paleoceanography* **7**:163–81.

Eshet, Y., Rampino, M.R., Visscher, H. (1995) Fungal event and palynological record of ecological crisis and recovery across the Permian–Triassic boundary. *Geology* **23**: 967–970.

Fedonkin, M. A. (2007) New data on *Kimberella*, the Vendian mollusc-like organism (White Sea region, Russia): palaeoecological and evolutionary implications. In: Vickers-Rich, P. and Komarower, P. (eds), *The Rise and Fall of the Ediacaran Biota.* Geological Society, London, Special Publications 2007; v. 286; 157–79.

Fedonkin, M. A. and Waggoner, B. M. (1997) The Late Precambrian fossil *Kimberella* is amollusc-like bilaterian organism. *Nature* **388**: 868–71.

Fike, D.A., Grotzinger, J. P., Pratt, L. M. *et al.* (2006) Oxidation of the Ediacaran ocean. *Nature* **444**:744–7.

Fortey, R. A. and Owens, R. M. (1999) Feeding habits in trilobites. *Palaeontology* **42**:429–65.

Fraiser, M.L., and Bottjer, D.J. (2004) The non-actualistic Early Triassic gastropod fauna: a case study of the Lower Triassic Sinbad Limestone Member. *Palaios* **19**: 259–275.

Fraiser, M.L., Twitchett, R.J., and Bottjer, D.J. (2005) Unique microgastropod biofacies in the Early Triassic: Indicator of long-term biotic stress and the pattern of biotic recovery after the end-Permian mass extinction. *Comptes Rendus Palevol* **4**: 475–484.

Fraiser, M.L., and Bottjer, D.J. (2007) When bivalves took over the world. *Paleobiology* **33**: 397–413.

Fraiser, M.L., Clapham, M.E., and Bottjer, D.J. (2010) Mass extinctions and changing taphonomic processes: Fidelity of the Guadalupian, Lopingian, and Early Triassic fossil records. *In*: Bottjer, D.J., and Allison, P. A. (Eds.), *Taphonomy: Process and Bias Through Time,* Plenum Press.

Galfetti, T., Bucher, H., Ovtcharova, M., et al. (2007) Timing of the Early Triassic carbon cycle perturbations inferred from new U-Pb ages and ammonoid biochronozones. *Earth and Planetary Science Letters* **258**: 593–604.

Ganeshram, R. S., Calvert, S. E., Pedersen, T. F. *et al.* (1999) Factors controlling the burial of organic carbon in laminated and bioturbated sediments off NW Mexico:

Implications for hydrocarbon preservation. *Geochimica et Cosmochimica Acta* **63**:1723–34.

Gehling, J. G. (1999) Microbial mats in terminal Proterozoic siliciclastics: Ediacaran death masks. *Palaios* **14**:40–57.

Gehling, J. G. (2000) Environmental interpretation and a sequence stratigraphic framework for the terminal Proterozoic Ediacara Member within the Rawnsley Quartzite, South Australia. *Precambrian Research* **100**:65–95.

Gehling, J. G., Droser, M. L., Jensen, S. R. *et al.* (2005) Ediacara Organisms: Relating Form to Function In: *Evolving Form and Function: Fossils and Development: Proceedings of a symposium honouring Adolph Seilacher for his contributions to paleontology in celebration of his 80th birthday.* Briggs, D.E.G. (ed).

Grard, A., François. L.M., Dessert, C., Dupré, B., and Goddéris, Y. (2005) Basaltic volcanism and mass extinction at the Permo-Triassic boundary: Environmental impact and modeling of the global carbon cycle. *Earth and Planetary Science Letters* **234**: 207–221.

Grasby, S. E., and Beauchamp, B. (2009) Latest Permian to Early Triassic basin-to-shelf anoxia in the Sverdrup Basin, Arctic Canada. *Chemical Geology* **264**: 232–246.

Grazhdankin, C. (2004) Patterns of distribution in the Ediacaran biotas: facies versus biogeography and evolution *Paleobiology*, **30**: 203–21.

Griffin, J.M., Marenco, P. J., Fraiser, M.L., and Clapham, M.E. (2010) Stromatolite- sponge-Tubiphites reefs in the Virgin Limestone Member of the Moenkopi Formation, Nevada: Implications for biotic recovery following the end-Permian mass extinction. *Geological Society of America Annual Meeting, Abstracts with Programs* **42**: 72.

Grotzinger, J. P., Bowring, S. A., Saylor, B. Z. *et al.* (1995) Biostratigraphic and geochronologic constraints on early animal evolution. *Science* **270**:598–604.

Guo, Q. F. (2007) The diversity-biomass-productivity relationships in grassland management and restoration. *Basic and Applied Ecology* **8**:199–208.

Haas, J., Demeny, A., Hips, K., *et al.* (2006) Carbon isotope excursions and microfacies changes in marine Permian-Triassic boundary sections in Hungary. *Palaeogeography, Palaeoclimatology, Palaeoecology* **237**: 160–181.

Haas, J., Demeny, A., Hips, K., et al. (2007) Biotic and environmental changes in the Permian- Triassic boundary interval recorded on a western Tethyan ramp in the Bukk Mountains, Hungary. *Global and Planetary Changes* **55**: 136–154.

Hays, L.E., Beatty, T., Henderson, C.M., Love, G.D., and Summons, R.E. (2007) Evidence for photic zone euxinia through the end-Permian mass extinction in the Panthalassic Ocean (Peace River Basin, Western Canada). *Palaeoworld* **16**: 39–50.

Hotinski, K.L. Bice, L.R. Kump, R.G. Najjar, and M.A. Arthur (2001) Ocean stagnation and end-Permian anoxia. *Geology* 29: 7–10.

Hu, Z.W., Huang S.J., Qing, H.R., Wang, C.M., and Gao, X.Y. (2008) Evolution and global correlation for strontium isotopic composition of marine Triassic from Huaying Mountains, eastern Sichuan, China. *Science in China Series D: Earth Sciences* 51: 540–549.

Huang, S.J., Qing, H.R., Huang, P.P., Hu, Z.W., Wand, Q.D., Zou, M.L., and Liu, H.N. (2008) Evolution of strontium isotopic composition of seawater from Late Permian to Early Triassic based on study of marine carbonates, Zhongliang Mountain, Chongqing, China. *Science in China Series D: Earth Sciences* 51: 528–539.

Ieno, E. N., Solan, M., Batty, P. *et al.* (2006) How biodiversity affects ecosystem functioning: roles of infaunal species richness, identity and density in the marine benthos. *Marine Ecology Progress Series* 311:263–71.

Isozaki, Y. (1994) Superanoxia across the Permo-Triassic boundary: Record in accreted deep-sea pelagic chert in Japan. *In*: Embry, A.F., Beauchamp, B., Glass, D.J. (Eds.), *Pangea: Global Environments and Resources*. Canadian Society of Petroleum Geologists Memoirs 17: 805–812.

Isozaki, Y. (1997) Permo-Triassic boundary superanoxia and stratified superocean; records from lost deep sea. *Science* 276: 235–238.

Ivantsov, A.Y. (2009) New reconstruction of *Kimberella*, problematic Vendian metazoan. *Paleontological Journal* 43: 601–11.

Jensen, S., Droser, M. L. and Gehling, J. G. (2006) A critical look at the Ediacaran trace fossil record. In: *Neoproterozoic Geobiology and Paleobiology*. Xiao, S. and Kaufman, A. J. (eds), Springer, The Netherlands, 115–57.

Jiang, H., Wu, Y., and Cai, C. (2008) Filamentous cyanobacteria fossils and their significance in the Permian-Triassic boundary section at Laolongdong, Chongqing. *Chinese Science Bulletin* 53: 1871–1879.

Kaiho, K., Chen, Z.Q., Ohashi, T., Arino, A., Swada, K., and Cramer, B.S. (2005) Negative carbon isotope anomaly associated with the earliest Lopingian (Late Permian) mass extinction. *Palaeogeography, Palaeoclimatology, Palaeoecology* 223: 172–180.

Kato, Y., Nakao, K., Isozaki, Y. (2002) Geochemistry of Late Permian to Early Triassic pelagic cherts from Southwest Japan; implications for an oceanic redox change. *Chemical Geology* 182: 15–34.

Kaufman, A. J., Hayes, J. M. and Knoll, A. H. *et al.* (1991) Isotopic compositions of carbonates and organic carbon from upper Proterozoic successions in Namibia: stratigraphic variation and the effects of diagenesis and metamorphism. *Precambrian Research* 49:301–27.

Kershaw, S., Li, Y., Crasquin-Soleau, S., Feng, Q., Mu, X., Collin, P.-Y., Reynolds, A., and Gou, L. (2007) Earliest Triassic microbialites in the South China block and other areas: Controls on their growth and distribution. *Facies* 53: 409–425.

Kowalewski, M., Goodfriend, G.A. and Flessa, K.W. (1998) High-Resolution Estimates of Temporal Mixing within Shell Beds: The Evils and Virtues of Time-Averaging. *Paleobiology* 24: 287–304.

Kidder, D.L., and Worsely, T.R. (2004) Causes and consequences of extreme Permo-Triassic warming to globally equable climate and relation to the Permo-Triassic extinction and recovery. *Palaeogeography, Palaeoclimatology, Palaeoecology* 203: 207–238.

Kidwell, S.M. (2002) Time-averaged molluscan death assemblages: Palimpsests of richness, snapshots of abundance. *Geology* 30:803–6.

Knoll, A.H., Bambach, R.K., Payne, J.L., Pruss, S., and Fischer, W.W. (2007) Paleophysiology and end-Permian mass extinction. *Earth and Planetary Science Letters* 256: 295–313.

Korte, C., Kozur, H.W., Bruckschen, P., and Veizer, J. (2003) Strontium isotope evolution of Late Permian and Triassic seawater. *Geochimica et Cosmochimica Acta* 67: 47–62.

Krassilov, V.A., Afonin, S.A., and Baranova, S.S. (1999) *Tympanicysta* and the terminal Permian events. *Permophiles* 35: 16–17.

Kump, L.R., Pavlov, A.A., and Arthur, M.A. (2005) Massive release of hydrogen sulfide to the surface ocean and atmosphere during intervals of oceanic anoxia. *Geology* 33: 397–400.

Laflamme, M. and Narbonne, G.M.. (2008a) Ediacaran fronds. *Palaeogeography, Palaeoclimatology, Palaeoecology*, 258:162–79.

Laflamme, M., and Narbonne, G.M.. (2008b) Competition in a Precambrian world: Palaeoecology and functional biology of Ediacaran fronds. *Geology Today*, 24:182–7.

Laflamme, M., Narbonne, G.M. and Anderson, M.M. (2004) Morphometric analysis of the Ediacaran frond *Charniodiscus* from the Mistaken Point Formation, Newfoundland. *Journal of Paleontology*, 78:827–37.

Laflamme, M., Narbonne, G.M. and Greentree, C. *et al.* (2007) Morphology and taphonomy of the Ediacaran frond: Charnia from the Avalon Peninsula of Newfoundland. In: *The Rise and Fall of the Ediacaran Biota*. Vickers-Rich, P. and Komarower, P. (eds), Geological Society, London, Special Publications, **286**: 237–57.

Laflamme, M., Schiffbauer, J.D. and Narbonne, G. M. (In press) Deep-Water Microbially Induced Sedimentary Structures (MISS) in Deep Time: The Ediacaran Fossil

Ivesheadia, In: *Microbial Mats in Sandy Deposits (Archean Era to Today)*, Noffke, N. K. and Chafetz, H. (eds), SEPM Special Publication.

Laflamme, M., Schiffbauer, J.D., Narbonne, G.M. *et al.* (2011) Microbial biofilms and the preservation of the Ediacara biota. *Lethaia*, **44**: 203–13.

Laflamme, M., Xiao, S. and Kowalewski, M. 2009. Osmotrophy in modular Ediacara organisms. *Proceedings of the National Academy of Sciences of USA* **106**: 14438–43.

Laschet, C. (1984) On the origin of cherts. *Facies* **10**: 257–289.

Lee, C. (1992) Controls on organic carbon preservation: The use of stratified water bodies to compare intrinsic rates of decomposition in oxic and anoxic systems. *Geochimica et Cosmochimica Acta* **56**:3323–35.

Lehrmann, D.J. (1999) Early Triassic calcimicrobial mounds and biostromes of the Nanpanjiang Basin, South China. *Geology* **27**: 359–362.

Liu, A.G., McIlroy, D., Antcliffe, J., Brasier, M.D. (2010) Effaced preservation in the Ediacara biota and its implications for the early macrofossil record. *Palaeontology* **54**: 607–630.

Liu, A.G., McIlroy, D., Antcliffe, J., Brasier, M.D. (2011) Effaced preservation in the Ediacara biota and its implications for the early macrofossil record. *Palaeontology* **54**: 607–630. Logan, G.A., Hayes, J.M., Hieshima, G.B. *et al.* (1995) Terminal Proterozoic reorganization of biogeochemical cycles. *Nature* **376**: 53–7.

Loreau, M., Naeem, S., Inchausti, P. *et al.* (2001) Biodiversity and ecosystem functioning: current knowledge and future challenges. *Science* **294**:804–8.

Marenco, K.N., and Bottjer, (2011) Quantifying bioturbation in Ediacaran and Cambrian rocks. *i*In: *Quantifying the Evolution of Early Life: Numerical Approaches to the Evaluation of Fossils and Ancient Ecosystems, Topics in Geobiology Series*, Laflamme, M., Schiffbauer, J.D. and Dornbos, S.Q. (eds), **36**:135–60, Springer, Dordrecht.

Marinelli, R. L., and Williams, T. J. (2003) Evidence for density dependent effects of infauna on sediment biogeochemistry and benthic-pelagic coupling in nearshore systems. *Estuary and Coastal Shelf Science* **57**:179–92.

McIlroy, D. and Logan, G. A. (1999) The impact of bioturbation on infaunal ecology and evolution during the Proterozoic–Cambrian transition. *Palaios* **14**: 58–72.

Mermillod-Blondin, F., François-Carcaillet, F. and Rosenberg, R. (2005) Biodiversity of benthic invertebrates and organic matter processing in shallow marine sediments: an experimental study. *Journal of Experimental Marine Biology and Ecology* **315**:187–209.

Mittelbach, G. G., Steiner, C. F., Scheiner, S. M. *et al.* (2001) What is the observed relationship between species richness and productivity? *Ecology* **82**:2381–96.

Naeem, S. (1998) Species redundancy and ecosystem reliability. *Conservation Biology*. **12**:39–45.

Naeem, S. (2002) Disentangling the impacts of diversity on ecosystem functioning in combinatorial experiments. *Ecology* **83**:2925–35.

Narbonne, G. M. (2004) Modular construction of Early Ediacaran complex life forms. *Science*, **305**:1141–4.

Narbonne, G. M. (2005) The Ediacara biota: Neoproterozoic origin of animals and their ecosystems. *Annual Review of Earth and Planetary Sciences* **33**:421–42.

Narbonne, G. M., Laflamme, M., Greentree, C. *et al.* (2009) Reconstructing a lost world: Ediacaran rangeomorphs from Spaniard's Bay, Newfoundland. *Journal of Paleontology* **83**:503–523.

Nielsen, J., and Shen, Y. (2004) Evidence for sulfidic deep water during the Late Permian in the East Greenland Basin. *Geology* **32**: 1037–1040.

Norling, K., Rosenberg, R., Hulth, S. *et al.* (2007) Importance of functional biodiversity and species-specific traits of benthic fauna for ecosystem functions in marine sediment. *Marine Ecology Progress Series* **332**:11–23.

Ogg, J.G., Ogg, G., and Gradstein, F.M. (2008) *The Concise Geologic Time Scale*. Cambridge University Press, 150 pp.

Payne, J. L., and Finnegan, S. (2006) Controls on marine animal biomass through geological time. *Geobiology* **4**: 1–10.

Payne, J.L., and Kump, L.R. (2007) Evidence for recurrent Early Triassic massive volcanism from quantitative interpretation of carbon isotope fluctuations. *Earth and Planetary Science Letters* **256**: 264–277.

Payne, J. L., and van de Schootbrugge, B. (2007) Life in Triassic oceans: Links between planktonic and benthic recovery and radiation. 165–189. *In* Falkowski, P.G., and Knoll, A.H. (Eds.), *Evolution of Primary Producers in the Sea*. Academic Press.

Payne, J.L., Lehrmann, D.J., Wei, J., and Knoll, A.H. (2006) The pattern and timing of biotic recovery from the end-Permian extinction of the Great Bank of Guizhou, Ghizhou Province, China. *Palaios* **21**: 63–85.

Paytan, A., and Griffith, E. M. (2007) Marine barite: recorder of variations in ocean export productivity. *Deep Sea Research Part II: Topical Studies in Oceanography* **54**:687–705.

Pell, S. D., McKirdy, D. M., Jansyn, J. *et al.* (1993) Ediacaran carbon isotope stratigraphy of South Australia—an initial study. *Transactions of the Royal Society of South Australia* **117**:153–61.

Petchey O. L., and Gaston, K. J. (2002) Functional diversity (FD), species richness and community composition. *Ecology Letters* **5**:402–11.

Petchey O. L. and Gaston, K. J. (2005) Extinction and the loss of functional diversity. *Proceedings of the Royal Society of London, B* **269**:1721–7.

Petchey, O. L., Hector, A. and Gaston, K. J. (2004) How do different measures of functional diversity perform? *Ecology* **85**:847–57.

Piper, D. Z., and Calvert, S. E. (2009) A marine biogeochemical perspective on black shale deposition. *Earth-Science Reviews* **95**:63–96.

Piper, D. Z., and Perkins, R. B. (2004) A modern vs. Permian black shale—the hydrography, primary productivity, and water-column chemistry of deposition. *Chemical Geology* **206**:177–97.

Plotnick, R.E., Dornbos, S.Q. and Chen, J.Y. (2010) Information landscapes and the sensory ecology of the Cambrian radiation. *Paleobiology* **36**: 303–17.

Pomeroy, L.R., Williams, P.J.I., Azam, F. *et al.* (2007) The microbial loop. *Oceanography* **20**: 28–33.

Pruss, S.B., Bottjer, D.J., Corsetti, F.A., and Baud, A. (2006) A global marine sedimentary response to the end-Permian mass extinction: Examples from southern Turkey and the western United States. *Earth-Science Reviews* **78**: 193–206.

Racki, G. (1999) Silica-secreting biota and mass extinctions: Survival patterns and processes. *Palaeogeography, Palaeoclimatology, Palaeoecology* **154**: 107–132.

Rampino, M.R., and Caldeira, K. (2005) Major perturbation of ocean chemistry and a 'Strangelove Ocean' after the end-Permian mass extinction. *Terra Nova* **17**: 554–559.

Raup, D.M., and Sepkoski, J.J. (1982) Mass extinctions in the marine fossil record. *Science* **215**: 1501–1503.

Riccardi, A., Kump, L.R., Arthur, M.A., and D'Hondt, S. (2007) Carbon isotopic evidence for chemocline upward excursions during the end-Permian event: *Palaeogeography, Palaeoclimatology, Palaeoecology* **248**: 73–81.

Riding, R. (2000) Microbial carbonates: The geological record of calcified bacterial-algal mats and biofilms. *Sedimentology* **47**: 179–214.

Rosenzweig, M. L. and Z. Abramsky. (1993) How are diversity and productivity related? Pp. 52–65 In: *Species Diversity and Ecological Communities*. Rickleffs, R. E. and Schluter, D. (eds), University of Chicago Press, Chicago.

Rothman, D.H., Hayes, J.M. and Summons, R.E. (2003) Dynamics of the Neoproterozoic carbon cycle: *Proceedings of the National Academy of Sciences* **100**:8124–9.

Sedlacek, A.R.C., Saltzman, M.R., and Linder, J.S. (2008) The Permian-Triassic boundary in the western United States. *Geological Society of America, Annual Meeting Abstracts and Programs*, Houston.

Seilacher, A. (1992) Vendobionta and Psammocorallia: lost constructions of Precambrian evolution. *Journal of the Geological Society, London* **149**: 607–13.

Seilacher, A. (1999) Biomat-related lifestyles in the Precambrian. *Palaios* **14**:86–93.

Seilacher, A., Buatois, L. A. and Mángano, M. G. (2005) Trace fossils in the Ediacaran-Cambrian transition: behavioral diversification, ecological turnover and environmental shift. *Palaeogeography, Palaeoclimatology, Palaeoecology* **227**:323–56.

Seilacher, A., and Pflüger, F. (1994) From biomats to benthic agriculture: a biohistoric revolution. *Biostabilization of sediments.* In: Krumbein, W. E., Paterson, D. M. and Stal, L. J. (eds), Bibliotheks und Informationssystem der Universitat Oldenburg, Oldenburg, 97–105.

Sephton, M.A, Looy, C.V., Brinkhuis, H., Wignall, P.B., de Leeuw, J.W., Visscher, H. (2005) Catastrophic soil erosion during the end-Permian biotic crisis. *Geology* **33**: 941–944.

Servais, T., Owen, A. W., Harper, D. A. T., Kröger, B. R., and Munnecke, A. (2010) The Great Ordovician Biodiversification Event (GOBE): The palaeoecological dimension. *Palaeogeography, Palaeoclimatology, Palaeoecology* **294**: 99.

Sheehan, P. (1996) A new look at Ecologic Evolutionary Units (EEUs). *Palaeogeography, Palaeoclimatology, Palaeoecology* **127**: 21–32.

Sheldon, N.D. (2006) Abrupt chemical weathering increase across the Permian-Triassic boundary. *Palaeogeography, Palaeoclimatology, Palaeoecology* **231**: 315–321.

Shen, B., Dong, L., Xiao, S.H. and Kowalewski, M. (2008) The Avalon explosion: evolution of Ediacara morphospace. *Science* **319**:81–4

Solan, M., P. Batty, Bulling, M. T. and Godbold, J. A. (2008) How biodiversity affects ecosystem processes: implications for ecological revolutions and benthic ecosystem function. *Aquatic Biology* **2**:289–301.

Solan, M., Cardinale, B. J., Downing, A. L. *et al.* (2004) Extinction and ecosystem function in the marine benthos. *Science* **306**:1177–80

Sperling, E.A., Ingle, J.C., Jr. (2006) A Permian-Triassic boundary section at Quinn River Crossing, northwestern Nevada, and implications for the cause of the Early Triassic chert gap on the western Pangean margin. *Geological Society of America Bulletin* **118**: 733–746.

Sperling, E.A., Peterson, K.J. and Laflamme, M. (2011) Rangeomorphs, *Thectardis* (Porifera?) and dissolved organic carbon in the Ediacaran ocean. *Geobiology*, **9**: 24–33.

Sperling, E.A., Pisani, D. and Peterson, K.J. (2007) Poriferan paraphyly and its implications for Precambrian palaeobiology. In: *The Rise and Fall of the Ediacaran Biota*, Vickers-Rich P, Komarower P (eds), *Geological Society of London, Special Publications* **286**: 355–68.

Sperling, E.A. and Vinther, J. (2010) A placozoan affinity for Dickinsonia and the evolution of late Proterozoic metazoan feeding modes. *Evolution and Development* **12**:201–9.

Suzuki, N., Ishida, K., Shinomiya, Y., and Ishiga, H. (1998) High productivity in the earliest Triassic ocean: Black shales, southwest Japan. *Palaeogeography, Palaeoclimatology, Palaeoecology* **141**: 53–65.

Takahashi, S., Oba, M., Kaiho, K., Yamakita, S., and Sakata, S. (2009) Panthalassic oceanic anoxia at the end of the Early Triassic: A cause of delay in the recovery of life after the end-Permian mass extinction. *Palaeogeography, Palaeoclimatology, Palaeoecology* **274**: 185–195.

Takemura, A., Aita, Y., Sakai, T., Kamata, V., Suzuki, N., Iiori, R.S., Famakita, S., Sakakibara, M., Campbell, H., Fujiki, T., Ogane, K., Takemura, S., Sakamoto, S., Kodama, A., and Nakamura, Y. (2003) Conodont-based age determination for a radiolarian-bearing Lower Triassic chert sequence in Arrow Rocks, New Zealand. *Tenth Meeting of the International Association of Radiolarian Palaeontologists, Abstracts and Programs*, University of Lausanne.

Thayer, C. W. (1979) Biological bulldozers and the evolution of marine benthic communities. *Science* **203**:458–61.

Thayer, C.W. (1983) Sediment-mediated biological disturbance and the evolution of marine benthos. *In*: Tevesz, M.J.S., McCall, P.L. (Eds.), *Biotic Interactions in Recent and Fossil Communities*. Plenum Press, New York: 479–625.

Thomas, B. M., Willink, R. J., Grice, K., Twitchett, R. J., Purcell, R. R., Archbold, N. W., George, A. D., Tye, S., Alexander, R., Foster, C. B. and Barber, C. J. (2004) Unique marine Permian–Triassic boundary section from Western Australia. *Australian Journal of Earth Sciences* **51**: 423–430.

Tilman, D. (1982) *Resource Competition and Community Structure*. Princeton University Press, Princeton.

Tilman, D., Knops, J., Wedin, D. *et al.* (1997) The influence of functional diversity and composition on ecosystem processes. *Science* **277**:1300–2.

Tomašových A., and Kidwell, S.M. (2010) Predicting the effects of increasing temporal scale on species composition, diversity, and rank-abundance distributions. *Paleobiology* **36**:672–95.

Trappe, J. (1994) Pangean phosphorites-ordinary phosphorite genesis in an extraordinary world? *Pangea: Global Environments and Resources, Canadian Society of Petroleum Geologists* **17**: 469–478.

van Ruijven, J., and Berendse, F. (2005) Diversity-productivity relationships: initial effects, long-term patterns, and underlying mechanisms. *Proceedings of the National Academy of Sciences, USA* **102**:695–700.

Waggoner, B. (2003) The Ediacaran biotas in space and time. *Integrative and Comparative Biology* **43**:104–13.

Wagner P.J., Kosnik, M.A., and Lidgard S. (2006) Abundance distributions imply elevated complexity of post-Paleozoic marine ecosystems. *Science* **314**: 1289–1292.

Waide, R. B., Willig, M. R., Steiner, C. F. *et al.* (1999) The relationship between productivity and species richness. *Annual Review of Ecology and Systematics* **30**:257–3000.

Wignall, P.B. and Twitchett, R.J. (1996) Oceanic anoxia and the end Permian mass extinction. *Science* **272**: 1155–1158.

Wilby, P.R., Carney, J.N. and Howe, M.P.A. (2011) A rich Ediacaran assemblage from eastern Avalonia: Evidence of early widespread diversity in the deep ocean. *Geology* **39**: 655–8.

Winguth, A.M.E., and Maier-Reimer, E. (2005) Causes of marine productivity and oxygen changes associated with the Permian-Triassic boundary: A re-evalution with ocean general circulation models. *Marine Geology* **217**: 283–304.

Whittington, H.B. and Briggs, D.E.G. (1985) The largest Cambrian animal, Anomalocaris, Burgess Shale, British Columbia. *Philosophical Transactions of the Royal Society of London* **309**: 569–609.

Wood, D. A., Dalrymple, R.W., Narbonne, G. M. *et al.* (2003) Palaeoenviromental analysis of the late Neoproterozoic Mistaken Point and Trepassey formations, southeastern Newfoundland. *Canadian Journal of Earth Sciences* **40**:1375–139.

Worm, B. and Duffy, J. E. (2003) Biodiversity, productivity, and stability in real food webs. *Trends in Ecology and Evolution* **18**:628–32.

Worm, B., Lotze, H. K., Hillebrand, H. *et al.* (2002) Consumer versus resource control of species diversity and ecosystem functioning. *Nature* **417**:848–51.

Xiao, S. and Laflamme, M. (2009) On the eve of animal radiation: phylogeny, ecology and evolution of the Ediacara biota. *Trends in Ecology and Evolution* **24**:31–40.

Xie, S., Pancost, R.D., Yin, H., Wang, H., and Evershed, R.P. (2005) Two episodes of microbial change coupled with Permo/Triassic faunal mass extinction. *Nature* **434**: 494–497.

Zhao, F., Caron, J.B., Hu, S.X. *et al.* (2009) Quantitative analysis of taphofacies and paleocommunities in the early Cambrian Chengjiang Lagerstätte. *Palaios* **24**: 826–39.

CHAPTER 6

The analysis of biodiversity–ecosystem function experiments: partitioning richness and density-dependent effects

Lisandro Benedetti-Cecchi and Elena Maggi

6.1 Introduction

Since early manipulations of species richness in the Ecotron facility and in grasslands (Naeem *et al.* 1994; Tilman *et al.* 1996; Hooper and Vitousek 1997; Hector *et al.* 1999), experimental analyses of biodiversity–ecosystem functioning relationships have become common in many terrestrial and aquatic environments (reviewed in Hooper *et al.* 2005; Cardinale *et al.* 2006; Stachowicz *et al.* 2007). This research has provided insights into the relative importance of species complementary versus species identity (Loreau and Hector 2001) in regulating key functional properties of ecosystems, including productivity, nutrient cycling, and stability. Complementarity occurs when increasing species richness results in the addition of new functional traits to a system, so that available resources are used more efficiently by the assemblage as a whole. Identity effects occur when only few species possess the appropriate traits that guarantee the efficient use of resources so that the presence of these species, more than richness *per se*, is important in driving the functioning of the system. Because diverse assemblages are likely to include particularly important species by chance—the sampling or positive selection effect (Huston 1997, Loreau 2000)—both complementary and identity effects may contribute to generate positive relationships between species richness and measures of ecosystem function.

A constructive debate in the biodiversity literature has fostered the development of experimental designs (Huston 1997; Allison 1999; Tilman 2001) and analytical techniques (Loreau and Hector 2001) to tease apart complementary from identity effects. Basic substitutive and additive replacement series designs (Jolliffe 2000), which underlie the majority of biodiversity experiments, have been implemented to include random assemblages nested within levels of species richness—e.g. Hooper and Vitousek 1997; Hector *et al.* 1999; Tilman 2001; Downing and Leibold 2002. The resulting hierarchical designs enable one to assess the role of species identity in terms of compositional effects within levels of richness—i.e. assemblage effects. The basic experimental design can be implemented to address more specific questions, such as the relative importance of species versus functional richness, or the analysis of biodiversity effects at multiple sites (reviewed in Schmid *et al.* 2002).

A general problem with replacement series designs is that either the relative abundance of individual species—substitutive designs—or total abundance—additive designs—changes with levels of species richness (Connolly 1988, Taylor and Aarsen 1989, Joliffe 2000). There is therefore the potential problem of confounding complementary or identity effects with density-dependent processes—we use this expression to indicate those processes such as population growth rate, and intra- and inter-specific interactions whose magnitude is affected by changes in species abun-

Marine Biodiversity and Ecosystem Functioning. First Edition. Edited by Martin Solan, Rebecca J. Aspden, and David M. Paterson.
© Oxford University Press 2012. Published 2012 by Oxford University Press.

dance—using replacement series designs. This problem has been debated in several areas of ecology where such designs are employed to discriminate between intra- and inter-specific interactions. Examples include experiments on competition—e.g. Underwood 1978—and predation experiments examining the combined effects of multiple predators on prey abundance—Aukema et al. 2004; Griffen 2006.

In the context of experimental biodiversity, ecologists have approached the problem of confounding abundance with richness effects from different angles. Confounding is often considered a minor problem in studies that focus on highly dynamic assemblages. In these circumstances, it is usually assumed that rapid population growth would quickly bring species abundances near carrying capacity, making any effects due to initial differences in established densities negligible by the end of the experiment (Douglass et al. 2008). This is akin to the 'law of constant final yield' in plant ecology (Harper 1977), which states that final stand biomass in monocultures is not affected by initial plant density. Another usual approach consists of deriving expected yields of species in mixtures as weighed averages of yields in monocultures (Loreau 1998; Griffin et al. 2009), then using expected and observed yields to separate complementary from identity effects (Loreau and Hector 2001). A third and much less common approach consists of incorporating abundance (density) explicitly as a factor in the experiment, crossed with one or more richness treatments. These experiments examine whether richness effects depend on total assemblage abundance (He et al. 2005; O'Connor and Crowe 2005; Griffin et al. 2008), or use abundance gradients of monocultures to derive expected yields of species in mixtures (Griffin et al. 2009). While enabling the analysis of abundance effects and richness × abundance interactions, these experiments cannot partition density-dependent from complementary or identity effects because richness is still manipulated as a replacement series within abundance levels.

The statistical significance of richness and, for those studies that include abundance as a factor richness × abundance interactions, are usually analysed by means of linear statistical models, including regression techniques and analysis of variance.

These techniques assume independent observations, a condition that may be easily violated if experimental units are sampled repeatedly through time, or if they are too close in space to be under the direct influence of spatially auto-correlated processes. Standard techniques also assume homogeneous variances, a condition that may be violated in biodiversity studies that include replicate assemblages within richness levels, and where species are sampled with replacement in generating the richness gradient (the variance-reduction effect: Huston 1997; Huston and McBride 2002). Biodiversity experiments may also include multiple sources of stochastic variation leading to complex variance–covariance structures. For example, a study may examine the relationship between richness and the focal response variable of interest at a random sample of sites. The intercepts and slopes of the response variable–richness relationship vary randomly across sites, and may display positive or negative covariance. These variance–covariance structures cannot be modeled with standard techniques.

Mixed-effect models are inferential statistical techniques that can deal with auto-correlated observations and complex variance-covariance structures (Singer and Willet 2003). These techniques are receiving increasing attention in ecology (McMahon and Diez 2007), but they have been applied only occasionally to the analysis of biodiversity experiments (see Godbold et al. 2008; Bulling et al. 2008). Here we use mixed-effect models to analyse a biodiversity experiment that includes the manipulation of algal richness (number of taxa) and abundance in factorial combinations, with replicate assemblages nested within richness levels, as described in Benedetti-Cecchi (2004). We implement the mixed-effect model with a test based on linear contrasts that enables one to assess the magnitude of richness effects relative to the strength of density-dependent processes. The logic behind our approach is that the importance of richness in the presence of density-dependent interactions emerges only if the effect of an increase in abundance due to the addition of new species to an assemblage is larger than the effect of an equivalent increase in abundance due to the addition of pre-existing species—i.e. by keeping richness constant.

6.2 Partitioning richness and abundance effects

Consider the outcome of a classical biodiversity experiment based on a substitutive design, in which there is a positive relationship between species richness and the specific response variable of interest. Because relative species abundances and number of species change simultaneously as richness increases (Figure 6.1a), one must assume that density-dependent processes are unimportant in order to unequivocally ascribe observed changes in the response variable to richness. This assumption may or may not be justified, depending on the environmental and biological context of the study.

How then can one examine and interpret richness and density-dependent effects in biodiversity studies? In principle, factorial manipulation of species (functional) richness and abundance should provide insights into the relative importance of these effects through significance tests on estimated parameters. This analysis, however, does not provide a direct comparison of the magnitude of richness versus density-dependent effects. Consider the hypothetical outcome of a factorial manipulation of species abundance and richness as illustrated in Figure 6.1b. With enough precision, the analysis of these data would reveal a significant richness × abundance interaction. This outcome would indicate that the relationship between richness and the response variable differs between levels of abundance, but it would not separate richness from density-dependent effects. Richness effects cannot be quantified directly from biodiversity experiments because changes in richness are inevitably associated with changes in relative or total species abundance in experimental treatments. Density-dependence, in contrast can be measured precisely by examining abundance effects within richness levels. A test to separate richness effects from density-dependent processes can therefore be performed by comparing the average effect of simultaneously changing richness and abundance—the average of Δ_{yHA} and Δ_{yLA} in Figure 6.1b—with the average effect of increasing abundance *within* levels of richness—the mean of the five Δ_y in Figure 6.1b. If the substitutive, factorial

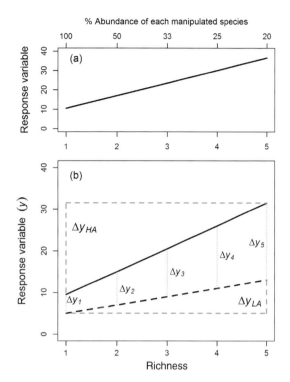

Figure 6.1 Interpretation of richness and abundance effects in biodiversity experiments. (a) Increasing level of richness is associated with a reduction in abundance of each manipulated species in replacement series designs. (b) Richness effects can be tested over and above density-dependent processes by comparing the average effect of simultaneously changing richness and abundance (the average of Δy_{HA} and Δy_{LA}) to the effect of an increase in abundance within richness levels (the average of Δ_{y1}–Δ_{y5}). These effects can be extracted from an experiment that includes manipulation of abundance and richness in factorial combinations (black-dashed and continuous lines reflect low and high abundance treatments, respectively). Further details are given in the text.

experiment is designed in such a way that the reductions in species abundance that occur with increasing richness (Figure 6.1b) match the differences between levels of the abundance factor, then under the null hypothesis of no richness effects one would expect:

$$H_0: \frac{1}{2}(\Delta y_{HA} + \Delta y_{LA}) = \overline{\Delta}_y \qquad \text{(Equation 6.1)}$$

This hypothesis can be tested as an *a priori* contrast using linear combinations of the coefficients obtained by fitting a linear regression model to the data. For the example in Figure 6.1b with five levels of species richness (and with richness centered over the first level), Δy_{HA} is equal to $4(\gamma_R + \gamma_{RA})$ and Δy_{LA} is equal to $4(\gamma_R)$, where γ_R and γ_{RA} are the regression coefficients associated with the main effects of richness and the richness × abundance interaction, respectively. Similarly, Δ_{y1} in Figure 6.1b is measured by γ_A, the coefficient associated with the main effect of abundance in the regression analysis—after centering richness over the first level—while Δ_{yj}, for any level of richness R_j between 2 and 5—between 1 and 4 if richness is centered over the first level—is equal to $\gamma_A + R_j\gamma_{RA}$. Hence, for this example with centered richness, $\frac{1}{2}(\Delta y_{HA} + \Delta y_{LA}) = \frac{1}{2}(4\gamma_R + 4\gamma_{RA} + 4\gamma_R)$ and $\overline{\Delta}_y = \frac{1}{5}(5\gamma_A + 10\gamma_{RA})$; setting the difference between these two equations equal to 0 enables one to identify the vector of constants that when multiplied by the vector of regression coefficients provides the linear contrasts corresponding to the hypothesis specified in Equation 6.1. The resulting vector of constants for this example is {0;4;-1;0} and the vector of regression coefficients would include the intercept, the main effect of richness, the main effect of abundance and the richness × abundance interaction.

Equation 6.1 assumes that the strength of density-dependent processes due to changes in species abundance across levels of richness (those captured by the left-hand side of the equation) are equal to those estimated by the comparison of abundance effects within levels of richness (given by the right-hand side of the equation). This may or may not be the case, and depends on the chosen levels of factors abundance and richness. In general, species will undergo large reductions in abundance in substitutive designs that include many levels of richness and where monocultures are established at high abundance. The potential for density-dependent processes to confound richness effects will be large under these conditions. The experimenter must therefore impose sufficiently large differences between levels of the abundance factor to ensure that the changes in abundance that occur between and within richness levels are comparable.

Alternatively, one may rescale Equation 6.1 with respect to the total change in species abundance associated with each term:

$$H_0: \frac{1}{2}\left(\frac{\Delta y_{HA}}{\Delta A_H} + \frac{\Delta y_{LA}}{\Delta A_L}\right) = \frac{\overline{\Delta}_y}{\Delta A} \qquad \text{(Equation 6.2)}$$

Here ΔA_H and ΔA_L reflect the overall reductions in species abundances associated with increasing levels of richness for the high and low abundance treatments, respectively. These quantities are calculated separately for each abundance level as:

$$\Delta A_{H,L} = \sum_{i=1}^{S} \sum_{j=R_k}^{R_{MAX}-1} (A_{ij} - A_{i,j+1}) \qquad \text{(Equation 6.3)}$$

where S is the number of species used to generate the $R_{MAX} - 1$ levels of factor richness, with R_{MAX} being the highest level of richness, R_k is the level of richness in which species i appears for the first time, and A_{ij} is the abundance of species i in richness level j. ΔA is simply the difference between levels of factor abundance, as imposed by design. Thus, the left-hand side of Equation 6.1 quantifies the combined effect of changes in species richness and abundance per unit of change in species abundance, while the right-hand side measures the effect of a unit of change in abundance.

As an example of the derivation of ΔA_H and ΔA_L through Equation 6.3, consider a hypothetical experiment with five levels of species richness (as in Figure 6.1), and with monocultures established with 100 and 60 individuals for the high and low abundance treatments, respectively. When a new species is added to the high abundance treatment, the abundance of the first species is halved (Figure 6.1a), contributing a 50 individuals change to ΔA_H. When a third species is added, the abundance of each of the two pre-existing species undergoes a reduction from 50 to 33 individuals (16 or 17 individuals per species), contributing an overall change of 33 individuals to ΔA_H. When richness is further increased with a fourth species, the collective reduction in abundance of the three pre-existing species contributes another change of 25 individuals to ΔA_H. Finally, when the fifth species is added, each of the four pre-existing species undergoes a reduction of five individuals contributing collectively a change of 20 individuals to ΔA_H. Adding up these values results in an overall change in abundance of 128 individuals, corresponding to ΔA_H. Repeating the same procedure for the low abundance treatment yields a series of changes in abundance (starting from a monoculture with 60 individuals) of 30, 20, 15, and 12 individuals. This would result in a value of ΔA_L of 77 individuals. ΔA is simply the difference between the high and low abundance treatments: $100 - 60 = 40$ individuals. Applying Equation 6.2 to the ΔAs just obtained yields $\frac{1}{2}\left(\frac{4\gamma_R + 4\gamma_{RA}}{128} + \frac{4\gamma_R}{77}\right)$ $= \frac{1}{5}\left(\frac{5\gamma_A + 10\gamma_{RA}}{40}\right)$, resulting in the vector of constants {0;0.0416;–0.05;–0.0344} that when multiplied by the vector of regression coefficients (including the intercept, the main effect of richness, the main effect of abundance, and the richness × abundance interaction), would provide the appropriate linear contrast to test the hypothesis in Equation 6.1.

6.3   Empirical example

6.3.1   Experimental layout

We illustrate the procedure of separating abundance (density-dependent) from richness effects using data from a biodiversity experiment undertaken on the rocky shores south of Livorno in Italy (43°30′N, 10°20′E), between July 2004 and December 2006. The experiment examined how changes in richness, composition, and abundance of algal taxa (morphological groups, hereafter MG) characterizing early stages of colonization of cleared surfaces in the low shore habitat (Benedetti-Cecchi and Cinelli 1994), affected subsequent patterns of recovery. Manipulated algae included the following species or MG: encrusting coralline algae (mostly *Lithophyllum orbiculatum*), encrusting cyanobacteria (*Rivularia* spp.), green filamentous algae (*Cladophora* spp. and *Chaetomorpha aerea*), red corticated filamentous algae (*Polysiphonia sertularioides*), and red uniseriate filamentous algae (*Ceramium* spp. and *Callithamnion granulatum*). Algal biodiversity was manipulated using stainless-steel plates of 12 × 8 cm as experimental units. Twenty four sandstone cubes (the same rock as the natural shore) of 2 × 2 × 2 cm each were screwed individually to each plate to form an individual surface for settlement. Preliminary studies indicated that the patterns of colonization of these plates did not differ in terms of species composition and abundance with respect to clearings produced on natural surfaces. In July 2004, we exposed 140 plates to natural colonization for 11 months. In June 2005, the individual cubes were disassembled and rearranged among plates to form appropriate experimental treatments, following the procedure described in Stachowicz *et al.* 1999. We manipulated numbers, identities, and abundances of algal taxa, following the experimental design described in Benedetti-Cecchi 2004. This design enabled the effect of increasing algal abundance within a given level of richness to be compared to the effect of an equivalent increase in abundance obtained by the addition of new taxa (Figure 6.2). Unfortunately, several plates went lost during the colonization period, preventing the use of 24 cubes on each experimental unit. Instead, we used 16 cubes per plate so that each experimental unit consisted of a surface of 8 × 8 cm. This was still deemed an appropriate size to manipulate the tiny algae that characterized the study site.

Richness levels consisted of experimental units with 1, 2, and 4 taxa (individual species or MG), and an additional level with 0 taxa was obtained by clearing cubes of all visible organisms with a paint scraper. Only encrusting coralline algae were established as a single taxon, since only these algae occurred in isolation from other groups on the settling plates. Additional richness levels of 2 and 4 taxa with nested replicated assemblages were generated by combining encrusting algae with other algal taxa that characterized early stages of recovery, selected randomly among those present on the plates. All treatments with 1, 2, and 4 taxa were established at each of two levels of total abundance, corresponding to 6 and 12 occupied cubes per experimental unit (37.5% and 75% of occupied substratum, respectively). Each treatment was replicated four times (Figure 6.2).

In addition to incorporating abundance as a factor, our design retained all the other key features that have been considered important to separate complementary from identity effects in biodiversity experiments (Hector *et al.* 1999; Downing and Leibold 2002), including: (1) the constrained random selection of taxa; this required that once a species or functional group was selected to create a low level of richness, these same organisms should also

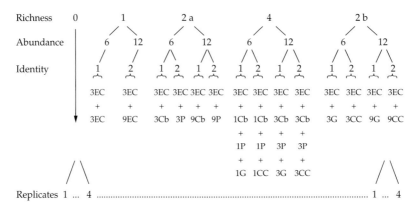

Figure 6.2 Schematic representation of the experimental design. Richness: number of manipulated taxa. Abundance: number of occupied sandstone cubes corresponding to 37.5% (6 cubes, low abundance) and 75% (12 cubes, high abundance) algal cover. Identity: replicated assemblages within richness levels 2 and 4. EC = encrusting corallines, Cb = encrusting cyanobacteria, P = *Polysiphonia sertularioides* (a red filamentous alga with corticated filaments), G = green filamentous algae (*Cladophora* spp., *Chaetomorpha aerea*), CC = *Ceramium* spp. and *Callithamnion granulatum* (red filamentous algae with uniseriate filaments).

occur in the more diversified treatments together with new, randomly drawn taxa; (2) the replication of random combinations of taxa within richness levels; and (3) the inclusion of additional treatments to ensure that the taxa characterizing intermediate and high richness levels also occurred in the less diversified treatments. These criteria resulted in a three-way mixed layout with richness, assemblage, and abundance as factors. Richness and abundance were fixed, crossed factors, while assemblage was random, nested within richness, and crossed with abundance (Figure 6.2).

The experimental units were sampled six times, after 3, 5, 8, 11, 14, and 18 months from start. At each time, the percentage cover of all visible algal species or functional-form groups was assessed with the aid of a plastic frame of 8 × 8 cm with 64 holes. The frame was overlaid onto the experimental units and the number of quadrats containing each of the focal species was counted. Species were lumped into MG when they could not be distinguished by visual sampling. We illustrate the analysis of this experiment and the use of Equations 6.1–6.3 to separate richness from abundance effects using Simpson diversity ($D = 1 - \sum_{i=1}^{S} p_i^2$, where p_i is the relative abundance of species i as the response variable. We calculated this diversity index for each panel at each sampling date by focusing on new colonizing

species (MG), therefore excluding manipulated algae form calculations. Simpson's diversity index provides an easily interpretable metric for this task reflecting, in the form used here, the probability that any two individuals (counts in our case) sampled randomly from an infinitely large community belong to different species (functional-form groups). We therefore tested the hypothesis that changes in richness, composition and abundance of early successional algae affected subsequent patterns and rates of assemblage development. Although our study does not focus on classical measures of ecosystem functioning such as productivity or nutrient cycling, it can be viewed as a study of resilience, which still reflects an important functional property of the system (see Maggi *et al.* 2011, for further details of this experiment).

6.3.2 Fitting the mixed-effect model and evaluating contrasts

Data were analysed with a multilevel mixed-effect model (Singer and Willet 2003; McMahon and Diez 2007). Plots of mean values of Simpson diversity against time for the different treatments indicated non-linear, often hump-shaped temporal trajectories of the response variable (Figure 6.3). We therefore modeled the random effects associated with individual experimental units as a linear and quadratic function of time. We also included assemblages as a random effect in the analysis to assess identity effects associated with individual species or MG. Examining differences among random combinations of species or other taxonomic groupings—we called these random combinations assemblages—nested within levels of richness was suggested as an alternative to including monocultures for all manipulated organisms to discriminate between identity and richness effects—e.g., Hooper and Vitousek 1997; Tilman 2001. The statistical model was:

$$Y_{ij} = \pi_{0i} + \pi_{1i}Time_{ij} + \pi_{2i}Time_{ij}^2 + \varepsilon_{ij} \quad \text{(Equation 6.4)}$$

where Y_{ij} is the observation from experimental unit i at time j, modeled as a function of the level 1 coefficients π_{0i}, π_{1i} and π_{2i}; π_{0i} is the true intercept of

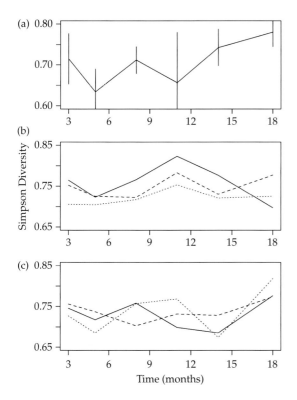

(a) 0.80
0.70
0.60

3   6   9   12   15   18

(b) 0.85
0.75
0.65

3   6   9   12   15   18

Simpson Diversity

(c) 0.85
0.75
0.65

3   6   9   12   15   18
Time (months)

Figure 6.3 Temporal changes of Simpson diversity in (a) cleared (zero taxa) panels, (b) panels established with low algal abundance and (c) panels established with high abundance. Dotted lines: 1 taxon; dashed lines: 2 taxa; continuous lines: 4 taxa. Standard errors are omitted from panels (b) and (c) for clarity of presentation.

experiment unit i, π_{1i} is the true linear trajectory of change of experimental unit i, π_{2i} is the true quadratic effect of time associated with unit i, and ε_{ij} is the level 1 residual reflecting the deviation between the observed and predicted outcome of observation i at time j. Equation 6.4 is further decomposed as:

$$\pi_{0i} = \gamma_{00} + \gamma_{01}R_i + \gamma_{02}A_i + \gamma_{03}RA_i + \phi_{0i} \quad \text{(Equation 6.4a)}$$

$$\pi_{1i} = \gamma_{10} + \gamma_{11}R_i + \gamma_{12}A_i + \gamma_{13}RA_i + \phi_{1i} \quad \text{(Equation 6.4b)}$$

$$\pi_{2i} = \gamma_{20} + \gamma_{21}R_i + \gamma_{22}A_i + \gamma_{23}RA_i + \phi_{2i} \quad \text{(Equation 6.4c)}$$

where γ_{00}, γ_{10} and γ_{20} are the level 2 intercepts describing population average values of level 1 coefficients π_{0i}, π_{1i} and π_{2i}, respectively, when the levels of predictor variables are at 0, with the other γ's describing the effects of richness (R_i), abundance (A_i) and their interactions on level 1 coefficients of observation i and ϕ_{0i}, ϕ_{1i} and ϕ_{2i} are the level 2 residuals allowing individual i's observations to deviate from population averages.

The stochastic component of the model assumed that level 1 residuals were normally distributed with mean 0 and variance σ_ε^2, while the level 2 residuals were assumed to be multivariate normal with unknown parameters σ_0^2, σ_1^2, σ_2^2, σ_{01}^2, σ_{02}^2, and σ_{12}^2 reflecting variance among intercepts, linear temporal trajectories, quadratic effects of time, and related covariances, respectively. An additional component was included to account for random variation among assemblages, σ_{Ass}^2. Variation among assemblages was assumed to be independent of variation within and among experimental units, so the covariance terms between σ_{Ass}^2 and the other stochastic components were set to 0.

Time, abundance, and richness were centered so that intercepts reflected data from the last sampling date at the low level of abundance and at the richness level of one taxon. Hence, we could examine how biodiversity manipulation affected mean values of Simpson diversity at the end of the experiment (Equation 6.4a) and its effects on linear (Equation 6.4b) and quadratic (Equation 6.4c) temporal changes.

Fixed and random parameters were estimated using Restricted Maximum Likelihood (REML) methods, and the significance of individual regression coefficients was tested using the Wald test

(Singer and Willet 2003). Plots of residuals versus fitted values and quantile-quantile plots were examined to assess visually whether the assumptions underlying the fitted model were reasonable.

The Wald test was also employed to test the significance of the linear contrasts used to examine the hypotheses in Equation 6.1 and Equation 6.2. The vector of constants corresponding to Equation 6.1 is {0;3;–1;0.167}, and the linear contrasts were obtained by multiplying this vector by the set of regression coefficients in Equation 6.4a–c. This resulted in three linear contrasts that tested the hypothesis defined by Equation 6.1 on mean values of Simpson diversity at the end of the experiment and on parameters describing linear and quadratic temporal changes, respectively. Equation 6.3 indicated that in our experiment, the difference in percentage cover between levels of factor abundance ($\Delta A = 37.5\%$) did not match the changes in cover occurring across levels of richness within the high and low abundance treatments ($\Delta A_H = 93.75\%$ and $\Delta A_L = 31.25\%$, mean change = 62.5%). Hence, we also derived rescaled contrasts using Equation 6.2 to account for difference between ΔA and the average of ΔA_H and ΔA_L. The vector of constants corresponding to Equation 6.2 is {0;0.064;–0.027;–0.02}.

6.4 Results

Treatment effects on Simpson diversity were generally weak (Figure 6.4, Table 6.1), although some patterns emerged. A non-significant trend ($0.05 < P < 0.09$) suggesting a negative effect of richness on mean diversity, was observed at the end of the experiment (Figure 6.4a). In contrast, there were significant richness × abundance interactions on linear and quadratic terms describing temporal changes in diversity (Table 6.1). Increasing richness had negative effects on these terms under low abundance, while the opposite occurred at high abundance (Figures 6.4b,c). These results clearly showed that the relationship between Simpson diversity and manipulated richness depended on the initial level of abundance at which assemblages were established. Hence, richness effects within levels of abundance were likely to be confounded by density-dependent processes, but the regression analysis could not discriminate between these two sources of variability.

Wald contrasts based on Equation 6.2 indicated that when scaled with respect to a unit of change in abundance, richness had a significant negative effect on linear temporal changes in Simpson diversity, over and above any effect due to density-

Table 6.1   Results of fitting a mixed effect model on Simpson diversity index. *, $P < 0.05$; ***, $P < 0.001$. Parameter estimates lower than 10^{-4} were set to zero.

(A) Fixed effects		Mean at last sampling date (MLD) (π_{0i})		Linear rate of change (π_{1i})		Quadratic rate of change (π_{2i})
		Coefficient (se)		Coefficient (se)		Coefficient (se)
Intercept	γ_{00}	0.7708 (0.0229)***	γ_{10}	0.0085 (0.0056)	γ_{10}	0.0004 (0.0004)
Richness = R	γ_{01}	–0.0251 (0.0143)	γ_{11}	–0.0080 (0.0035)*	γ_{11}	–0.0004 (0.0002)
Abundance = A	γ_{02}	–0.0006 (0.0008)	γ_{12}	–0.0001 (0.0002)	γ_{12}	0.0000 (0.0000)
R × A	γ_{03}	0.0002 (0.0005)	γ_{13}	0.0003 (0.0001)*	γ_{13}	–0.0002 (0.0000)*

(B) Random effects

Variances			Covariances		
Among assemblages	σ^2_{Ass}	0.0000	MLD, linear	σ^2_{01}	–0.0004
Within units	σ^2_{ϵ}	0.0075	MLD, quadratic	σ^2_{02}	0.0000
Among MLD	σ^2_0	0.0040	linear, quadratic	σ^2_{12}	0.0000
In linear rate of change	σ^2_1	0.0001			
In quadratic rate of change	σ^2_2	0.0000			

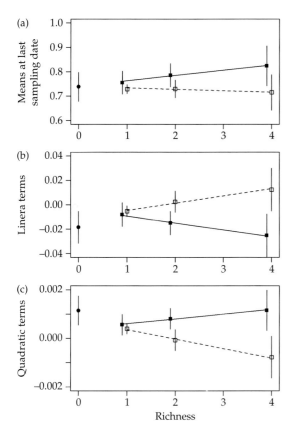

Figure 6.4 Effects of richness and abundance on Simpson diversity. Shown are the prototypical trajectories and fitted means from the mixed-effect model (parameters in Table 6.1) for (a) means at last sampling date, (b) linear, and (c) quadratic terms describing temporal changes in diversity. Treatments included experimental units with 0 (filled circles), 1, 2, and 4 taxa established at low (open squares) and high (filled squares) abundance. The zero treatment was included in the analysis, but it was not used to fit prototypical trajectories. Error bars are 1 standard deviation of the mean.

dependent processes ($W = -0.0005$, S.E. $= 0.0002$, $P < 0.05$). The same result was obtained using Equation 6.1 ($W = -0.0244$, S.E. $= 0.0106$, $P < 0.05$). In contrast, the analysis on the quadratic term failed to reject the null hypothesis of a richness effect over and above density-dependent processes (Equation 6.2: $W = -0.0003$, S.E. $= 0.00001$; Equation 6.1: $W = -0.0012$, S.E. $= 0.0006$; $0.08 < P < 0.05$ in both cases). Contrasts were not evaluated for mean values of Simpson diversity at the end of the experiment because treatment had no significant effect on the last sampling date (Table 6.1).

On balance, these analyses indicated that when manipulated at low abundance, algal richness was negatively related to the linear and quadratic terms describing temporal changes in Simpson diversity. The relationship between Simpson diversity and time shifts from positive to negative with increasing richness (see coefficients in Figures 6.4b,c). Simpson diversity was a concave-down function of time in these circumstances, as also shown in Figure 6.3b. The pattern observed for linear terms reflected a true effect of richness, whereas the pattern displayed by quadratic terms was driven by density-dependent effects. When manipulated at high abundance, richness had slightly positive effects on linear and quadratic terms (positive linear terms in Figure 6.4b). These results indicated that increasing total algal cover outbalanced the negative effects of algal richness at low abundance.

The stochastic part of the fitted model indicated that variances in intercepts and slopes, the variance among assemblages (reflecting identity effects) and covariance terms were negligible (Table 6.1). A reduced model retaining only random intercepts was significantly undistinguishable from the full model (log-likelihood ratio test: $\chi 2 = 4.5$, d.f. $= 6$, $P > 0.5$). This reduced model would provide a more parsimonious construct than the full model for predictive purposes.

6.5 Conclusions

We have shown how richness effects can be partitioned from density-dependent processes in biodiversity experiments that include factors richness and abundance in factorial combinations. The mixed-effect model revealed significant richness × abundance interactions on the linear and quadratic terms describing temporal changes in Simpson diversity. These interactions indicated that richness effects depended on the abundance at which assemblages were established (and *vice versa*), but they did not provide a direct measure of the relative magnitude of these effects. The analysis was implemented using the linear contrasts defined in Equation 6.1 and Equation 6.2, which enabled richness effects to be assessed relative to density-dependent processes.

It is important to note that these contrasts complement the mixed-effect model analysis of richness and abundance effects and by no means should they be viewed as an alternative to this analysis, or a way to force the interpretation of a main effect of richness when it interacts with abundance. In the presence of richness × abundance interactions, as observed in our study, the linear contrasts provide a means to assess whether richness or density-dependent processes prevail in driving the interaction. Our results indicated that richness effects prevailed in determining negative linear temporal trends in Simpson diversity, whilst density-dependent effects prevailed in determining quadratic trends. Unfortunately, we do not have a solid explanation for this result. We note, however, that failure to reject the null hypothesis of no richness effects for the quadratic terms originated from strict adherence to a Type I error rate of $\alpha = 0.05$. Having obtained $0.08 < P < 0.05$ as the outcome of the linear contrast, it may well be the case that richness effects are not as unimportant as the dichotomous nature of decisions in hypothesis testing implies.

We can now focus on the mechanisms accounting for the negative effect of richness on temporal trends of diversity when algal taxa are established at low abundance, and why increasing total algal cover reversed this effect. As shown in Figure 6.3b, increasing richness under low abundance resulted in greater peaks and more drastic declines of Simpson diversity. This pattern was not evident under high levels of algal abundance where Simpson diversity tended to increase by the end of the experiment in all richness treatments (Figure 6.3c). Hence the negative effect of richness on linear and quadratic terms estimated by the mixed-effect model reflected increasing fluctuations in Simpson diversity with increasing algal richness, for assemblages established at low abundance.

The low abundance treatments had more than 60% of the substratum unoccupied at the beginning of the experiment, while only 30% of the substratum was free in high abundance treatments. There was therefore a greater potential for richness effects to operate more effectively under low than high levels of abundance. Manipulated filamentous algae might have provided a suitable environment for

propagule settlement of later colonists either by maintaining a moist microhabitat at low tide, or by physically entrapping spores and larvae (Santelices 1990). Similarly, encrusting algae might have enhanced propagule settlement by increasing the topographic complexity of the substratum at the micro-scale or by releasing chemical cues (Hadfield & Paul 2001). These complementary, facilitative effects might account for the peak in Simpson diversity observed under high richness and low algal abundance after ten months from the start of the experiment. Positive effects, however, fade away afterwards, leading to a marked decline in diversity. The similarity of effects of encrusting and filamentous algae may also explain why identity effects were unimportant in our study, in contrast to what observed in many other biodiversity experiments (Stachowicz et al. 2007).

Assemblages established at high algal abundance underwent different temporal dynamics in diversity compared to those faced by low abundance treatments. With most of the substratum already occupied by manipulated algae, colonists establishing in high abundance treatments required relatively long periods to become established. Hence, high algal abundance delayed the build-up of diversity that peaked by the end of the experiment for all levels of algal richness (Figure 6.3). Other studies on marine sessile organisms have shown inhibitory effect due to space monopolization on species colonization—e.g. Stachowicz et al. 1999. In our case, however, space monopolization by early successional algae delayed, but did not prevent the build-up of diversity. The key aspect of our results was, however, that the effect of richness on temporal dynamics of diversity changed in relation to the total abundance at which early assemblages were established (see Maggi et al. 2011 for further results and responses of individual taxa from this experiment).

In synthesis, we believe that the analysis of biodiversity experiments can be advanced by paying more attention to separating richness from density-dependent processes. These processes can and should be scrutinized because they may operate in concomitance with richness effects—so that observing one effect does not exclude the other—and they may contribute to explain some of the currently unexplained variability in biodiversity-ecosystem functioning experiments. We have proposed a test based on linear contrasts that works in this direction. Our test can be easily extended to other study designs that include abundance as a factor, and can be used to partition biodiversity effects among complementary, identity, and density-dependent mechanisms in a wide range of aquatic and terrestrial systems.

References

Allison, G.W. (1999) The implications of experimental design for biodiversity manipulations. *American Naturalist*, **153**, 26–45.

Aukema, B.H., Clayton, M.K. and Raffa, K.F. (2004) Density-dependent effects of multiple predators sharing a common prey in an endophytic habitat. *Oecologia*, **139**, 418–26.

Benedetti-Cecchi, L. (2004) Increasing accuracy of causal inference in experimental analyses of biodiversity. *Functional Ecology*, **18**, 761–8.

Benedetti-Cecchi, L. and Cinelli, F. (1994) Recovery of patches in an assemblage of geniculate coralline algae: variability at different successional stages. *Marine Ecology Progress Series*, **110**, 9–18.

Bulling, M.D., Solan, M., Dyson, K.E. *et al.* (2008) Species effects on ecosystem processes are modified by faunal responses to habitat composition. *Oecologia*, **158**, 511–20.

Cardinale, B.J., Srivastava, D.D., Duffy *et al.* (2006) Effects of biodiversity on the functioning of trophic groups and ecosystems. *Nature*, **443**, 898–992.

Connolly, J. (1988) What is wrong with replacement series. *Trends in Ecology and Evolution*, **3**, 24–6.

Douglass, J.G., Duffy, J.E. and Bruno, J.F. (2008) Herbivore and predator diversity interactively affect ecosystem properties in an experimental marine community. *Ecology Letters*, **11**, 598–608.

Downing, A.L. and Leibold, M. A. (2002) Ecosystem consequences of species richness and composition in pond food webs. *Nature*, **416**, 837–41.

Godbold, J.A., Solan, M. and Killham, K. (2008) Consumer and resource diversity on marine macroalgal decomposition. *Oikos*, **118**, 77–86.

Griffen, B.D. (2006) Detecting emergent effects of multiple predators species. *Oecologia*, **148**, 702–9.

Griffin, J.N., De La Haye, K.L., Hawkins, S.J. *et al.* (2008) Predator diversity and ecosystem functioning: density modifies the effect of resource partitioning. *Ecology*, **89**, 298–305.

Griffin, J.N., Méndez, V., Johnson, A.F. *et al.* (2009) Functional diversity predicts overyielding effects of species combination on primary productivity. *Oikos*, **118**, 37–44.

Hadfield, M.G. and Paul, V.J. (2001) Natural chemical cues for settlement and metamorphosis of marine-invertebrate larvae. In: *Marine chemical ecology*, McClintock, J. B. and Baker, J. B. (eds), CRC Press, Boca Raton, 431–61.

Harper, J. L. (1977) *The population biology of plants*. Academic Press, London.

He, J., Wolfe-Bellin, K.S., Schmid, B. *et al.* (2005) Density may alter diversity-productivity relationships in experimental plant communities. *Basic and Applied Ecology*, **6**, 505–17.

Hector, A., Schmid, B., Beierkuhnlein, C. *et al.* (1999) Plant diversity and productivity experiments in European grasslands. *Science*, **286**, 1123–7.

Hooper, D.U., Chapin, F.S., III, Ewell, J.J. *et al.* (2005) Effects of biodiversity on ecosystem functioning: a consensus of current knowledge. *Ecological Monographs*, **75**, 3–35.

Hooper, D.U. and Vitousek, P.M. (1997) The effects of plant composition and diversity on ecosystem processes. *Science*, **277**, 1302–5.

Huston, M.A. (1997) Hidden treatments in ecological experiments: re-evaluating the ecosystem function of biodiversity. *Oecologia*, **110**, 449–60.

Huston, M.A. and McBride, A. C. (2002) Evaluating the relative strengths of biotic versus abiotic controls on ecosystem processes. In: *Approaches to Understanding Biodiversity and Ecosystem Function*, Loreau, M., Naeem, S. and Inchausti, P. (eds), Oxford University Press, Oxford, 47–60.

Joliffe, P. A. (2000) The replacement series. *Journal of Ecology* **88**, 371–85.

Loreau, M. (1998) Separating sampling and other effects in biodiversity experiments. *Oikos*, **82**, 600–2.

Loreau, M. (2000) Biodiversity and ecosystem functioning: recent theoretical advances. *Oikos* **91**, 3–17.

Loreau, M. and Hector, A. (2001) Partitioning selection and complementarity in biodiversity experiments. *Nature*, **412**, 72–6.

Maggi, E., Bertocci, I., Vaselli, S. *et al.* (2011) Connell and Slatyer's models of succession in the biodiversity era. *Ecology* **92**, 1399–1406.

McMahon, S.M. and Diez, J.M. (2007) Scales of association: hierarchical linear models and the measurement of ecological systems. *Ecology Letters*, **10**, 437–52.

Naeem, S., Thompson, L.J., Lawler, S.P. *et al.* (1994) Declining biodiversity can alter the performance of ecosystems. *Nature*, **368**, 734–7.

O'Connor, N.E. and Crowe, T.P. (2005) Biodiversity loss and ecosystem functioning: distinguishing between number and identity of species. *Ecology*, **86**, 1783–96.

Santelices, B. (1990) Patterns of reproduction, dispersal and recruitment in seaweeds. *Oceanography and Marine Biology Annual Review*, **28**, 177–276.

Schmid, B., Hector, A., Huston, H.A. *et al.* (2002) In: Loreau, M., Naeem, S. and Inchausti, P. (eds), *Biodiversity and Ecosystem Functioning*, Oxford University Press, Oxford, 61–78.

Singer, J.D. and Willett, J.B. (2003) *Applied longitudinal data analysis: modeling change and event occurrence.* Oxford University Press, New York.

Stachowicz, J.J., Bruno, J. and Duffy, J.E. (2007) Understanding the effects of marine biodiversity on community and ecosystems. *Annual Review of Ecology, Evolution and Systematics*, **38**, 739–66.

Stachowicz, J.J., Whitlatch, R.B. and Osman, R.W. (1999) Species diversity and invasion resistance in a marine ecosystem. *Science*, **286**, 1577–9.

Taylor, D.R. and Aarssen, L.W. (1989). On the density dependence of replacement-series competition experiments. *Journal of Ecology*, **77**, 975–88.

Tilman, D. (2001) Distinguishing between the effects of species diversity and species composition. *Oikos*, **80**, 185.

Tilman, D., Wedin, D. and Knops, J. (1996) Productivity and sustainability influenced by biodiversity in grassland ecosystems. *Nature*, **379**, 718–20.

Underwood, A.J. (1978) An experimental evaluation of competition between three species of intertidal prosobranch gastropods. *Oecologia*, **33**, 185–202.

CHAPTER 7

The importance of body size, abundance, and food-web structure for ecosystem functioning

Mark C. Emmerson

7.1 Introduction

Marine ecosystems are astoundingly complex, often with hundreds of plant and animal species co-existing together. In an attempt to describe this complexity, Darwin, over one 150 years ago, evoked, rather provocatively, the concept of a 'tangled bank' of species. He recognized explicitly the importance of species interactions. We now recognize that these interactions are important for the maintenance of ecological stability in the face of natural and anthropogenic perturbations. Yet the mechanisms which underlie ecosystem stability in the face of environmental change remain elusive. Arguably, over the last 150 years, community ecologists have failed to deliver a predictive science. Yet the need for understanding, and for predictions over the likely consequences of natural and anthropogenic perturbations, has never been more pressing, especially as the world's ecosystems undergo unprecedented changes with species being lost from a wide range of ecosystems and trophic levels (Duffy *et al.* 2007). The drivers of these biodiversity changes are known to include the processes of habitat loss and fragmentation (Fahrig 2003), the introduction of alien species (Seabloom *et al.* 2006), and the cascading effects produced by the internal dynamics of natural systems (Estes *et al.* 1998).

Biodiversity loss is pervasive, does not always result from the loss of species, but also of local populations, and is ultimately driven by the scale of human endeavour and success. Whilst we have excelled in many ways, a limiting step in our ability to predict how ecosystems work and function has been a lack of understanding over what determines the strength of ecological interactions among species. The complexity that emerges from predator–prey interactions can be considered in the context of food-webs, which provide a way of visualizing the feeding links among species, and represent a unifying framework within which to consider the drivers of biodiversity change in a wide range of ecosystem types. Food-webs have been a central organizing concept in ecology since the early 1920s (Elton 1927), and provide a useful tool for representing the complexity of communities. Patterns of body size, interaction strength, energy and material flows, and indirect effects can all be analysed using the food-web framework. In the face of continued biodiversity loss, food-webs therefore provide one way to identify the basic constraints underpinning the dynamics and functioning of natural and managed ecosystems.

Whilst the food-web framework is well established, challenges remain, and a major limitation of the food-web approach has been a poor mechanistic understanding of processes driving patterns of interaction strength, the basic terms used to create parameters for food-web models. In conjunction, the computational facilities required to simulate complex dynamical systems have historically been limited, and finally, whilst the ecological literature is replete with data, these data are lacking from a food-web perspective. Arguably, a predictive food-web framework requires good quantitative descriptions of: 1) species interactions (who eats whom), 2) the population sizes of species, 3) the body sizes of component species, and 4) descriptions of how all of these change over time (see Figure 7.1). Whilst parts of this trivariate data structure (Cohen *et al.* 2003) are routinely collected—i.e. body mass, abundance, or food-web structure—there are still relatively few studies that are able to present all three (Jonsson *et al.* 2005; Woodward *et al.* 2005; Layer *et al.* 2010; McLaughlin *et al.* 2010; O'Gorman and Emmerson 2010). To date, no studies are available that quantify how these properties change over at least two generations of the largest organism in the food-web. The lack of such fundamental data hampers the development of a predictive science: firstly, because the basic building blocks of food webs are missing—e.g. body mass, who eats whom, and the strength of species interactions—and secondly, because the multispecies time series describing food-web dynamics are missing, without such data it is impossible to validate the predictions of food-web models. Food-webs therefore represent

a valuable tool, but one which still lacks any predictive credibility.

This chapter will focus on the building blocks of food-webs, body mass, abundance, and predator–prey interactions, and the interrelationships between body mass and trophic position—where trophic position emerges as a function of species specific predator–prey interactions when embedded within a community—between body mass and abundance, between abundance and trophic position, and between body mass and the strength of trophic interactions. The examples used are not restricted to marine ecosystems, reflecting the possibility that an understanding of the drivers and consequences of biodiversity change, in a wide range of ecosystems, might be gained if ecologists are able to synthesize information from disparate study systems. Working beyond traditional discipline boundaries focused on marine, freshwater, or terrestrial ecology may enable us to identify some simple basic constraints that govern food-web, community, and ecosystem dynamics.

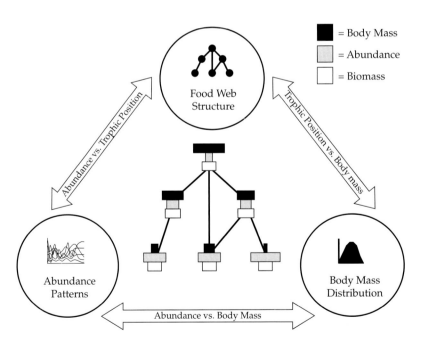

Figure 7.1 Conceptual relationships between body mass, abundance and food web structure. Combining the measurement of body mass, abundance, and predator–prey interactions provides a way of integrating consistently measured components of communities (after Cohen *et al.* 2003).

7.2 Historical context and the evolution of an idea

The field of Biodiversity and Ecosystem Functioning (BEF) research has emerged from studies in the early 1970s that were largely focused on relationships between diversity and stability in communities and food-webs (Gardner and Ashby 1970; May 1973; Pimm and Lawton 1977; Pimm and Lawton 1978; Yodzis 1981). These studies focused on the stability properties of mathematical models that described the dynamical behaviours of model and real-world food-webs close to equilibrium. These studies quantified either the probability of stability occurring as a function of species richness within the food-webs, or they focused on a measure of stability, here resilience, which described how the food-webs would recover and return to equilibrium after a perturbation. It was generally concluded that increased complexity of food-webs would reduce the probability of finding stable food-webs when the predator–prey interaction strengths in these webs were parameterized at random (May 1973; Pimm and Lawton 1977). Increasing food-chain length and the presence of omnivory also reduced stability by increasing the time it took for food-webs to recover to equilibrium following a perturbation (Pimm and Lawton 1978). In a study of 40 real food-webs, these findings were subsequently reconciled with the observation that complex systems exist in nature (Yodzis 1981). Yodzis assigned the strengths of predator–prey interactions based on the traits of the predator in each interaction. Specifically, Yodzis (1981) assumed that interaction strengths could be defined using the basal metabolic rates of each predatory species, and that these in turn could be estimated using the body masses of species in the food-webs.

A limitation of these theoretical explorations was that the specific measure of stability, resilience, or the return time to equilibrium was difficult to measure empirically. The populations of species in most natural communities fluctuate either around a long-term time average (e.g. Hunt *et al.* 1987), or show pronounced non-equilibrium behaviour (e.g. Stenseth *et al.* 1997). Against this backdrop of variability, it is empirically difficult to determine when a community has finally settled at equilibrium. Consequently, dissatisfaction with the mathematical equilibrium-based approaches in community ecology, which were largely focused on small food-web modules, led to the development of conceptual relationships between the number of species (biodiversity) and ecosystem functioning; e.g. the rivet (Ehrlich and Ehrlich 1981), redundancy (Walker 1992), and idiosyncratic hypotheses (Lawton 1994). Throughout the early 1990s, research efforts flourished and considerable advances were made in exploring the relationships between biodiversity and ecosystem functioning (see McCann 2000 for a comprehensive review). These advances helped to identify issues over the interpretation of biodiversity and ecosystem functioning experiments—e.g. hidden treatments (Huston 1997), pseudo-replication (Hurlbert 1984; Huston 1997), and portfolio effects (Doak *et al.* 1998; Tilman 1999)—and mechanisms producing positive biodiversity–ecosystem functioning relationships—e.g. complementary resource use (Loreau and Hector 2001) and insurance effects (Yachi and Loreau 1999).

The vast majority of work carried out during the 1990s focused on biodiversity changes within single trophic levels, and dealt almost exclusively with plant communities or their invertebrate consumers (Duffy *et al.* 2007). Recently, there has been considerable interest in the consequences of biodiversity change across and within multiple trophic levels (Duffy *et al.* 2001; Duffy 2002; Dulvy *et al.* 2004; Hillebrand and Cardinale 2004; Bascompte *et al.* 2005; Bruno and O'Connor 2005; Finke and Denno 2005; Gamfeldt *et al.* 2005 Byrnes *et al.* 2006; Dobson *et al.* 2006; Duffy *et al.* 2007; Long *et al.* 2007; O'Connor and Bruno 2007; Bruno *et al.* 2008; Griffin *et al.* 2008; O'Gorman *et al.* 2008; Byrnes and Stachowicz 2009; Hillebrand *et al.* 2009; O'Connor and Bruno 2009; Moran *et al.* 2010), placing biodiversity and ecosystem functioning research firmly in a food-web context.

From a food-web perspective, diversity–stability research has also continued, following two broad paths, either the study of food-web dynamics or food-web topologies. Research focused on food-web dynamics has been largely confined to the study of relatively small model food-webs (see Figure 7.2a), and has been based on the exploration of the mechanisms that result in feasibility; that is, species have positive equilibrium population densities (e.g. Emmerson and Yearsley 2004); persistence, whereby species populations co-exist in the long term (e.g. McCann *et al.* 1998); and stability, i.e. species populations return to equilibrium following a small disturbance (e.g. Jonsson and Ebenman 1998). There are many other studies too numerous to cite here. In contrast, the study of food-web topology has been based largely on the empirical description of large and complex food-webs from a range of ecosystem types (see Figure 7.2b) (Riede *et al.* 2010; Yvon-Durocher *et al.* 2010). These studies have quantified empirical patterns of energy and material flow (Layman *et al.* 2007), food-web structure and topology (Dunne *et al.* 2002a; Montoya and Sole 2002; Riede *et al.* 2010),

and the pattern of species specific traits within food webs, e.g. how body mass and abundance change between ecosystems and across trophic levels (Cohen *et al.* 2003; Jonsson *et al.* 2005; Yvon-Durocher *et al.* 2010).

In a food-web context, the dynamic and topological lines of research have been reunited by combining the study of patterns in body mass, abundance, and food-web structure (e.g. Brose *et al.* 2006). The emphasis on these lines of research has yet to translate into the exploration of multitrophic-level biodiversity and ecosystem functioning-focused questions, and this clearly represents an avenue for future research. In the following sections, I highlight some of the opportunities and issues associated with exploiting our growing understanding of how body mass, abundance, and food-web structure interplay to affect stability, and how this in turn might be used to address multitrophic level biodiversity and ecosystem functioning questions.

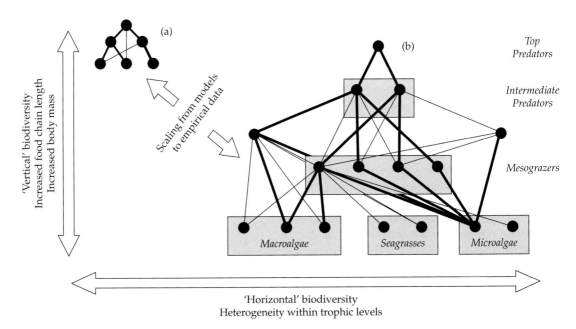

Figure 7.2 Conceptual relationships between vertical (food-chain length) and horizontal (heterogeneity) components of biodiversity. Scaling dynamics from (a). simple model food-webs, to (b). complex empirical descriptions of food-webs is a major challenge. Changes in biodiversity may influence ecosystem processes via altered horizontal biodiversity within trophic levels, or vertical biodiversity reflecting impacts on food chain length (e.g. loss of predators), or altered body masses (e.g. loss or harvesting of predators). Loss of predator species, functional groups (grey rectangles), or body mass size classes might alter the strength of top-down control by either altering food-web structure or the relative strengths of trophic interactions as predator and prey body masses change. The complex web reflects a partial food-web from an eelgrass bed (*Zostera marina*) in Chesapeake Bay, USA (after Duffy 2006).

7.2.1   Integrating body mass, abundance, and food-web structure into biodiversity and ecosystem functioning studies

The importance of body size, hereafter body mass, has been recognized for many years in ecology (Peters 1986; Calder 1996). These studies focused on the use of allometric relationships to explore the dependency of physiological rates on body mass, exploring how physiological processes might scale to influence and determine community level patterns. The mechanisms underlying regularities in these patterns at the community or macro-ecological scale have also received much attention (Cyr et al. 1997a; Kerr and Dickie 2001; White et al. 2007). Recently, the study of body mass within food-webs has provided fruitful avenues for research, with clear patterns of increasing body mass with trophic height (Jonsson et al. 2005), increased risk of extinction for large bodied species due to anthropogenic activity (Solan et al. 2004), and increased susceptibility of large-bodied, high trophic-level species to climate change impacts (Petchey et al. 1999). Experimental manipulations of consumer body masses at higher trophic levels would provide clear insights into the consequences of climate change, specifically warming, and harvesting.

The focus on body-mass relationships has stimulated theories about how species population dynamics, interaction strengths, and food-web structure could depend on predator–prey body-size ratios (Yodzis and Innes 1992; Emmerson and Raffaelli 2004; Brose et al. 2006). Since the distribution of trophic links and interaction strengths in a community may reflect size constraints on who eats whom, the ratio of body sizes between predators and prey may play an important role in explaining regularities in food-web structure (Brose et al. 2005) and stability (Neutel et al. 2002; Brose et al. 2006). The link between predator and prey body mass and the strength of trophic interactions (Jonsson and Ebenman 1998; Aljetlawi et al. 2004; Emmerson and Raffaelli 2004; Emmerson et al. 2005; Brose et al. 2008; Vucic-Pestic et al. 2010) therefore provides an opportunity to integrate dynamic and topological food-web approaches using body mass-based parameter estimates for the strength of trophic interactions. Empirical descriptions of food-webs that detail patterns of body mass across trophic levels, and which, essentially, detail the body masses of predators and prey, therefore facilitate the development of simple models, using simple rules, that could describe complex ecosystem dynamics.

The pattern of predator and prey body masses is known to vary across habitats and predator and prey types; e.g. predator–prey body-size ratios are on average significantly higher (1) in freshwater habitats than in marine or terrestrial habitats, (2) for vertebrate than for invertebrate predators, and (3) for invertebrate than for ectotherm vertebrate prey (Brose et al. 2006a). These patterns of predator and prey body mass mean that the predator–prey body-mass ratio will vary in a systematic and predictable way between ecosystem types and taxonomic groups, suggesting that the pattern of interaction strengths will vary in an equally predictable way. Such predictions remain to be formally tested, and represent an avenue of interesting research.

From a biodiversity and ecosystem-functioning perspective, the study of large complex food-webs represents a major challenge (Stachowicz et al. 2007). BEF studies tend to be experimental in nature, testing specific hypotheses about the consequences of biodiversity change (Duffy et al. 2001; Emmerson et al. 2001; Byrnes and Stachowicz 2009), or they examine the basic mechanisms leading to diversity effects (Bruno et al. 2008; Griffin et al. 2008). Manipulations of diversity in a food-web context are difficult for a number of reasons: 1) food-webs span very large spatial scales beyond the scope of most research groups to manipulate experimentally in a replicated fashion; and 2) a change in predator species richness affects the horizontal diversity within trophic levels and the vertical diversity across trophic levels; i.e. how many trophic levels exist in the system. In addition to these horizontal and vertical components of biodiversity, the trophic complexity of the system will also be affected. To provide some context, here I provide a brief overview of some basic food-web concepts and measures.

Trophic complexity can be described in a number of ways: at the food-web scale, connectance (C) describes the probability of any two species interacting and is calculated as $C = L/(S^2 – S)$, where L is the number of links observed in the food-web and S is the number of species. Connectance can be thought of as the number of possible interspecific links in the food-web under study that are realized, and provides a measure of trophic complexity that is comparable across food-webs. At the level of individual species within food-webs, complexity can also be described in a number of ways; these include: *generality*, which describes the number of prey for each species; *vulnerability* describes the number of predators for a given species; and *connectedness* describes how many predators and prey each species has, which provides a measure of how highly connected the species is within the food-web (Digel *et al.* 2010). Connectance is important for food-web stability, where stability can also be measured in a range of ways. For example, early studies recognized that there was an interplay between the number of species (S), the average strength of species interactions (A), and connectance (C), so that if $A(SC)^{1/2} < 1$ then the resulting food-web would be locally stable (May 1973); that is, it would return to equilibrium following a small perturbation. Here, if the diversity (S) of a food-web were to increase, and the average strength of the trophic interactions remained the same, then connectance (C) would have to decline for the web to remain stable. The probability of stability is often used as a measure of a systems stability. There are other measures including *resilience*, which describes how long it takes for the community to recover after a perturbation; *resistance*, describing how much the community changes following a perturbation; *persistence*, which describes how long component species populations are able to persist together before species become extinct; and *variability*, where more variable populations or community metrics, e.g. biomass, are considered less stable. Variability is often measured using the coefficient of variation. More recently the term 'robustness', has been used to describe the number of secondary extinctions that occur after an initial species becomes extinct within a community or food-web (Dunne *et al.* 2002b, 2004). See McCann (2000) and Montoya *et al.* (2006) for comprehensive reviews.

How the complexity of food-webs affects their stability has been the focus of a range of theoretical studies. For example, McCann *et al.* (1998) found that the pattern of weak and strong trophic interactions were important for reducing variation in the population sizes of species embedded in simple model food-webs. When trophic interactions were just strong, species populations within the food-webs fluctuated dramatically, but when weak interactions were also included, then the weak interactions had a stabilizing effect, damping the large oscillations in population size caused by strong predator–prey interactions. Emmerson and Yearsley (2004) also found that the arrangement of weak interactions within omnivorous loops in model food-webs conferred local and global stability onto the resulting food -webs. Using empirically described complex soil food-webs, Neutel *et al.* (2002) developed the concept of loop weight, calculated as the geometric mean of interaction strengths in omnivorous loops. They showed that the heaviest loops, containing sets of strong interactions, tended to occur at low trophic levels in food-webs and were of only three links in length. In contrast, loops that spanned many trophic levels from the bottom to the top of a food-web tended to be less heavy, and so contained combinations of weaker interactions. Therefore the pattern and arrangement of weak interactions was vitally important for conferring stability to these food-webs.

Food-web stability is therefore not a simple function of how many species are present within a trophic level or across trophic levels. Rather, stability emerges as a function of the interplay between species richness, complexity (measured in a number of ways), and the strength of predator–prey interactions. In food-webs, these interaction strengths depend on the body mass of consumers and their prey, and their respective population sizes. The building blocks of body mass and population size could therefore provide the basis with which to explore the consequences of species loss in food-webs.

It seems obvious that there should be a relationship between the stability of an ecosystem or food-web and the functioning of that same system. Yet until recently, ecologists have largely failed to make explicit links between the many measures of stability and ecosystem functioning. A notable exception

is the use of variability, or the coefficient of variation, as a measure of stability (Tilman 1996). In
empirical studies, variability over time is typically
quantified from aggregate measures of biomass at
the community level. In turn, the component species level measures are derived from organismal
mass and population size, again, two measures of
importance at the food-web scale. Many studies have
now shown that in more diverse communities, ecosystem processes measured at the aggregate scale
are less variable over time. This phenomena has been
termed the 'portfolio effect' (Doak *et al.* 1998). The
portfolio effect suggests that even in the absence of
ecological interactions, statistical effects can cause
more diverse communities to have lower oscillations
in aggregate ecosystem measures such as biomass. It
was subsequently argued, using empirical data from
a large scale BEF study, that reduced variability
could occur due to species competitive interactions
producing compensatory dynamics that maintained,
and hence stabilized, community biomass (Tilman
et al. 1998). It is now recognized that asynchronous
population dynamics brought about by interspecific
interactions stabilize aggregate measures of ecosystem functioning, and that such asynchronies largely
depend on the way that different species have independent and different tolerances to the environment,
which varies at spatial and temporal scales (Yachi
and Loreau 1999; Loreau 2010).

In a food-web context, therefore, BEF studies
which employ factorial experimental designs will
typically replicate for either species or functional
group richness within or across trophic levels. These
studies must also account for how food-web complexity varies as species richness changes in experimental food-webs. This is not trivial and can rapidly
increase levels of replication, scale, and the logistics
of implementing this type of experimental design.
The accepted view for many years has been that
food-web loops featuring omnivory were rare in
many systems; however, recent highly resolved
descriptions of food-webs show this not to be the
case (see Figure 7.2). These food-web loops, and/or
food chains, would most likely form the basis of
model experimental units requiring replication and
control for species identity effects. Omnivory is
present in food-webs, but often omnivorous inter-

actions are associated with ontogenetic shifts in
diet, as small consumers grow into large consumers, and consequently feed across a wide range of
prey body mass, and hence trophic levels as they
grow. Replicating these types of food-web structures, not just diversity within and across trophic
levels, may represent a major stumbling block in
experimental biodiversity and ecosystem functioning studies.

7.3 The relevance of body mass to biodiversity–ecosystem functioning research

Body size is one of the most fundamental traits of
an organism. It is linked to biological rates such as
growth, respiration, reproduction, and mortality
(Peters 1986; Brown *et al.* 2004), and hence is fundamental to our understanding of dynamical biological systems. The role of body mass in structuring
communities has been recognized for many years
(Peters 1986). These early studies made extensive
use of allometric relationships to predict a wide
range of species traits and system properties from
energetics to abundance. Subsequent early studies
also tried to relate body mass and energetics to the
dynamical properties of food-webs (Yodzis and
Innes 1992), while more recently the development
of a Metabolic Theory of Ecology has met with some
controversy (Brown *et al.* 2004). The use of emergent
features, such as food-web structure, abundance,
and body mass, and the relationship between body
mass and the energetics of the species in a food-
web, provide a basis for studying the consequences
of biodiversity change using body mass as a trait
that might underpin the exploration of food-web
dynamics and ecosystem processes in a wide range
of different situations.

Recently, it has been shown that across a wide
range of ecosystems, including marine, standing
freshwaters, and terrestrial systems, as predator
mass (M_j) increased, so too did prey body mass (M_i)
(Riede *et al.* 2011). Using a database of 35 food-webs
including 1313 predators and their associated interactions, these authors showed comprehensively that
predator body mass increased with trophic height. It
was found that average prey body mass increased

disproportionately as predator mass increased, so that at higher trophic levels, there were systematic declines in predator–prey body mass ratios (M_j/M_i), showing that predators at the top of food-webs are, on average, more similar in size to their prey than predators at the base of food-webs.

A number of recent studies have explored the relationship between predator and prey body mass and interaction strength, measured empirically as either a *per capita* effect (Emmerson and Raffaelli 2004), or as components of a functional response (Brose *et al.* 2008; Vucic-Pestic *et al.* 2010). These studies show that the relationship between predator and prey body mass ratio and interaction strength is positive and linear at small to intermediate body mass ratios (see Figure 7.3a), but above some critical threshold, either predators do not perceive prey, or the predator makes a trade-off against

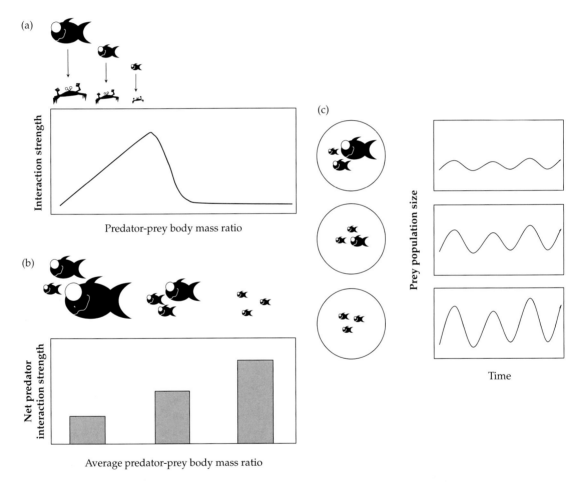

Figure 7.3 Conceptual relationships between predator and prey body mass, interaction strength, and prey population dynamics. (a) The relationship between predator (*j*) and prey (*i*) body mass ratio (M_j/M_i) and *per capita* interaction strength (a_{ij}) is linear at low to intermediate body mass ratios. When the body mass ratio is too large, the predator fails to perceive the prey and interaction strengths decline, producing a hump-shaped relationship between predator and prey body mass ratio and interaction strength. Large-bodied predators tend to have small predator–prey body mass ratios, whilst small predators have large predator–prey body mass ratios (Riede *et al.* 2011); (b) As body mass is reduced in harvested systems, average population level interaction strengths ($\sum a_{ij} X$) will increase, where *X* is the abundance of the predator in the population; (c) Intact predator assemblages will have a balance of weak and strong interactions reducing average interaction strength, which in turn will result in more stable prey population dynamics. In contrast, in harvested systems where large-bodied consumers have been removed, body mass ratios and hence interaction strengths increase, destabilizing prey population dynamics.

time spent pursuing a small prey item versus the energetic gain from such an activity. This trade-off leads to a hump-shaped relationship between the predator–prey body mass ratio and interaction strength. These results suggest that large-bodied predators at higher trophic levels will interact relatively weakly with their prey (See Figure 7.3a). The biological basis for this is that a large predator which kills and consumes a large prey may not need to kill again quickly, fulfilling its nutritional requirements with one prey item. In contrast, small predators at the base of webs will have relatively large body mass ratios, resulting in stronger trophic interactions with their prey (see Figure 7.3a). Again, the biological basis for this is that as the prey are relatively small compared to the predator, then the predator will need to consume more small prey items to fulfil its nutritional requirements.

These results are not intuitive and suggest that in harvested systems, as size structure is lost due to over-exploitation of large bodied individuals (e.g. Jennings *et al.* 2002), then body mass ratios will become larger and hence trophic interactions will become stronger (Figure 7.3bB). For example, Jennings and Warr (2003) found that in harvested marine systems, body mass ratios between predators and their prey were smaller in longer food chains, suggesting that interactions are on average weaker in longer food chains. This result is analogous to the findings of Neutel *et al.* (2002) who found that long omnivorous loops in food-webs were composed of weak links and conferred stability onto the resulting food-webs. McCann *et al.* (1998) found that when all interactions in a food-web were strong, then large oscillations in prey population dynamics resulted, but if there was a balance of weak and strong interactions within the system, then fluctuations in prey population size were reduced. There is a direct analogy with the overharvesting of marine systems, which has an effect on size structure, and consequently on the pattern and arrangement of trophic interactions determined by predator and prey body mass ratios. The consequence of harvesting large-bodied organisms and changing the size structure of the community is that as large, yet weak, interactors are removed from a system, then average interaction

strengths will increase. In turn, prey variability will increase and stability will decline (Figure 7.3c).

These patterns raise important questions in the context of the BEF debate. Body mass is a key trait of many species. It is easy to manipulate and measure, both between and within species, and there are many ways in which size versus biodiversity effects on prey population dynamics could be quantified. Simple manipulations of predator body size within treatments could alter the pattern of interaction strengths, destabilizing prey population dynamics. In an applied context, there are important implications for the management and predictability of fished stocks that have had their size structure altered. Fishing down these webs is likely to increase the variability of component populations, making predictions over their future population size much harder. There are limitations of these size-based approaches: size-based predator–prey interactions do not hold for all types of consumer resource interaction. The provision of biocontrol services in agro-ecosystems, where parasitoids prey on pest aphids, may not be well reflected using these size-based interactions.

7.4   Abundance, body mass, and species diversity patterns

Here, it has been argued that body mass is a critical functional trait that contributes to many of the properties that confer stability on ecological networks, and therefore to the stable provision of the ecosystem services that these networks in turn deliver. Body mass is believed to scale with metabolic rate as approximately $M^{3/4}$ (Brown *et al.* 2004), suggesting that larger bodied species have greater energetic demands. It has also been shown that body mass scales with abundance on a global scale as $M^{-3/4}$ (Enquist *et al.* 1998), suggesting that larger bodied species are less abundant than smaller bodied species. Combining these two relationships $(M^{3/4} \times M^{-3/4} = M^0)$ results in the suggestion of constant energy use for each species. This concept is known as the energy equivalence hypothesis (Brown and Gillooly 2003), and suggests that population-level energy use is roughly equivalent for each species—at least when species share a common source of energy. Such predictions suggest that in a multitrophic-level context, all species are equal in their use of energy, and consequently species identity is not important. However, it is well known that identity effects are very important in biodiversity and ecosystem functioning studies (O'Connor *et al.* 2008).

Whilst the slopes of body mass–abundance relationships may vary from local to global scales (White *et al.* 2007), they appear to be remarkably conserved across a wide range of habitats and ecosystems (Cyr *et al.* 1997b). A limited number of studies suggest that body mass–abundance relationships may be robust to major food-web perturbations. For example, in Tuesday Lake, Michigan, the replacement of three species of planktivorous fish with one species of piscivorous fish caused widespread changes in invertebrate composition, an increase in zooplankton biomass, and a reduction in algal biomass (Carpenter *et al.* 1987). Despite this large-scale perturbation and the occurrence of a trophic cascade within the system, many characteristics of the Tuesday Lake food-web remained relatively unchanged, including the slope of the average species body mass–abundance relationship (Jonsson *et al.* 2005). Similarly, Marquet *et al.* (1990) showed that the average species body mass–abundance relationships from inside and outside two marine reserves in Chile—Las Cruces and Montemar—were not affected by ecosystem-wide trophic cascades. Inside both reserves, the absence of human harvesting caused an increase in the density of predatory gastropods, leading to cascading effects on the body mass and abundance of invertebrate and algal species (Castilla and Duran 1985). These human-induced trophic cascades had no noticeable effect on the body mass–abundance relationships at either reserve (Marquet *et al.* 1990).

The number of studies that report both the slope of the body mass–abundance relationship, and the existence of a trophic cascade, are limited, but the robustness of the body mass–abundance relationships in the face of such dramatic changes may have its basis in the insurance provided by biodiversity at regional scales. Widely documented relationships suggest that in intact ecosystems, body mass–abundance relationships should scale with a negative slope of ¾ (Figures 7.4a and b). If large-bodied

species are removed from the top of a food-web (Figure 7.44c), as they are in harvested systems, then one might predict a trophic cascade (Figure 7.44d). As intermediate sized species increase in abundance and biomass, they in turn have a cascading effect on smaller producers and consumers within the system. The consequence of a trophic cascade should be an increase in intermediate-sized consumers and a reduction in the smaller consumers; this would result in a reduced body mass–abundance slope (Figure 7.4c). By extension, loss of intermediate consumers would lead to an increase in the smallest size classes of animals and plants in the system, and so we might predict an overall increase in abundance of the smallest individuals with no overall net change in the slope of the body mass–abundance relationship (Figures 7.44e and f). Such predictions are not well supported by the few examples available in the literature (Carpenter *et al.* 1987; Marquet *et al.* 1990; Jonsson *et al.* 2005), but present an interesting avenue for research. If these effects fail to manifest, then this suggests an alternative mechanism responsible for maintaining these wide ranging macro-ecological relationships.

Following the introduction of piscivorous fish in Tuesday Lake, a trophic cascade occurred, yet there were no significant effects on the slopes of the body mass–abundance relationships (Jonsson *et al.* 2005). Two alternative, but not mutually exclusive, hypotheses might explain the robustness of these relationships. Firstly, species in the absence of predation might grow larger, leading to an increase in average species body mass for some species. These species could grow to fill the size classes that were previously occupied by predators, and essentially undergo an increase in body mass and a decrease in abundance (Figures 7.5a and 7.5b). Second, with a change in prevailing environmental conditions— here I include the presence or absence of consumers or competitors as contributing to a press perturbation in the system that could change the prevailing conditions— new species could colonize the vacant body-size niches from nearby habitats and therefore maintain the body mass–abundance relationships (Figures 7.5a and 7.5c). A combination of these two mechanisms might also prevail. In Tuesday Lake, following the introduction of the piscivorous fish,

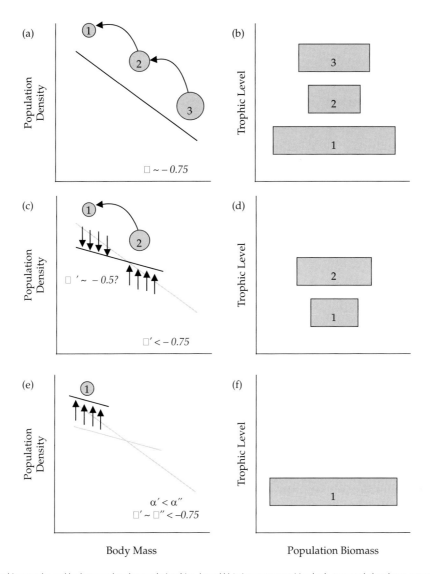

Figure 7.4 Trophic cascades and body mass abundance relationships. (a and b) In intact communities, body mass and abundance across a wide range of body masses scales as $M^{-3/4}$; (c and d) If large-bodied species are removed by harvesting, then intermediate trophic levels are released from predation resulting in a trophic cascade: intermediate consumers increase in abundance, which in turn reduces the abundance of primary consumers, and there is a reduction in the slope of the body mass abundance relationship and $\beta' < -0.75$; (e and f) If a system is further harvested, leaving just primary consumers in the system, then these species are released from predation, their abundances increase and there is an overall increase in the standing stock of these species and the intercept of the relationship between mass and abundance may increase so that $\alpha' < \alpha''$.

there was a 40% turnover in the composition of the food-web between 1984 and 1986, suggesting a large-scale reorganization of the community that maintained the body mass–abundance relationships. The identity of these species is lost at the macro-ecological level, yet the patterns that they define are maintained, ensuring ecological structure and function in the face of human activities. At large spatial scales—landscape, seascape, or regional scales—there is a sufficiently large species pool—gamma diversity (γ)—and a diversity of local habitats, to provide a source of colonists (Figure 7.5d)

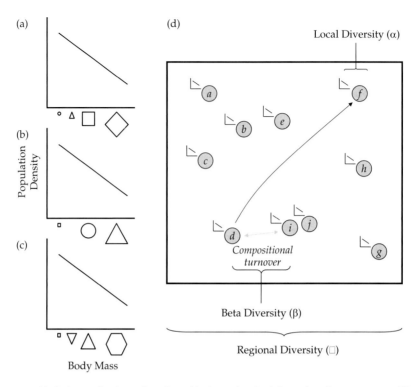

Figure 7.5 Maintenance of the body mass abundance relationship and biodiversity loss. (a–c) Shapes along the *x*-axis represent different species. The size of the shape reflects its body mass; (a) An intact ecosystem and the resulting body mass abundance relationship; (b) A disturbed ecosystem, showing that previously small species have been able to grow larger in the absence of the larger predators (triangles), and that their abundances have declined in the process. In contrast, some species may have reduced in body mass either through predation or competition for resources (squares). The resulting body mass abundance relationship is not significantly different to the undisturbed system in a; (c) A disturbed ecosystem, showing the presence of new species that have been able to colonize following the removal of larger bodied species, as existing small species increase in size, they create gaps in the size spectrum and new novel species are able to colonize the system here too; (d) A conceptual diagram showing how a disturbance occurring in patch *f*, which changed local conditions, could be buffered by dispersal from similar patches at a regional scale. If the new conditions in patch *f* are similar to conditions in a patch at the landscape scale, e.g. patch *d*, then there is some capacity for species to colonize *f* and maintain function. If, however, patches *a* to *j* become more homogenous then there is a reduced capacity to insure composition, structure, and hence functioning in the disturbed patch. There is a critical ratio between regional diversity (γ) and local diversity (α), below which there is no capacity to maintain macro-ecological functions if local conditions change.

that may maintain patterns in perturbed patches. Environmental heterogeneity provides a diversity of patches which vary in environmental conditions, and provides the capacity to also maintain high levels of beta-diversity (β) at regional scales. If at these large spatial scales biodiversity is reduced, the capacity to ensure function at local scales (α) can be eroded (Yachi and Loreau 1999), so that there will be a critical local–regional ratio of biodiversity (α/γ), below which macro-ecological relationships such as body mass and abundance cannot be maintained.

Tackling all of these issues will be particularly difficult, but a promising avenue for research will

be through the use of microcosm communities, based on bacteria, algae, and protists. These simplistic laboratory-based communities provide a model analogue to real-world ecosystems, where body mass, abundance, and food-web structure can potentially be manipulated, and where population dynamics can be measured at very fine temporal scales. The advent of modern imaging technologies may soon enable us to quantify population dynamics of protist-based systems in real time, enabling us for the first time to quantify real-time changes in abundance, size structure, and the nature of predation.

7.5   Conclusions

Using a food-web framework and the composite descriptions of body mass, abundance, and trophic interactions, it is possible to generate testable hypotheses over the relationships between biodiversity, and the structure and functioning of ecological systems. Over the last decade, we have seen major advances in our understanding of the mechanisms underpinning relationships between biodiversity and ecosystem functioning. Incorporating an understanding of food-web dynamics into the work on biodiversity and ecosystem functioning remains a major challenge for the next decade. Some small steps have been made in this direction, using body mass, abundance, and community and network structure.

Major challenges still remain in integrating the predictions from theory with empirical tests developed in the context of biodiversity and ecosystem functioning research. The provision of relevant data to move community and food-web models from a hypothesis generating tools to predictive frameworks represents a very significant challenge. Using novel data descriptions, including body mass, abundance, and food-web structure, marine ecologists can address these issues and lead the way, integrating our understanding of food-webs and biodiversity using novel experimental designs and statistical techniques.

Acknowledgements

Many thanks to Eoin O'Gorman, Marion Twomey, Catherine Palmer, Aurélie Aubry, Orla McLaughlin, and Anna Eklöf for useful and insightful discussions. The ideas and concepts in this manuscript have benefited from filtering in the ESF funded research networking program SIZEMIC.

References

Aljetlawi, A. A., Sparrevik, E.and Leonardsson, K. (2004) Prey-predator size-dependent functional response: derivation and rescaling to the real world. *Journal of Animal Ecology* 73: 239–52.

Bascompte, J., Melian, C. J. and Sala, E. (2005) Interaction strength combinations and the overfishing of a marine food web. *Proceedings of the National Academy of Sciences of the USA* 102: 5443–7.

Brose, U., Cushing, L., Berlow, E. L. *et al.* (2005) Body sizes of consumers and their resources. *Ecology* 86: 2545–2545.

Brose, U., Ehnes, R. B., Rall, B. C. *et al.* (2008) Foraging theory predicts predator-prey energy fluxes. *Journal of Animal Ecology* 77: 1072–8.

Brose, U., Williams, R. J. and Martinez, N. D. (2006) Allometric scaling enhances stability in complex food webs. *Ecology Letters* 9:1228–36.

Brown, J. H. and Gillooly, J. F. (2003) Ecological food webs: high-quality data facilitate theoretical unification. *Proceedings of the National Academy of Sciences, USA* 100: 1467–8.

Brown, J. H.,Gillooly, J. F., Allen, A. P. *et al.* (2004) Toward a metabolic theory of ecology. *Ecology* 85: 1771–89.

Bruno, J. F., Boyer, K. E., Duffy, J. E. *et al.* (2008) Relative and interactive effects of plant and grazer richness in a benthic marine community. *Ecology* 89: 2518–28.

Bruno, J. F. and O'Connor, M. I. (2005) Cascading effects of predator diversity and omnivory in a marine food web. *Ecology Letters* 8: 1048–56.

Byrnes, J., Stachowicz, J. J., Hultgren, K. M. *et al.* (2006) Predator diversity strengthens trophic cascades in kelp forests by modifying herbivore behaviour. *Ecology Letters* 9: 61–71.

Byrnes, J. E. and Stachowicz, J. J. (2009) The consequences of consumer diversity loss: different answers from different experimental designs. *Ecology* 90: 2879–88.

Calder, W. (1996) Size, Function and Life History. Dover Publications, New York.

Carpenter, S. R., Kitchell, J. F., Hodgson, J. R. *et al.* (1987) Regulation of lake primary productivity by food web structure. *Ecology* 68: 1863–76.

Castilla, J. C. and Duran, L. R. (1985) Human exclusion from the rocky intertidal zone of central Chile: the effects on *Concholepas concholepas* (Gastropoda). *Oikos* 45: 391–9.

Cohen, J. E., Jonsson, T. and Carpenter, S. R. (2003) Ecological community description using the food web, species abundance, and body size. *Proceedings of the National Academy of Sciences of the USA* 100: 1781–6.

Cyr, H., Downing, J. A. and Peters, R. H. (1997a) Density body size relationships in local aquatic communities. *Oikos* 79: 333–46.

Cyr, H., Peters, R. H. and Downing, J. A. (1997b) Population density and community size structure: Comparison of aquatic and terrestrial systems. *Oikos* 80: 139–49.

Digel, C., Riede, J. and Brose, U. (2011) Body sizes, cumulative and allometric degree distributions across natural food webs. *Oikos* 120: 503–9.

Doak, D. F., Bigger, D., Harding, E. K. *et al.* (1998) The sta-tistical inevitability of stability-diversity relationships in community ecology. *American Naturalist* **151**: 264–76.

Dobson, A., Lodge, D., Alder, J. *et al.* (2006) Habitat loss, trophic collapse, and the decline of ecosystem services. *Ecology* **87**: 1915–24.

Duffy, J. E. (2002) Biodiversity and ecosystem function: the consumer connection. *Oikos* **99**: 201–19.

Duffy, J. E. (2006) Biodiversity and the functioning of seagrass ecosystems. *Marine Ecology-Progress Series* **311**: 233–50.

Duffy, J. E., Cardinale, B. J., France, K. E. *et al.* (2007) The functional role of biodiversity in ecosystems: incorporating trophic complexity. *Ecology Letters* **10**: 522–38.

Duffy, J. E., Macdonald, K. S., Rhode, J. M. *et al.* (2001) Grazer diversity, functional redundancy, and productivity in seagrass beds: An experimental test. *Ecology* **82**: 2417–34.

Dulvy, N. K., Freckleton, R. P. and Polunin, N. V. C. (2004) Coral reef cascades and the indirect effects of predator removal by exploitation. *Ecology Letters* **7**: 410–16.

Dunne, J. A., Williams, R. J. and Martinez, N. D. (2002a) Food-web structure and network theory: The role of connectance and size. *Proceedings of the National Academy of Sciences of the USA* **99**: 12917–22.

Dunne, J. A., Williams, R. J. and Martinez, N. D. (2002b) Network structure and biodiversity loss in food webs: robustness increases with connectance. *Ecology Letters* **5**: 558–67.

Dunne, J. A., Williams, R. J. and Martinez, N. D. (2004) Network structure and robustness of marine food webs. *Marine Ecology-Progress Series* **273**: 291–302.

Ehrlich, P.R., Ehrlich, A. (1981) Extinction: The Causes and Consequences of the Disappearance of Species. Random House. 305 pp.

Elton, C. 1927. *Animal Ecology*. Sidgwick and Jackson, London.

Emmerson, M., Montoya, J. M. and Woodward, G. (2005) Body size, interaction strength and food web dynamics. In: *Dynamic food webs: Multispecies assemblages, ecosystem development, and environmental change.* de Ruiter, P. C., Wolters, V. and Moore, J. C. (eds), Academic Press, Burlington, MA, 167–78.

Emmerson, M. and Yearsley, J. M. (2004) Weak interactions, omnivory and emergent food-web properties. *Proceedings of the Royal Society of London Series B—Biological Sciences* **271**: 397–405.

Emmerson, M. C. and Raffaelli, D. (2004) Predator-prey body size, interaction strength and the stability of a real food web. *Journal of Animal Ecology* **73**: 399–409.

Emmerson, M. C., Solan, M., Emes, C. *et al.* (2001) Consistent patterns and the idiosyncratic effects of biodiversity in marine ecosystems. *Nature* **411**: 73–7.

Enquist, B. J., Brown, J. H. and West, G. B. (1998) Allometric scaling of plant energetics and population density. *Nature* **395**: 163–5.

Estes, J. A., Tinker, M. T., Williams, T. M. *et al.* (1998) Killer whale predation on sea otters linking oceanic and nearshore ecosystems. *Science* **282**: 473–6.

Fahrig, L. (2003) Effects of habitat fragmentation on biodiversity. *Annual Review of Ecology Evolution and Systematics* **34**: 487–515.

Finke, D. L. and Denno, R. F. (2005) Predator diversity and the functioning of ecosystems: the role of intraguild predation in dampening trophic cascades. *Ecology Letters* **8**: 1299–306.

Gamfeldt, L., Hillebrand, H. and Jonsson, P. R. (2005) Species richness changes across two trophic levels simultaneously affect prey and consumer biomass. *Ecology Letters* **8**: 696–703.

Gardner, M. R. and W. R. Ashby. (1970) Connectance of large dynamic (cybernetic) systems: critical values for stability. *Nature* **228**: 784.

Griffin, J. N., De la Haye, K. L., Hawkins, S. J. *et al.* (2008) Predator diversity and ecosystem functioning: Density modifies the effect of resource partitioning. *Ecology* **89**: 298–305.

Hillebrand, H. and Cardinale, B. J. (2004) Consumer effects decline with prey diversity. *Ecology Letters* **7**: 192–201.

Hillebrand, H., Gamfeldt, L., Jonsson, P. R. *et al.* (2009) Consumer diversity indirectly changes prey nutrient content. *Marine Ecology-Progress Series* **380**: 33–41.

Hunt, H. W., Coleman, D. C., Ingham, E. R. *et al.* (1987) The detrital food web in a shortgrass prairie. *Biology and Fertility of Soils* **3**: 57–68.

Hurlbert, S. H. (1984) Pseudoreplication and the design of ecological field experiments. *Ecological Monographs* **54**: 187–211.

Huston, M. A. (1997) Hidden treatments in ecological experiments: re-evaluating the ecosystem function of biodiversity. *Oecologia* **110**: 449–60.

Jennings, S., Greenstreet, S. P. R., Hill, L. *et al.* (2002) Long-term trends in the trophic structure of the North Sea fish community: evidence from stable-isotope analysis, size-spectra and community metrics. *Marine Biology* **141**: 1085–97.

Jennings, S. and Warr, K. J. (2003) Smaller predator-prey body size ratios in longer food chains. *Proceedings of the Royal Society of London Series B-Biological Sciences* **270**: 1413–17.

Jonsson, T., Cohen, J. E. and Carpenter, S. R. (2005) Food webs, body size, and species abundance in ecological community description. *Advances in Ecological Research* **36**: 1–84.

Jonsson, T. and Ebenman, B. (1998) Effects of predator-prey body size ratios on the stability of food chains. *Journal of Theoretical Biology* **193**: 407–17.

Kerr, S. R. and Dickie, L. M. (2001) *The biomass spectrum: A predator prey theory of aquatic production.* Columbia University Press, New York.

Lawton, J. H. (1994) What do species do in ecosystems? *Oikos* **71**: 367–74.

Layer, K., Riede, J. O., Hildrew, A. G. *et al.* (2010) Food web structure and stability in 20 streams across a wide pH gradient. *Advances in Ecological Research* **42**: 265–99.

Layman, C. A., Arrington, D. A., Montana, C. G. *et al.* (2007) Can stable isotope ratios provide for community-wide measures of trophic structure? *Ecology* **88**: 42–8.

Long, Z. T., Bruno, J. F. and Duffy, J. E. (2007) Biodiversity mediates productivity through different mechanisms at adjacent trophic levels. *Ecology* **88**: 2821–9.

Loreau, M. (2010) *From populations to ecosystems: Theoretical foundations for a new synthesis.* Princeton University Press, Princeton, USA.

Loreau, M. and Hector, A. (2001) Partitioning selection and complementarity in biodiversity experiments. *Nature* **412**: 72–6.

Marquet, P. A., Navarrete, S. A. and Castilla, J. C. (1990) Scaling population density to body size in rocky intertidal communities. *Science* **250**: 1125–7.

May, R. M. (1973) *Stability and complexity in model ecosystems.* 3rd edition. Princeton University Press, Princeton, USA.

McCann, K., Hastings, A. and Huxel, G. R. (1998) Weak trophic interactions and the balance of nature. *Nature* **395**: 794–8.

McCann, K. S. (2000) The diversity-stability debate. *Nature* **405**: 228–33.

McLaughlin, O., Jonsson, T. and Emmerson, M. C. (2010) Temporal variability in predator–prey relationships of a forest floor food web. *Advances in Ecological Research* **42**: 171–264.

Montoya, J. M., Pimm, S. L. and Sole, R. V. (2006) Ecological networks and their fragility. *Nature* **442**: 259–64.

Montoya, J. M. and Sole, R. V. (2002) Small world patterns in food webs. *Journal of Theoretical Biology* **214**: 405–12.

Moran, E. R., Reynolds, P. L., Ladwig, L. M. *et al.* (2010) Predation intensity is negatively related to plant species richness in a benthic marine community. *Marine Ecology-Progress Series* **400**: 277–82.

Neutel, A. M., Heesterbeek, J. A. P. and de Ruiter, P. C. (2002) Stability in real food webs: Weak links in long loops. *Science* **296**: 1120–3.

O'Connor, M. I. and Bruno, J. F. (2009) Predator richness has no effect in a diverse marine food web. *Journal of Animal Ecology* **78**: 732–40.

O'Connor, N. E. and Bruno, J. F. (2007) Predatory fish loss affects the structure and functioning of a model marine food web. *Oikos* **116**: 2027–38.

O'Connor, N. E., Grabowski, J. H., Ladwig, L. M. *et al.* (2008) Simulated predator extinctions: Predator identity affects survival and recruitment of oysters. *Ecology* **89**: 428–38.

O'Gorman, E. and Emmerson, M. C. (2010) Manipulating interaction strengths and the consequences for trivariate patterns in a marine food web. *Advances in Ecological Research* **42**: 301–419.

O'Gorman, E., Enright, R. and Emmerson, M. (2008) Predator diversity enhances secondary production and decreases the likelihood of trophic cascades. *Oecologia* **158**: 557–67.

Petchey, O., McPhearson, P., Casey, T. *et al.* (1999) Environmental warming alters food-web structure and ecosystem function. *Nature* **402**: 69–72.

Peters, R. 1986. *The ecological implications of body size.* Cambridge University Press, UK.

Pimm, S. L. and Lawton, J. H. (1977) Number of trophic levels in ecological communities. *Nature* **268**: 329–31.

Pimm, S. L. and Lawton, J. H. (1978) On feeding on more than one trophic level. *Nature* **275**: 342–4.

Riede, J. O., Brose, U. Ebenman, B. *et al.* (2011) Stepping in Elton's footprints: a general scaling model for body masses and trophic levels across ecosystems. *Ecology Letters* **14**: 169–78.

Riede, J. O., Rall, B. C., Banasek-Richter, C. *et al.* (2010) Scaling of food-web properties with diversity and complexity across ecosystems. *Advances in Ecological Research* **42**: 139–70.

Seabloom, E. W., Williams, J. W., Slayback, D. *et al.* (2006) Human impacts, plant invasion, and imperiled plant species in California. *Ecological Applications* **16**: 1338–50.

Solan, M., Cardinale, B. J., Downing, A. L. *et al.* (2004) Extinction and ecosystem function in the marine benthos. *Science* **306**: 1177–80.

Stachowicz, J. J., Bruno, J. F. and Duffy, J. E. (2007) Understanding the effects of marine biodiversity on communities and ecosystems. *Annual Review of Ecology Evolution and Systematics* **38**: 739–66.

Stenseth, N. C., Falck, W., Bjørnstad, O. N. *et al.* (1997) Population regulation in snowshoe hare and Canadian lynx: Asymmetric food web configurations between hare and lynx. *Proceedings of the National Academy of Sciences of the USA* **94**: 5147–52.

Tilman, D. (1996) Biodiversity: Population versus ecosystem stability. *Ecology* **77**:350–63.

Tilman, D. (1999) The ecological consequences of changes in biodiversity: A search for general principles. *Ecology* **80**: 1455–74.

Tilman, D., Lehman, C. L., and Bristow, C. E. (1998) Diversity-stability relationships: Statistical inevitability or ecological consequence? *American Naturalist* **151**: 277–82.

Vucic-Pestic, O., Rall, B. C., Kalinkat, G. *et al.* (2010) Allometric functional response model: body masses constrain interaction strengths. *Journal of Animal Ecology* **79**: 249–56.

Walker, B. H. (1992) Biodiversity and ecological redundancy. *Conservation Biology* **6**: 18–23.

White, E. P., Ernest, S. K. M., Kerkhoff, A. J. *et al.* (2007) Relationships between body size and abundance in ecology. *Trends in Ecology & Evolution* **22**: 323–30.

Woodward, G., Speirs, D. C. and Hildrew, A. G. (2005) Quantification and resolution of a complex, size-structured food web. *Advances in Ecological Research* **36**: 85–135.

Yachi, S. and Loreau, M. (1999) Biodiversity and ecosystem productivity in a fluctuating environment: the insurance hypothesis. *Proceedings of the National Academy of Sciences of the USA* **96**: 1463–8.

Yodzis, P. (1981) The stability of real ecosystems. *Nature* **289**: 674–6.

Yodzis, P. and Innes, S. (1992) Body size and consumer-resource dynamics. *American Naturalist* **139**: 1151–75.

Yvon-Durocher, G., Reiss, J., Blanchard, J. *et al.* (2011) Across ecosystem comparisons of size structure: Methods, approaches, and prospects. *Oikos* **120**: 550–63.

CHAPTER 8

Effects of biodiversity–environment conditions on the interpretation of biodiversity–function relations

Jasmin A. Godbold

8.1 Introduction

Concerns over the ecosystem consequences of anthropogenic activities on biodiversity loss and changes in community composition led to the emergence of what it now a major sub-discipline within the field of ecology (*sensu* Caliman *et al.* 2010), focusing on the relationship between biodiversity and ecosystem functioning (Schmid *et al.* 2009; Solan *et al.* 2009). The main reasoning behind this area of research is that the hitherto unprecedented rates of change observed in global biodiversity (Dirzo and Raven 2003) may significantly endanger fluxes of energy and matter provided by ecosystems which underpin important services that humans rely on for food, shelter, economic prosperity, and well-being (MEA 2005; Fischlin *et al.* 2007; White *et al.* 2010).

Over the past decade, biodiversity–ecosystem function hypotheses have been successfully articulated and examined, largely through experimental manipulation of species richness under controlled conditions (Loreau *et al.* 2001; Covich *et al.* 2004; Hooper *et al.* 2005). These studies have, collectively, substantially advanced our understanding of the links between changes in biodiversity and levels of various ecosystem functions, including primary productivity, decomposition, and nutrient cycling (Duffy 2009). Since 2006, a number of influential meta-analyses (Balvanera *et al.* 2006; Cardinale *et al.* 2006, 2007, 2011; Worm *et al.* 2006; Stachowicz *et al.* 2007; Schmid *et al.* 2009) have summarized the available evidence for the importance of biodiversity in maintaining a variety of ecosystem functions and processes across taxa, trophic levels, habitats, and ecosystems. Although these studies independently conclude that biodiversity is vital for ecosystem process and functioning, this research area continues to be subject to controversy (e.g. Huston and McBride 2002; Wardle and Jonsson 2009) and debate over the applicability of research findings to natural systems that are subject to environmental variation, and their ability to inform policy-relevant issues including conservation management (Srivastava and Vellend 2005; Thompson and Starzomski 2007). In response, and recognizing that experimental manipulations are inherently constrained in the number of additional driver variables that can be tested simultaneously, ecologists have diversified and broadened empirical approaches by developing increasingly complex experimental designs that allow a greater degree of biological and environmental realism across a range of contexts (Naeem 2008). Experimental biodiversity–ecosystem function studies in aquatic ecosystems now include a variety of *in-situ* manipulations of species diversity (O'Connor and Crowe 2005; Lecerf *et al.* 2007; Godbold *et al.* 2009; Griffin *et al.* 2010), the incorporation of habitat or resource heterogeneity (Dyson *et al.* 2007; Weis *et al.* 2008; Godbold *et al.* 2011), two or more trophic levels (e.g. Duffy *et al.* 2007; Bruno *et al.* 2008; Douglass *et al.* 2008) as well as larger spatial and longer temporal scales (O'Connor and Crowe 2005; Stachowicz *et al.* 2008; Weis *et al.* 2008) that are supported by observational studies in natural systems (e.g. Ruesink *et al.* 2006; Danovaro *et al.* 2008; Godbold and Solan 2009).

The magnitude, direction, and shape of the relationship between species richness and ecosystem function varies within and between manipulative and observational studies, and responses appear to be dependent on a variety of additional biotic and/or abiotic conditions (Emmerson *et al.* 2001; Bolam *et al.* 2002; O'Connor and Crowe, 2005; Danovaro *et al.* 2008; Langenheder *et al.* 2010). The large variability between studies remains a contentious issue, and initial reviews of the biodiversity–ecosystem function literature have suggested that effects are significantly weaker under less well-controlled conditions (Balvanera *et al.* 2006), implying that in natural systems, the effect of biodiversity on ecosystem properties may be of secondary importance to the effects of spatio-temporal variability of other abiotic and biotic drivers of change (Lecerf *et al.* 2007; Grace *et al.* 2007; Wardle and Jonsson 2009; but see Godbold and Solan 2009 for a counter-example). In fact, it has been argued that ecosystem functions such as productivity are primarily mediated by components of the abiotic environment, such as climate, soils, and resource availability, making diversity effects only detectable in highly controlled experimental systems in which the influence of the abiotic environment is controlled or has been excluded altogether (e.g. Thompson *et al.* 2005; Grace *et al.* 2007). Whilst it is not surprising that biodiversity effects vary with context, because ecosystems are structured by multiple, inextricably linked, and simultaneously operating abiotic and biotic factors, remarkably only a few attempts have been made to specifically distinguish the relative importance of biodiversity and environmental factors in modifying ecosystem properties (e.g. Healy *et al.* 2008; Godbold and Solan 2009).

In this chapter, I review biodiversity–ecosystem function studies in aquatic (marine and freshwater) ecosystems that have incorporated and statistically accounted for additional abiotic or biotic drivers in mediating ecosystem responses. The aim is to assess whether these studies give credence to the view that species richness is of minor importance relative to abiotic or biotic drivers of change or other biodiversity measures (e.g. species

composition, species evenness) across the full suite of ecosystem properties that have been measured. In order to avoid repetition of results in previous meta-analyses, hypotheses relating to the influence of, for example, experimental design and duration, geographical location, or ecosystem versus population level responses, have been omitted. I will show how the strength of the relationship between species richness and ecosystem properties, and other drivers of change and ecosystem properties are affected by (a) *model systems,* that differ in the degree of experimental control of species richness and environmental conditions (including laboratory experiments, where species richness and abiotic conditions are highly controlled; field and in-situ experiments, where species richness is manipulated, but experimental fauna and flora are subject to ambient abiotic conditions; and, observational studies that focus on natural, un-manipulated communities) and, (b) *ecosystem type of the experimental study* (e.g. rocky shore, stream, or seagrass). I will conclude by illustrating how ecosystem properties are mediated across aquatic ecosystems by considering the effects of abiotic, biodiversity (distinguishing between species richness, species evenness, species identity) and other biotic drivers of change combined. My approach is to deliberately consider various driver variables (abiotic, biotic, and biodiversity), ecosystem responses (e.g. nutrient fluxes, primary production, resource use efficiency, decomposition, bioturbation) and ecosystems simultaneously in order to detect over-arching patterns in ecosystem properties across aquatic (marine and freshwater) systems.

8.2 Methods of analysis

8.2.1 Compilation of publications

Publications were initially identified from the *ISI Web of Knowledge* using the *Science Index Expanded* and *Social Sciences Citation Index* databases. A general search using the search term *"('biodiversity' OR 'species diversity' OR 'species richness') AND ('ecosystem function' OR 'ecosystem proce*') AND ('marine' OR 'aquatic' OR 'freshwater')* in the titles and keywords of all document types, in all languages, was performed. Additional publications that were not identified using the initial search term were obtained from citations contained within the search return. I also contacted authors to obtain additional information and request further relevant publications. Where records from an experiment were contained within multiple scientific contributions, care was taken to avoid repeat counting. Where repeated measures were taken over time within an experiment, and time was not specifically included within the subsequent analysis, only the final time point was recorded.

This search resulted in 76 original, peer-reviewed studies published between 1997 and 2009, in which a total of 218 relationships between species richness and ecosystem functioning or process (following Balvanera *et al.* 2006, hereafter collectively referred to as ecosystem properties) in aquatic (marine and freshwater) systems. The database also includes a subset of 148 relationships between abiotic and/or biotic drivers and/or alternative components of biodiversity (e.g. species composition, functional richness, see below). These were incorporated as additional explanatory variables in studies ($n = 51$) investigating the relationship between species richness and ecosystem properties.

8.2.2 Calculation of effect sizes

The correlation coefficient r was used to estimate the strength of the relationship between species richness and additional abiotic or biotic explanatory variables on ecosystem properties. All records which either included the effect size r, or incorporated the information from which the effect size r could be reliably calculated, were included in the analysis.

In models with multiple explanatory variables, the frequently used formula for converting F values to r yields a *partial r* correlation that can be substantially larger than the required *non-partial r*, depending upon the number and relevance of additional factors in the model (Levine and Hullett 2001; Hullett and Levine 2003; Nakagawa and Cuthill 2007). A preferential method for calculating non-partial r for single terms and their interactions within a multivariate model, is the effect sum of

squares (SS_{effect}) and the total sum of squares (SS_{total}), as SS_{total} will not vary according to the number of factors included in a model (Hullett and Levine 2003). *Non-partial r* ($n = 132$) were obtained using the F statistic (Rosenthal 1991): $r = \sqrt{\dfrac{F}{F + df_{error}}}$ from studies in which species richness was the only variable included in the model and d.f.$_{effect}$ = 1. In models with multiple explanatory variables, *non-partial r* values for each variable were calculated from SS_{effect} and SS_{total} of the whole model (Hullett and Levine 2003): $r = \sqrt{\dfrac{SS_{effect}}{SS_{total}}}$, thereby standardizing the effect size irrespective of the number of variables in the model. In publications where information on the sum of squares was not available, or could not be obtained, *partial r* for each variable was calculated using the *F statistic* ($n = 86$ effect sizes).

It was possible to pool *non-partial* and *partial* effect sizes within an analysis because the correlation between the partial effect size and the number of variables in a model (Hullett and Levine 2003) was insignificant (Pearson's product moment correlation, t = −1.797, d.f. = 131, p = 0.075, $r = -0.155$). The relationships between explanatory variables and ecosystem properties were categorized as positive, negative, or zero, based on the direction of trend reported in the original study. When added explanatory variables were expressed in relative terms (e.g. high or low current regime), or in a binary format (presence or absence of a predator), the r and its direction was recorded relative to the 'non-reference' or 'usual' conditions. For example, if the presence of a predator had a negative effect on resource use, the effect size was attributed a minus ('−') sign. Following Underwood (1998), only the effect sizes of the highest order terms were recorded. Thus, if species richness effects on an ecosystem property occurred as part of an interaction, then the effect size of the interaction was recorded rather than the effects size of species richness as a single term. Additional explanatory variables were only recorded if their effects, either as a single term or as part of an interaction, on ecosystem properties were significant. Following the nomenclature of

Cohen (1988) effect sizes are classified as small ($> 0.1\ r < 0.3$), medium ($> 0.3\ r < 0.5$) or large ($r > 0.5$).

8.2.3 Extraction of data

The present analysis is based on 76 peer-reviewed publications that focused on the relationship between species richness and ecosystem properties. Any *additional driver variables* were combined into the following categories: a) *abiotic*, including variables such as environmental disturbance, temperature, or nutrients; b) *biotic*, including variables such as predation, presence/absence of another focal species; c) *biodiversity*, here the effects of alternative measures (not species richness) such as species identity/composition, density, biomass, or evenness were pooled into one variable; d) *location*, where different sites or 'block' within a site were incorporated as a variable; e) *time*, where time was incorporated as an explanatory variable; and, f) any interactions between these groupings (denoted by a '×' between groups, e.g. abiotic × biodiversity). Studies were distinguished between marine ($n = 40$) and freshwater ($n = 36$) *ecosystem types* and subdivided into the specific habitats of the individual studies (i.e. bentho-pelagic, pond, stream, mangrove, rocky shore, soft benthos, seagrass, and wetland). Finally, studies were distinguished by *Model System* as being experimental manipulations of species richness that were carried out under controlled conditions in the laboratory (hereafter, *laboratory*), experimental manipulations of species richness that were carried out in the field or in outdoor mesocosms where experimental subjects were exposed to variation in natural abiotic conditions e.g. atmospheric and/or water temperature, wave action, tidal cycles (hereafter collectively termed *field*), or as *observational* studies investigating natural levels/changes in species richness and associated ecosystem properties (hereafter, *observational*).

8.2.4 Statistical Analysis

In models which had more than one additional driver, the effect size of the variable or interaction with the strongest r was ascribed to a new variable r_{A1}, whilst the variable with the second largest r was ascribed to r_{A2}. Initially, the mean effect sizes of the relationship between species richness and ecosystem properties, and between the additional explanatory variables (r_{A1} or r_{A2}) and ecosystem properties, were compared using T-test.

One-way analysis of variance was used to determine the effects of *model system* or *ecosystem type* as nominal explanatory variables on r_{SR} (the strength of the relationship between species richness and ecosystem properties) and r_A (the strength of the relationship between additional drivers and ecosystem properties). In addition, the data was analysed to determine the relative importance of the individual groups of abiotic, biotic, and biodiversity variables in mediating ecosystem properties. As there was a significant, but weak negative relationship between the total sample size of individual studies and effect size (r_{SR}, Pearson's product moment correlation, $t = -3.002$, d.f. = 179, $p < 0.05$, $r = -0.219$; r_A, $t = -3.631$, df = 85, $p < 0.01$, $r = -0.366$) linear mixed-effects (LME) statistical models were developed that incorporated the variable 'Total N' as a 'random effect' (Pinheiro and Bates 2000). Graphical exploratory techniques were used to check for outliers, normality, and homogeneity of variance (Quinn and Keough 2002). When model validation indicated normality but heterogeneity of variances, additional variance–covariate terms were incorporated as 'weighted random effects' to model the variance structure, thereby avoiding the need for data transformation (Godbold *et al.* 2009; Bulling *et al.* 2010). The optimal structure in terms of random components was determined using restricted maximum-likelihood (REML) estimation, whilst the fixed effects structure was then determined using maximum-likelihood (ML) estimation. The optimal random structure was determined by comparing the full analysis of variance model to the equivalent GLS model incorporating a specific variance structure using Aikaike information criteria (AIC) and inspection of model residual patterns. The importance of individual explanatory variables was assessed by comparing a reduced model (with the factor of interest removed) with the full model, using likelihood ratio test (West *et al.* 2007). Interactions between the explanatory variables (species richness and additional driver variable, and model system and ecosystem type) were not included because not all combinations were possible.

8.3   Are alternative drivers of change more important than species richness for ecosystem properties?

8.3.1   Summary of studies focusing on relationship between species richness and ecosystem properties

Of the 218 effect sizes reflecting the strength relationship between changes in species richness and

ecosystem properties, the majority of effect sizes ($n = 125$) are based on experiments carried out in marine ecosystems either from field studies (59.2%), laboratory experiments (32.8%) or observational studies (8%). Of the 93 effect sizes from freshwater ecosystems, 20.4% are from field experiments, 69.9% are from laboratory experiments, and 9.7% from observational studies.

Overall, 56% of the models include at least one additional driver of change when investigating the effects of species richness on ecosystem properties (Figure 8.1). Of these, more than half of the laboratory and field experiments (58.5% of marine studies, 56.7% of freshwater studies) included a

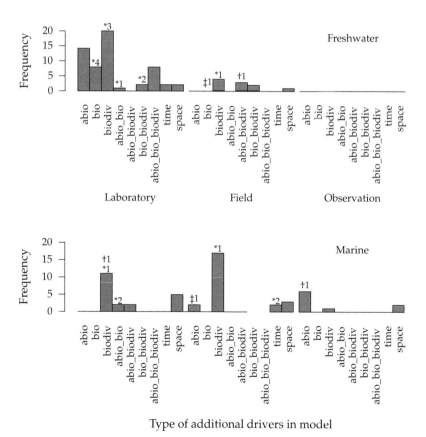

Figure 8.1   The total number of models in which additional abiotic, biotic, or biodiversity variables either alone or in combination were included in analyses investigating the effects of species richness on ecosystem properties. Abbreviations: abio = abiotic driver; bio = biotic driver; biodiv = biodiversity variable (species identity, biomass or density); '_' represents the presence of two additional variables within a model (e.g. abiotic and biotic variable is 'abio_bio'); * indicates number of models that also included time as an explanatory variable; † indicates number of models that also included a spatial explanatory variable, and ‡ indicates the number of models that incorporated both a temporal and spatial variables in the model.

measure of biodiversity other than species richness (species identity/combination, evenness, biomass, or density) either alone or in combination with additional abiotic or biotic drivers (Figure 8.1). Nearly one-third (28.4%) of laboratory experiments conducted in freshwater systems, also include either abiotic (e.g. nutrients, disturbance, temperature, etc.) or biotic (e.g. predation, presence/absence of other species, etc.) drivers, but only a few studies consider spatial and/or temporal aspects of the biodiversity–ecosystem function relationship (10.5% of freshwater studies, 24.5% of marine studies) (Figure 8.1).

The number of driver variables included in addition to species richness varies from one to 10 (median = 1) in freshwater field experiments, and from one to three (median = 0) in marine laboratory and field experiments (Figure 8.2). Irrespective of the system under study, 43% of studies do not measure

any additional driver variables, whilst about 39% of studies include only one additional driver to explain the observed ecosystem response. The number of significant additional drivers in these models varied from 0 (no significant variables) to 4 (marine) and 7 (freshwater) variables, either as single terms or as part of interactions, suggesting that ecosystem properties in aquatic environments are mediated by a variety of biotic and abiotic factors in addition to species richness. Just under half (43% marine, 41% freshwater) of the studies that investigated the ecosystem effects of species richness in isolation found that species richness effects were insignificant. In marine studies, the percentage of insignificant species richness effects increases to 55%, but decreases to 32% in freshwater when ecosystem responses were investigated in the presence of at least one additional driver variable. In those models in which species

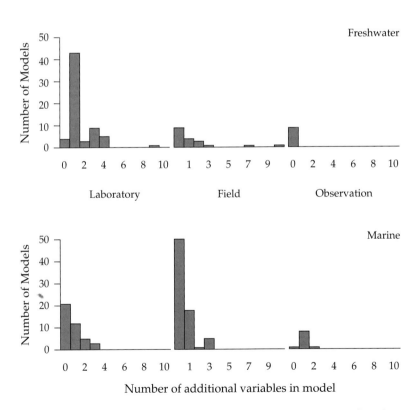

Figure 8.2 The total number of additional abiotic and biotic drivers (1–7) included in experiments investigating the effects of species richness on ecosystem properties. *Number of additional variables in the model* = 0 represents those models that considered species richness when acting in isolation.

richness effects were found to be insignificant and included an alternative measure of biodiversity as an additional driver (21 studies for marine, 18 studies for freshwater), species identity/composition is the driver that was most often added and found to have significant effects on ecosystem properties.

8.3.2   Effects of species richness and/or additional drivers of change on ecosystem properties

The mean effects size r for the relationship between species richness (alone or as part of an interaction) and ecosystem properties is small ($r < 0.3$), but significantly positive (mean $r_{SR} = 0.204 \pm 0.355$, t-value = 8.48, d.f. = 217, p < 0.0001), ranging from –0.991 to 0.933. The mean effect size r for the strongest additional driver is also significantly positive (mean $r_{A1} = 0.376 \pm 0.397$, t-value = 8.411, d.f. = 78, p < 0.0001, ranging from –0.830 to 0.984) and significantly stronger than mean r_{SR} (t-value = –4.357, d.f. = 78, p < 0.0001). In contrast, there is no difference between r_{SR} and the mean effect size for the second largest additional driver, r_{A2} (t-value = 1.475, d.f. = 19, p = 0.156). If only significant effects of species richness are considered, then the strength of the relationship between species richness and ecosystem properties increases (mean $r_{SR} \pm$ s.d., 0.271 ± 0.339, t-value = 7.834, d.f. = 125, p < 0.0001). The mean effect size, r_{SR}, is lower than r_{A1} (mean $r_{A1} = 0.288 \pm 0.315$, t-value = 4.792, d.f. = 38, p < 0.0001–although this difference is not significant (t-value = –1.468, d.f. = 38, p = 0.150). Thus, when species richness effects on ecosystem properties are significant, additional driver variables are not necessarily more important in mediating ecosystem properties.

Species richness effects vary depending on whether effects occur as part of a single species richness effect, or as part of a significant or insignificant interaction (L.ratio = 6.681, d.f. = 2, p < 0.05). Results reveal that species richness effects are similar (t-value = –1.059, p = 0.291) irrespective of whether effects of species richness occur in isolation (mean ± s.d. (number of effect

sizes), 0.251 ± 0.362 (131)) or as part of an interaction (mean ± s.d. (number of effect sizes), 0.203 ± 0.184 (33)). Consequently, the following analyses on species richness effects are based on both significant single and interaction effects combined.

The strength of the relationship between species richness and ecosystem properties (r_{SR}) does not vary significantly between model systems (L-ratio = 4.673, d.f = 2, p = 0.097) (Figure 8.3a). In contrast, the strength of the relationship between additional drivers and ecosystem properties was found to differ significantly between model systems (L-ratio = 7.683, d.f. = 2, p < 0.05) and is significantly stronger and less variable in observational studies (mean ± s.d., 0.562 ± 0.272, t = 2.974, p < 0.05) and field experiments (mean ± s.d., 0.399 ± 0.211, t = 2.691, p < 0.05) in comparison to laboratory experiments (mean ± s.d., 0.201 ± 0.411, Figure 8.3b).

The effects of species richness on ecosystem properties does not differ significantly between the ecosystems in which the studies were carried out (L-ratio = 7.625, d.f. = 5, p = 0.178) (Figure 8.4a). Nevertheless, r_{SR} in marine intertidal ecosystems (soft benthos, saltmarsh and rocky shore) are strong, positive, and less variable in comparison to those conducted in freshwater ecosystems (e.g. pond and stream). In contrast, effects of additional drivers on ecosystem properties differed significantly between habitat types (L-ratio = 26.743, d.f. = 6, p < 0.001) (Figure 8.4b). Particularly in studies conducted in rocky shore, seagrass, and bentho-pelagic ecosystems, r_A were high suggesting a very strong influence on ecosystem properties.

8.3.3   Distinguishing the effects of biodiversity, the abiotic and/or biotic environment on ecosystem properties

In order to determine what type of driver (abiotic, biotic, or any of the biodiversity measures) has the strongest effects on ecosystem properties overall, all significant driver variables were combined into one analysis. No significant difference

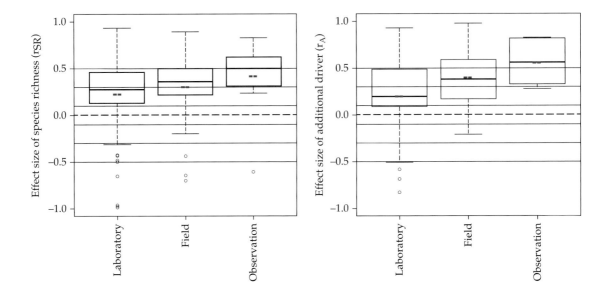

Figure 8.3 The effects of laboratory, field, and observational studies on the strength of the relationship between a) species richness and ecosystem properties, and b) additional abiotic, biotic or biodiversity variables and ecosystem properties. In each box, the median is indicated at the midpoint (thick, solid line in each box), the upper and lower quartiles are indicated by the hinges, vertical dashed lines represent the spread, and the open circles indicate outliers. In addition, '– –' represent the mean model predictions from the linear mixed effects analysis incorporating the factor *Model System* as a variance covariate. Dashed horizontal line represents no effect (for $0 > r < 0.1$), whilst the grey horizontal lines show small (> 0.1 $r < 0.3$), medium (> 0.3 $r < 0.5$), or large ($r > 0.5$) effect size classifications following the nomenclature of Cohen (1988).

was found between the effect sizes of the different driver variables and ecosystem properties (L-ratio = 19.826, d.f. = 14, p = 0.136). The results indicate species identity effects on ecosystem properties are strongest ($r = 0.348$), largely positive, and less variable in comparison to the effects of species richness (Figure 8.5). The average effects of the abiotic environment on ecosystem properties, although highly variable, are overall positive and similar in strength to the effects of identity and species richness. The strength of the interactive effect of species richness and additional explanatory variables on ecosystem properties is comparatively low, but much less variable in comparison to the single term effects of species identity and species richness.

8.4 Conclusions

Biodiversity–ecosystem function research has continually expanded over the past 20 years, with research focusing largely on the experimental manipulation of species richness to test new theory and develop a mechanistic understanding of the ecosystem consequences of biodiversity loss (Solan *et al.* 2009). Whilst the importance of species diversity for ecosystem properties has been well studied and is now well established (Duffy 2009), some of the strongest remaining controversies result from the assertion that most studies manipulate species richness in the absence of the appropriate environmental context, thereby reducing the relevance of results to processes in natural complex ecosystems (Bracken *et al.* 2008; Naeem 2008).

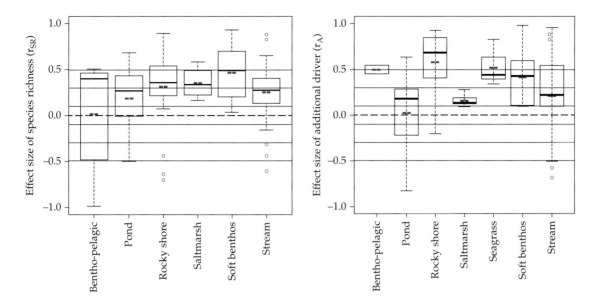

Figure 8.4 The effects of ecosystem type on the strength of the relationship between a) species richness, and b) additional abiotic, biotic, or biodiversity drivers and ecosystem properties. In each box, the median is indicated at the midpoint (thick, solid line in each box), the upper and lower quartiles are indicated by the hinges, vertical dashed lines represent the spread, and the open circles indicate outliers. In addition, '– –'represent the mean model predictions from the linear mixed effects analyses incorporating the factor *Ecosystem type* as a variance covariate for both models. Dashed horizontal line represents no effect (for 0 > r < 0.1), whilst the grey horizontal lines show small (> 0.1 r < 0.3), medium (> 0.3 r < 0.5), or large (r > 0.5) effect size classifications following the nomenclature of Cohen (1988).

Whilst increasing the complexity of experiments by manipulating additional abiotic or biotic variables may be desirable, it also reduces the tractability and interpretability of results. We are, arguably, able to explain how two or even three driver variables interact mechanistically to affect ecosystem properties (e.g. Dyson *et al.* 2007; Stachowicz *et al.* 2008; Bulling *et al.* 2010), but our ability to explain fully the mechanisms causing a particular net response is strongly reduced as more abiotic and biotic variables are introduced into a study (e.g. Godbold *et al.* 2009; Godbold *et al.* 2011; Leuzinger *et al.* 2011). In addition, despite the fact that increasing the realism of experimental manipulations and their context may be crucial for accurately predicting the ecological significance of biodiversity loss, the more system-driven and realistic an experi-

mental manipulation is, the less generic any predictions are likely to be (Balvanera *et al.* 2006; Naeem 2008).

Although the applicability of results to natural systems has been frequently questioned (Srivastava and Velland 2005), due to the often assumed predominance of experiments with single-factor responses, the present analysis reveals that more than half the studies carried out in aquatic ecosystems manipulate and consider at least one additional abiotic, biotic, or biodiversity-related driver variable. Whilst the inclusion of two experimental variables is far from accurately representing the complexities of the natural environment, it does allow for the generality of diversity effects under different abiotic and biotic conditions to be established. Crucially the results here suggest that the

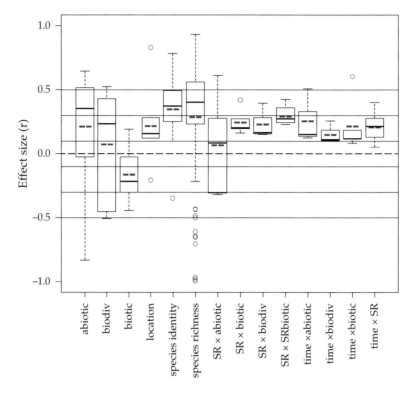

Figure 8.5 Effects of driver variable type, alone and in combination, on ecosystem properties. '×' represents the interaction between two explanatory variable, SR = species richness. In each box, the median is indicated at the midpoint (thick, solid line in each box), the upper and lower quartiles are indicated by the hinges, vertical dashed lines represent the spread, and the open circles indicate outliers. In addition, '—'represents the mean model predictions from the generalised least squares analysis incorporating the factor *driver variable type* as a variance covariate for both models.

strength of biodiversity effects obtained from laboratory and field/in-situ manipulations is, although overlapping, generally lower and more variable, in comparison to observational studies of biodiversity loss in natural systems. This supports the view that experimental results are likely to underestimate the influence of biodiversity on ecosystem properties in natural systems (Duffy 2009). In addition, effect sizes of interacting drivers are generally less variable in comparison to the effects of single drivers on ecosystem properties, which suggest that studies incorporating only a single driver variable may be overestimating the impacts on ecosystem properties. In their analysis of global change experiments in terrestrial systems,

Leuzinger *et al.* (2011) also found that the magnitude of the responses declined in higher order interactions and suggested that global scale impacts on terrestrial ecosystems may be dampened as a result of positive and negative interactions between individual drivers of change.

Although the analyses reported here provide further evidence of the presence and varying strength of species richness effects for ecosystem properties, it is important to consider how these effects compare to the effects of abiotic and/ or biotic variables, or the importance of other biodiversity-related variables in mediating ecosystem properties. The present analysis is the first to explicitly quantify and compare the effect sizes of additional variables that have either been explicitly manipulated and/or statistically accounted for in studies aiming to introduce abiotic and/or biotic context. Overall, the effects of the other explanatory variables in mediating ecosystem properties are slightly stronger than the effects of species loss, with decreasing, but largely overlapping effects from studies conducted in natural systems to laboratory experiments. In contrast to previous suggestions, these results do not conclusively support the view that ecosystem properties will be influenced more by the effects of abiotic and biotic forcing than by species loss per se (e.g. Huston and McBride 2002; Grace et al. 2007; Wardle and Jonsson 2009; Romanuk *et al.* 2009). In fact, it appears that significant species richness effects have largely similar effects to other abiotic and biotic drivers, irrespective of experimental system, highlighting the importance of abiotic–biotic coupling for functioning in natural systems. This result is not surprising, as it is well known that the component parts that make up the complexities of natural ecosystems are inextricably linked (Hughes *et al.* 2007), but it does provide compelling support for adopting a holistic philosophy to ecosystem management (White *et al.* 2010). That said, although effect sizes are overlapping, it does appear that the relative importance of species richness versus that of other drivers of change will vary with context. The argument that species richness effects on ecosystem properties in aquatic ecosystems are strongly dependent on abiotic and biotic context (e.g. Cardinale *et al.* 2000; Emmerson *et al.* 2001; Duffy *et al.* 2005; France and Duffy 2006; Lecerf *et al.* 2007; Mckie *et al.* 2009), the spatial and long-term extent of the study (e.g. O'Connor and Crowe 2005; Cardinale *et al.* 2007; Stachowicz *et al.* 2008), and the importance of species traits in affecting ecosystem functions (e.g. Solan *et al.* 2004) is well rehearsed. How the relative contributions of abiotic and biotic drivers to ecosystem properties change with species loss, however, has received considerably less attention (e.g. Godbold and Solan 2009).

Whilst it is true that the empirical manipulations of biodiversity, under a wealth of experimental and natural conditions for a range of ecosystem properties, have resulted in a variety of outcomes, criticisms over their non-applicability for the long-term management and conservation of our ecosystems are not warranted. The evidence base, despite its variability, points not only to the importance of conserving biodiversity but also to the importance of the abiotic and biotic environment to which it is inextricably linked for maintaining ecosystem properties and services in the long term. Only by considering species loss (or change) alongside other anticipated major ecosystem changes can the full consequences of impending environmental problems be considered. An immediate challenge in achieving this goal will be to identify which components of an ecosystem influence the provision of a variety of ecosystem properties, and understand how, and when, the interactions between these influencing variables become uncoupled and lead to negative ecological consequences. Such a holistic approach will need to dominate the next phase of biodiversity-ecosystem function research if we are to develop appropriate environmental policy to secure the needs of future generations.

References

Balvanera, P., Pfisterer, A.B., Buchmann, N. *et al.* (2006) Quantifying the evidence for biodiversity effects on ecosystem functioning and services. *Ecology Letters*, **9**, 1146–56.

Bolam, S., Fernandes, T., and Huxham, M. (2002) Diversity, biomass, and ecosystem processes in the marine benthos. *Ecological Monographs*, **72**, 599–615.

Bracken, M.E.S., Friberg, S.E., Gonzales-Dorantes, C.A. *et al.* (2008) Functional consequences of realistic biodiversity changes in a marine system. *Proceedings of the National Academy of Sciences USA*, **105**, 924–8.

Bruno, J.F., Boyer, K.E., Duffy, J.E. *et al.* (2008) Relative and interactive effects of plant and grazer richness in a benthic marine community. *Ecology*, **89**, 2518–28.

Bulling, M.T., Hicks, N., Murray, L. *et al.* (2010) Marine biodiversity-ecosystem functions under uncertain environmental futures. *Philosophical Transaction of the Royal Society—B Biological Sciences*, **365**, 2107–446.

Caliman, A., Pires, A.F., Esteves, F.A. *et al.* (2010) The prominence of and biases in biodiversity and ecosystem functioning research. *Biodiversity and Conservation*, **19**, 651–64.

Cardinale, B.J., Matulich, K.L., Hooper, D.U. *et al.* (2011) The functional role of producer diversity in ecosystems. *American Journal of Botany*, **98**, 572–92.

Cardinale, B. J., Nelson, K. and Palmer, M. A. (2000) Linking species diversity to the functioning of ecosystems: on the importance of environmental context. *Oikos*, **91**, 175–83.

Cardinale, B.J., Srivastava, D.S. Duffy, J.E. *et al.* (2006) Effects of biodiversity on the functioning of trophic groups and ecosystems. *Nature*, **443**, 989–92.

Cardinale, B.J., Wright, J.P., Cadotte, M.W. *et al.* (2007) Impacts of plant diversity on biomass production increase through time because of species complementarity. *Proceedings of the National Academy of Sciences of the USA*, **104**, 18123–8.

Cohen, J. (1988) *Statistical Power Analysis for the Behavioral Sciences* (2nd Ed). Lawrence Erlbaum Associates

Covich, A.P., Austen, M., Bärlocher, F. *et al.* (2004) The role of biodiversity in the functioning of freshwater and marine benthic systems. *Bioscience*, **54**, 767–75.

Danovaro, R., Gambi, C., Dell'Anno, A. *et al.* (2008) Exponential decline in deep-sea ecosystem functioning linked to benthic biodiversity loss. *Current Biology*, **18**, 1–8.

Dirzo, R., and Raven, P.H. (2003) Global state of biodiversity and loss. *Annual Review of Environment and Resources*, **28**, 137–67.

Douglass, J.G., Duffy, J.E. and Bruno, J.F. (2008) Herbivore and predator diversity interactively affect ecosystem properties in an experimental marine community. Ecology Letters, **11**, 598–608.

Duffy, J.E. (2009) Why biodiversity is important to the functioning of real-world ecosystems. *Frontiers in Ecology and the Environment*, **13**, 437–44.

Duffy, J.E., Cardinale, B.J., France, K.E. *et al.* (2007) The functional role of biodiversity in ecosystems: incorporating trophic complexity. *Ecology Letters*, **10**, 522–38.

Duffy, J. E., Richardson, J. P. and France, K. E. (2005) Ecosystem consequences of diversity depend on food chain length in estuarine vegetation. *Ecology Letters*, **8**, 301–9.

Dyson, K.E., Bulling, M.T., Solan, M. *et al.* (2007) Influence of macrofaunal assemblages and environmental heterogeneity on microphytobenthic production in experimental systems. *Proceedings of the Royal Society B—Biological Sciences*, **274**, 2547–54.

Emmerson, M.C., Solan, M., Emes, C. *et al.* (2001) Consistent patterns and the idiosyncratic effects of biodiversity in marine ecosystems. Nature 411, 73–7.

Fischlin, A., Midgley, G., Price, J. *et al.* (2007) Ecosystems, their properties, goods, and services. In IPCC (ed.) Climate Change 2007: Impacts, Adaptation and Vulnerability.

France, K.E. and Duffy, J.E. (2006) Diversity and dispersal interactively affect predictability of ecosystem function. *Nature*, **441**, 1139–43.

Godbold, J. A., Bulling, M. T. and Solan, M. (2011) Habitat structure mediates biodiversity effects on ecosystem properties. *Proceedings of the Royal Society B—Biological Sciences*, **278**, 2510–18.

Godbold, J.A. and Solan, M. (2009) Relative importance of biodiversity and the abiotic environment in mediating an ecosystem process. *Marine Ecology Progress Series*, **396**, 273–82.

Godbold, J.A., Solan, M. and Killham, K. (2009) Consumer and resource diversity effects on marine macroalgal decomposition. *Oikos*, **118**, 77–86.

Grace, J.B., Anderson, T.M., Smith, M.D. *et al.* (2007) Does species diversity limit productivity in natural grassland communities? *Ecology Letters*, **10,** 680–9.

Griffin, J.N., Noël, L.M.L.J, Crowe, T.P. *et al.* (2010) Consumer effects on ecosystem functioning in rock pools: roles of species richness and composition. *Marine Ecology Progress Series*, **420**, 45–56.

Healy, C., Gotelli, N.J., Potvin, C. (2008) Partitioning the effects of biodiversity and environmental heterogeneity for productivity and mortality in a tropical tree plantation. *Journal of Ecology*, **96**, 903–13.

Hooper, D.U., Chapin, F.S., and Ewel, J.J. *et al.* (2005) Effects of biodiversity on ecosystem functioning: a consensus of current knowledge. *Ecological Monographs*, **75**, 3–35.

Hughes, A.R., Byrnes, J.E., Kimbro, D.L. *et al.* (2007) Reciprocal relationships and potential feedbacks between biodiversity and disturbance. *Ecology Letters*, **10**, 849–64.

Hullett, C.R. and Levine, T.R. (2003) The overestimation of effect sizes from *F* values in Meta-Analysis: The cause and a solution. *Communication Monographs*, **70**, 52–67.

Huston, M.A. and McBride, A.C. (2002) Evaluating the relative strengths of biotic versus abiotic controls on ecosystem process. In: Loreau, M., Naeem, S., Inchausti, P. (eds), *Biodiversity and ecosystem functioning: synthesis and perspectives.* Oxford University Press, Oxford, 47–60.

Langenheder, S., Bulling, M.T., Solan, M. *et al.* (2010) Bacterial biodiversity-ecosystem functioning relations are modified by environmental complexity. *PLoS ONE*, **5**, e10834.

Lecerf, A., Risnoveanu, G., Popescu, C. *et al.* (2007) Decomposition of diverse litter mixtures in streams. *Ecology*, **88**, 219–27.

Leuzinger, S., Luo, Y., Beier, C. *et al.* (2011) Do global change experiments overestimate impacts on terrestrial ecosystems? *Trends in Ecology and Evolution*, **6**, 236–41.

Levine, T.R. and Hullett, C.R. (2001) Eta-squared, partial eta-squared, and misreporting of effect size in communication research. *Human Communication Research*, **28**, 612–25.

Loreau, M., Naeem, S., Inchausti, P. *et al.* (2001) Biodiversity and ecosystem functioning: current knowledge and future challenges. *Science*, **294**, 804–8.

Mckie, B.G., Schindler, M., Gessner, M.O. *et al.* (2009) Placing biodiversity and ecosystem functioning in context: environmental perturbations and the effects of species richness in a stream field experiment. *Oecologia*, **160**, 757–70.

Millennium Ecosystem Assessment (2005) Ecosystems and human well-being: biodiversity synthesis. Washington, DC: World Resources Institute.

Naeem, S. (2008) Advancing realism in biodiversity research. *Trends in Ecology and Evolution*, **23**, 414–16.

Nakagawa, S. and Cuthill, I.C. (2007) Effect size, confidence interval and statistical significance: a practical guide for biologists. *Biological Reviews*, **82**, 591–605.

O'Connor, N.E. and Crowe, T.P. (2005) Biodiversity loss and ecosystem functioning: Distinguishing between number and identity of species. Ecology 86, 1783–96.

Pinheiro, J. C. and Bates, D. M. (2000) Mixed-effects models in S and S-plus. Springer Verlag, New York.

Quinn, Q.P. and Keough, M.J. (2002) Experimental design and data analysis for biologists. Cambridge University Press, Cambridge.

Romanuk, T.N., Vogt, R.J. and Kolasa, J. (2009) Ecological realism and mechanisms by which diversity begets stability. *Oikos*, **118**, 819–28.

Rosenthal, R. (1991) Meta-analysis procedures for social research. Sage Publications, Newbury Park.

Ruesink, J.L., Feist, B.E., Harvey, C.J. *et al.* (2006) Changes in productivity associated with four introduces species: ecosystem transformation of a 'pristine' estuary. *Marine Ecology Progress Series*, **311**, 203–15.

Schmid, B., Balvanera, O., Cardinale, B.J. *et al.* (2009) Consequences of species loss for ecosystem functioning: meta-analyses of data from biodiversity experiments. In: S. Naeem, Bunker, D.E., Hector, A. *et al. Biodiversity, Ecosystem Functioning, and Human Wellbeing*, Oxford University Press, Oxford, 14–29.

Solan, M., Cardinale, B.J., Downing, A.L. *et al.* (2004) Extinction and ecosystem function in the marine benthos: *Science* 306, 1177–1180.

Solan, M. Godbold, J.A., Symstad, A. *et al.* (2009) Biodiversity—ecosystem function research and biodiversity futures: early bird catches the worm or a day late and a dollar short? In: Naeem, S., Bunker, D.E., Hector, A. *et al.* 'Biodiversity, Ecosystem Functioning, and Human Wellbeing', Oxford University Press, Oxford, 30–46.

Srivastava, D.S. and Vellend, M. (2005) Biodiversity–ecosystem function research: Is it relevant to conservation? *Annual Review of Ecology, Evolution, and Systematics*, **36**, 267–94.

Stachowicz, J.J., Best, R.J., Bracken, M.E.S. *et al.* (2008) Complementarity in marine biodiversity manipulations: reconciling divergent evidence from field and mesocosm experiments. Proc. Natl. Acad. Sci. USA, **105**, 18842–7.

Stachowicz, J.J., Bruno, J.F. and Duffy, J.E. (2007) Understanding the effects of marine biodiversity on communities and ecosystems. *Annual Review of Ecology, Evolution and Systematics*, **38**, 739–66.

Thompson, K., Askew, A.P., Grime, J.P. *et al.* (2005) Biodiversity, ecosystem function and plant traits in mature and immature plant communities. *Functional Ecology*, **19**, 355–8.

Thompson, R. and Starzomski B.M. (2007) What does biodiversity actually do? A review for managers and policy makers. *Biodiversity and Conservation*, **16**, 1359–78.

Underwood, A. J. (1998) Experiments in ecology: their logical design and interpretation using analysis of variance. Cambridge University Press, Cambridge, UK.

Wardle, D.A. and Jonsson, M. (2009) Biodiversity effects in real ecosystems—a response to Duffy. *Frontiers in Ecology and the Environment*, **8**, 10–11.

Weis, J.J., Madrigal, D.S. and Cardinale, B.J. (2008) Effects of algal diversity on the production of biomass in homogeneous and heterogeneous nutrient environments: a microcosm experiment. PLoS ONE, **3**, e2825.

West, B. T., Welch, K. B. and Gatecki, A. T. (2007) *Linear mixed models: a practical guide using statistical software*. London: Chapman and Hall, 28–9.

White, P.C.L., Godbold, J.A., Solan, M. *et al.* (2010) Ecosystem services and policy: A review of coastal wetland ecosystem services and an efficiency-based framework for implementing the ecosystem approach. In: Hester, R.E. and Harrison, R.M. 'Eco-system Services' *Issues in Environmental Science and Technology*, **39**, 29–46.

Worm, B., Barbier, E.B., Beaumont, N. *et al.* (2006) Impacts of biodiversity loss on ocean ecosystem services. *Science*, **314**, 787–90.

CHAPTER 9

Extending the approaches of biodiversity and ecosystem functioning to the deep ocean

Roberto Danovaro

9.1 Deep-sea ecosystems: characteristics, biodiversity, and functioning

The average depth of the global oceans is ca. 3850 metres (m). Deep-sea sediments (> 1000 m depth) cover 65% of the Earth's surface, and the dark areas of the oceans account for > 95% of the volume of the biosphere. With the exception of the oxygen minimum zones (Levin 2003), these deep waters are typically well oxygenated, and they are characterized by low temperatures at depths > 2000 m, from −1.9 °C (in deep Antarctic waters) to 2.0 °C to 4.0 °C in most oceanic regions. Some deep-sea systems have much higher temperatures—e.g. > 14.0 °C in the deep Mediterranean and the Red Sea, (Sardà *et al.* 2004). The deep sea floor is typically covered by fine sediments (medium sands to clays). However, hard substrates are not uncommon, and are associated with fault scarps, mid ocean ridges, seamounts, polymetallic nodules, and block of concrete sediment at the basis of the continental margins interested by landslides. Most of the abyssal sea floor is highly stable for long periods of time—i.e. low current speed and re-suspension—although in some regions, and especially along continental margins—e.g., in deep-sea canyons or on basin slopes—sediment instability, strong bottom currents, dense water cascading, and benthic storms can be relatively frequent (Gage and Tyler 1991).

In contrast to terrestrial ecosystems, the ocean interior lacks primary photosynthetic production. In-situ chemosynthetic primary production characterizes specific deep-sea habitats, such as hydrothermal vents and cold seeps. There is also increasing evidence that Archaea in the deep water column and the surface and subsurface sediments is responsible for primary chemosynthetic production; however, overall, one of the key ecological characteristics of deep-sea ecosystems is that they are note sites of de novo production. Rather, deep-sea assemblages are highly food-limited because their benthic production depends on the input of the energy supplied as the 'rain' of organic matter that is produced in the surface photic layer (Witte *et al.* 2003). In most cases, the extent of this limitation increases with increasing water depth. Indeed, most of the organic flux arrives as an attenuated rain of small particles—typically, only 0.5–2.0% of net primary production in the euphotic zone—which decreases inversely with increasing water depth (Buesseler *et al.* 2007), and varies regionally with the levels of primary production in the upper ocean (Yool *et al.* 2007). This minimal particle flux can be augmented by the fall of larger carcasses and the down-slope transport of organic material near continental margins (Smith and Demoupolos 2003; Tyler 2003).

Despite their limited food availability, deep-sea ecosystems are by far the largest reservoirs of biomass on Earth owing to the size of the deep sea and the sub-seabed biosphere. With increasing depth there is a notable decrease in the larger benthic components (i.e. megafauna), and a progressively less-evident decrease in the abundance and biomass of macrofauna, meiofauna, and microbes (Wei *et al.* 2010). Thus at bathyal-abyssal depths, the large

Marine Biodiversity and Ecosystem Functioning. First Edition. Edited by Martin Solan, Rebecca J. Aspden, and David M. Paterson.
© Oxford University Press 2012. Published 2012 by Oxford University Press.

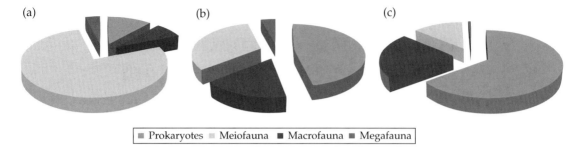

(a)   (b)   (c)

■ Prokaryotes ■ Meiofauna ■ Macrofauna ■ Megafauna

Figure 9.1 Pie charts showing the relative levels of the prokaryotic, meiofauna, macrofauna, and megafauna biomass over the total living benthic biomass inhabiting deep-sea sediments: (a) from 200 to 2000 m; (b) from 2000 to 4000 m; (c) from 4000 to 6000 m. The sequence shows the increasing relevance of the benthic prokaryotic biomass with increasing water depth.

majority of the biomass is accounted for by microbial components—i.e. mostly Bacteria and Archaea (Danovaro 2003; Figure 9.1). The microbial processes occurring there provide essential services, driving nutrient regeneration and global biogeochemical cycles, which are essential to sustain the primary and secondary production of the oceans (Dell'Anno and Danovaro 2005).

Despite the strong reduction in abundance and biomass of animals with increasing depth, the deep-sea ecosystems host an extremely high biodiversity (Danovaro *et al.* 2008a) that remains poorly understood. Deep-sea seabed ecosystems have been poorly studied owing to their vast size and remoteness (Figure 9.2). There is increasing evidence that deep-sea ecosystems host the largest yet-to-be-discovered metazoan biodiversity (McIntyre 2010). Most of this biodiversity is likely to be accounted for by small invertebrates. In considering this, it has been estimated that the deep-sea benthos will host from 0.3 to 8.3×10^6 species that are new to science (Gage and Tyler 1991; Grassle and Maciolek 1992), and it is not uncommon to gather deep-sea samples at abyssal depths where more than 80% of the invertebrate species are new to science (Glover *et al.* 2002; Martinez and Schminke 2005; Snelgrove and Smith 2003).

There are several ways in which deep-sea benthic biodiversity—i.e., a higher functional diversity— can promote ecosystem processes (Danovaro *et al.* 2004; Petchey and Gaston 2006):

i) A higher benthic diversity might increase bioturbation, with a consequent increase in benthic fluxes

(Lohrer *et al.* 2004; Meysman *et al.* 2006) and redistribution of foods within the sediment.

ii) Despite their small size, meiofauna and foraminiferans are responsible for the crypto-bioturbation for their numerical importance and activity (Giere 2009; Pike *et al.* 2001).

iii) A higher species richness can stimulate prokaryote carbon (C) production to a greater extent than selective grazing by a few species (De Mesel *et al.* 2004).

iv) Higher biodiversity levels can also promote higher rates of detritus processing, digestion, and reworking, thus resulting in faster rates of organic matter re-mineralization.

v) Higher numbers of predator species might influence the structural and functional diversity of meio-, macro- and megafauna assemblages, by preying selectively on the larvae of organisms displaying lower mobility (Giere 2009).

The global scale of the biodiversity crisis has stimulated explorations into the relationships between biodiversity, productivity, stability, and services in different ecosystems on Earth. The effects of biodiversity loss on ecosystem functioning have been the focus of an explosion of research over the past decade, and several studies have predicted that species loss will impair the functioning and sustainability of terrestrial ecosystems (Naeem *et al.* 1994; Loreau *et al.* 2001; Hooper *et al.* 2005; Worm *et al.* 2006); however, this research field has focused only very recently on the deep sea. Now we have evidence that climate change and human activities can also have severe impacts on

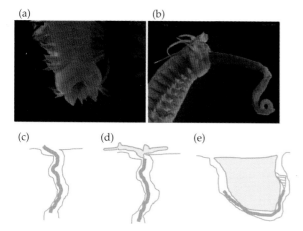

(a) (b)

(c) (d) (e)

Plate 1. Functional plasticity in the feeding behaviour of the polychaete ragworm, *Hediste diversicolor*. *H. diversicolor* is an omnivore which may feed on vegetative material (a), become a carnivore (b), or use other methods of deposit feeding. It may forage over the local sediment surface from its tube (c), or if surface algal material is present, the worm pulls material into its burrows to consume it (d). However, when submerged and where the overlying water contains sufficient suspended organic material, the worm retreats into its tube and spins a mucilage net. The worm then undulates to create a water flow which is filtered through the mucilage net which is later consumed by the feeding worm (d). Images C. Wood and M. Cholonkney, St Andrews (see Paterson 2005). See also Figure 3.2, page 29.

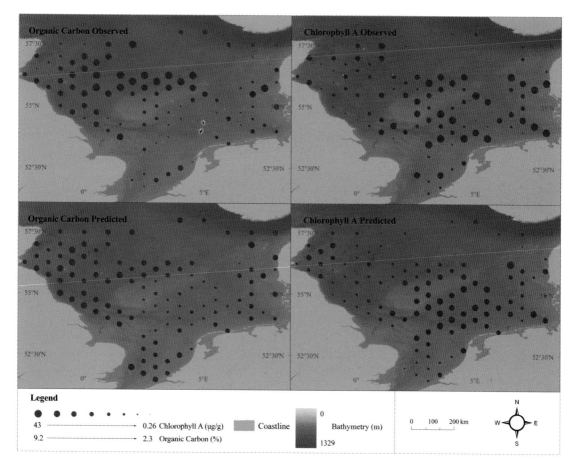

Plate 2. Observed and predicted levels of ecosystem functioning for sediment Org C (left-hand panels) and Chl *a* (right-hand panels) for each station in the North Sea. Predictions were fitted empirically using standard linear regression (Equations 10.2–10.5 in main text). See also Figure 10.2, page 135.

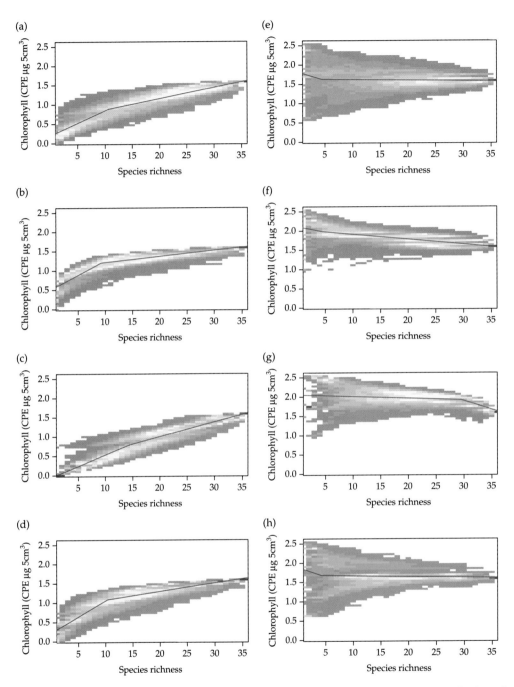

Plate 3. Representative predicted changes in sediment Chl *a* following benthic invertebrate extinctions at station 8 in the North Sea (predictions for the other 108 stations are available from the authors). Each panel within a station shows the results of 2000 simulations, each representing one trajectory of species loss from the full complement of species to a single species. The probabilistic order of species extinction was either random (panels a and e), or assumed to be proportional to specific traits: mean abundance within a station (rare species first, lowest to highest, panels b and f); mean abundance across the regional species pool (rare species first, lowest to highest, panels c and g); or related to body size (largest to smallest, panels d and h). Simulations (a), (b), (c), and (d) are for a non-interactive model of community assembly assuming no compensation by surviving species. Simulations (e), (f), (g), and (h) are for an interactive model that assumes full compensation by the surviving community following extinction. Lines represent segmented linear regression with the break-point indicating the level of species richness at which the effect on ecosystem process changes abruptly. See also Figure 10.3, page 137.

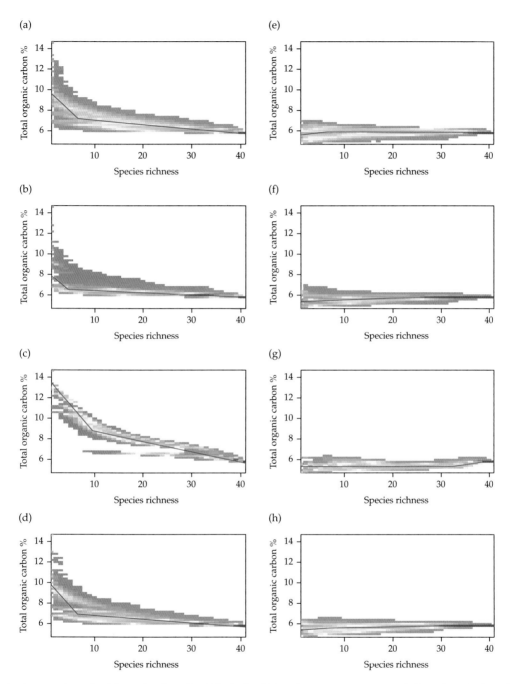

Plate 4. Representative predicted changes in sediment Org C following benthic invertebrate extinctions at station 54 in the North Sea (predictions for the other 108 stations are available from the authors). Each panel follows the probabilistic order of species extinctions as listed in Plate 3, Figure 10.3 (page 137). Lines represent segmented linear regression with the break-point indicating the level of species richness at which the effect on ecosystem process changes abruptly. See also Figure 10.4, page 138.

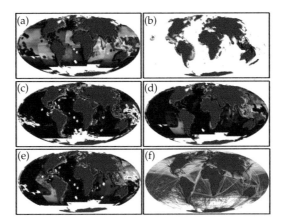

Plate 5. Global maps of selected human impacts on marine ecosystems. (a) pelagic fishing resulting in low bycatch (collateral damage to non-target species), (b) pelagic, high-bycatch fishing (taking substantial numbers of non-target species), (c) demersal (bottom) fishing that modifies habitat (such as reefs and seagrass beds), (d) demersal non-habitat modifying, low-bycatch fishing, (e) demersal non-habitat-modifying, high-bycatch fishing, (f) shipping. Blue and red denote low and high impacts, respectively, while white denotes no data. From Halpern *et al*. (2008), used with permission from AAAS. See also Figure 12.1, page 165.

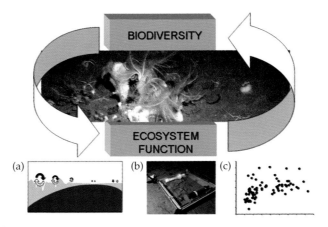

Plate 6. BEF relationships are not simple cause—effect relationships. They should be viewed as feedback loop relationships. This chapter argues for developing a process of iterating between: (a) observation of patterns, (b) local measurement and manipulation, and (c) formulation of general relationships. See also Figure 14.1, page 202.

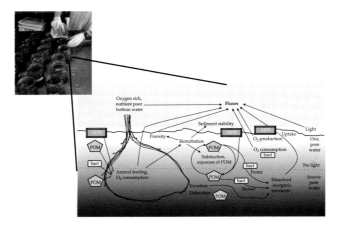

Plate 7. BEF processes are complex and it may not be easy to isolate individual mechanisms even in small-scale enclosure experiments. There are many mechanisms involved in the way in which animals, such as burrowing urchins, interact with biogeochemical processes to influence nutrient flux and productivity. See also Figure 14.2, page 205.

Deep-sea bottoms

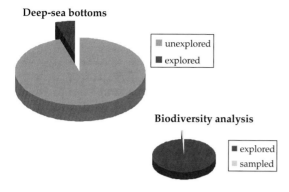

Figure 9.2   Pie charts showing overall estimates of relative portions of: (a) oceans explored by detailed topographic and remote or direct video (or equivalent) analyses; (b) sediments sampled, retrieved on-board and analysed in detail for their biodiversity.

deep-sea ecosystems (Glover and Smith 2003). Understanding the relationships between biodiversity and deep-sea ecosystem functioning is therefore crucial for understanding the functioning of our biosphere.

9.2   Approaches to the investigation of deep-sea biodiversity and ecosystem functioning

To date, manipulative experiments in aquatic systems have involved a very limited number of species (Sih *et al.* 1998), and there is a need for integrating these experiments with large-scale observational studies (Griffin *et al.* 2008). The inaccessibility and extreme conditions of deep-sea environments are not conducive to manipulative experiments similar to those that have been conducted in terrestrial ecosystems or in intertidal/coastal habitats (Bolam *et al.* 2002; Lohrer *et al.* 2004). Thus, with some notable exceptions (Micheli *et al.* 2005), the analysis of the relationships between biodiversity and ecosystem functioning are currently based on correlative approaches—especially when large spatial scales are considered—and/or on modelling approaches (Danovaro *et al.* 2008b). Despite their intrinsic limitations, comparative studies have great value in revealing patterns that would otherwise remain out of the reach of experimental studies when these patterns involve large

spatial scales, long timescales, or not very accessible ecosystems, such as in the deep sea.

Benthic-fauna diversity provides an ideal tool for exploring the relationships between biodiversity and ecosystem functioning in marine ecosystems (Snelgrove 1999). Among the deep-sea benthic-fauna taxa, nematodes are an excellent model organism (Bonger and Ferris 1999), as they account for 90% or more of the multicellular animals in the deep sea. Nematodes are also the most abundant metazoans on Earth, and in terrestrial ecosystems, they account for 80% of multicellular animals (Bonger and Ferris 1999). This Phylum is also characterized by:

i) very high species richness—i.e. the largest biodiversity among the animal Phyla living in the oceans (Lambshead 2004);
ii) well-established feeding habits (Wieser 1953; Jensen 1987) and known life strategies, which make it possible to identify functional diversity traits (Danovaro *et al.* 2001).

In addition, due to their size and abundance in all deep-sea systems (typically > 50 000 individuals m^{-2}), nematodes can potentially be used in manipulative experiments (Witte *et al.* 2003). Although comparative studies are rare, deep-sea nematode diversity appears to be related to that of other benthic components, including Foraminifera (Gooday *et al.* 1998), macrofauna (Levin *et al.* 2001), and the richness of the higher meiofauna taxa (a group which includes 22 of the 35 modern animal Phyla; Figure 9.3).

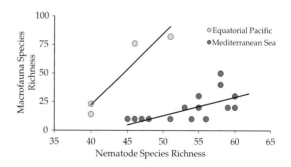

Figure 9.3   Relationships between nematode and macrofauna diversities (species richness) in deep-sea samples collected synoptically from the Mediterranean Sea (600–3055 m) and the equatorial Pacific Ocean (4305–4994 m).

Methods and protocols for the sampling, preservation and separation of deep-sea organisms from the sediments have been described in detail elsewhere (Danovaro 2010).

9.2.1 Biodiversity metrics

A complete census to species level of all of the benthic animals encountered in deep-sea sediments is almost impossible at present, not only because of the huge fraction of under-described species, but also for the extreme difficulty in pooling together the taxonomic expertise needed to classify all of the taxa. Of the 22 meiofaunal phyla encountered in deep-sea sediments—i.e. organisms of size < 0.5 mm—five occur exclusively in this benthic component—e.g. Loricifera, Gnatostomulida, Tardigrada, Gastrotricha, Kinorhyncha; (Higgins and Thiel 1988). The size of a large fraction of the benthic organisms inhabiting deep-sea sediments is often in the order of less than 200 µm. In certain cases, a proxy of the benthic diversity can be obtained through the census of the higher taxa, which can be identified as nematodes, harpacticoid copepods (including their naupliar stages), polychaetes, oligochaetes, isopods, cumaceans, tardigrades, amphipods, acarians, ostracods, tanaidaceans, cnidarians, kinorhynchs, turbellarians, gastrotriches, nemerteans, bivalves, priapulids, cladocerans, decapods (larvae), and loriciferans.

The results can be reported as the number of species present in a sample (the species richness) or using widely accepted biodiversity indices. The Shannon-Wiener diversity (H′) is calculated as $H' = -p_i \sum \log_2 p_i$, where $p_i = n_i / N$, n_i is the number of individuals of the ith-species, and N is the total number of individuals. Since most indices of species diversity are sample-size dependent, the rarefaction method is applied to reduce all of the samples to the same size, with ES(51) as the expected number of species in a hypothetical random sample of 51 individuals. Previous studies have shown that this approach provides robust data on species richness in the deep sea, and the expected species number is the best independent density index for comparing areas with non-standardized sample sizes (Lambshead et al. 2002; Danovaro et al. 2008a,b). The evenness of benthic assemblages can be calculated as the Pielou index (J), which indicates how the abundance is partitioned among the species, using the for-

mula $J = H'/H_{max}$, where H′ is the Shannon index and H_{max} is the maximum diversity—i.e. $\log_2 H$.

9.2.2 Functional diversity

The functional diversity is the range of functions that are performed by the organisms in a system (Cardinale et al. 2002, Stuart et al. 2003). Three descriptors can be used:

i) The number of different functional (trophic) traits based on an analysis of the feeding types according to the classical literature (Heemsbergen et al. 2004), and updated according to the most recent approaches.

ii) The diversity of the morpho-functional traits, assuming that different morphologies, buccal sizes, and other traits reflect a diverse ecological role—e.g. selection of food items within the same feeding guild (Petchey and Gaston 2006).

iii) The number of predator species, which depends upon the assumption that the number of species at the top of the benthic food web reflects a higher functional diversity of the entire benthic assemblage.

9.2.3 Deep-sea ecosystem functioning

Terrestrial ecologists have related biodiversity to ecosystem functioning through analyses of ecosystem processes that are estimated by measuring the rates of energy and material flow between the biotic and abiotic compartments—e.g. biomass production, organic matter decomposition, nutrient regeneration, or other measures of material production, transport, or loss. Applying the same approach through a series of independent and synoptic measures, it is possible to investigate the relationships between deep-sea biodiversity and ecosystem functioning. Deep-sea ecosystem functioning reflects the collective activities of animals, protists, and prokaryotes in exploiting and recycling the inputs of material from the photic zone. However, as they lack primary photosynthetic production, to a certain extent, these systems have simplified functioning, when compared to coastal marine ecosystems.

The following key processes can be identified for the characterization of deep-sea ecosystems:

i) the prokaryotic production in the sediment;

ii) the fauna biomass and/or production—a measure of the production of renewable resources by ecosystems;

iii) the rates at which organic matter is decomposed and recycled.

These three independent indicators of ecosystem functioning represent key variables of deep-sea ecosystems, as they regulate:

i) the transfer of mobilized organic matter into microbial biomass;

ii) the ability of the ecosystem to transfer microbial biomass to higher trophic levels, thus providing additional indications of the heterotrophic production of the ecosystem;

iii) nutrient regeneration processes, which reflect the ability of ecosystems to sustain their functions over time.

The biodiversity and functional measures should be derived from the same samples, so as to ensure contextual consistency.

9.2.4 Variables used for measuring ecosystem efficiency

The efficiency of an ecosystem can be defined as the available energy resources that the deep-sea biota can channel into biomass. There are three independent indicators of ecosystem efficiency that have been used to date (Danovaro *et al.* 2008b):

i) the ratio of benthic fauna biomass or production to the organic C input;

ii) the ratio of prokaryote C production to the organic C flux;

iii) the ratio of the benthic fauna biomass to the amount of food available in the sediment.

Organic C fluxes are typically measured using sediment traps, in which the settling particles are fixed in-situ with buffered and pre-filtered formalin, to minimize prokaryotic activity. Sediment traps can be deployed for several months, thus allowing the detection of changes in the food supply over the long term. The organic C concentrations and the presence of specific bioavailable organic particles are determined using standard protocols (Heussner *et al.* 1990).

The amount of food available to benthic consumers can be estimated through analysis of the biochemical composition of the organic matter in the sediment. The sum of the carbohydrate, protein, and lipid C equivalents is defined as the biopolymeric C fraction (Pusceddu *et al.* 2004), which typically accounts for approximately 30% of the total organic C pools. Unlike the geo-polymeric C fraction, which consists of black C, and humic and fulvic substances, the biopolymeric C fraction represents a more readily available food source for the deep-sea benthos than the bulk of the organic C. It includes a readily available (digestible) fraction and a semi-labile fraction. The protein in sediment sub-samples has been analysed according to standard protocols (Pusceddu *et al.* 2004).

9.3 Relationships between biodiversity and ecosystem functioning in the deep sea

The mechanisms involved in the biodiversity effects on ecosystem functioning are typically grouped into two main classes of biodiversity effects (Loreau 2008):

i) Complementarity: this occurs when species in a mixed community perform better on average than expected from their performances in monoculture, thereby contributing to enhanced ecosystem processes.

ii) Selection: this occurs when specific species tend to dominate, thereby contributing to either enhancing or deteriorating the ecosystem processes, depending on whether a better-performing or a lesser-performing species dominates.

The relative contributions of these two effects towards the results obtained in biodiversity experiments have been highly contentious, because they each have very different implications for the mechanisms that maintain the diversity in natural assemblages. However, studies that have been conducted in deep-sea ecosystems have revealed, for the first time, that the functioning of these ecosystems is positively and exponentially related to their biodiversity in all deep-sea regions investigated (Figure 9.4). While different mechanisms can contribute to the relationships between saturating

(a)

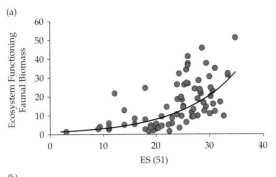

(b)

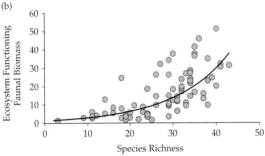

Figure 9.4 Relationships between biodiversity and ecosystem functioning (total fauna biomass; mg C g⁻¹), with biodiversity measures of: (a) expected species number, ES(51); (b) species richness.

biodiversity and ecosystem functioning, only positive species interactions are known to yield accelerating relationships, and thus to generate exponential relationships (Bruno *et al.* 2003).

Facilitative, or positive, interactions are encounters between organisms that benefit at least one of the participants and do not cause harm to either. Such interactions are considered 'mutualisms', when both species derive benefit from the interaction. These interactions can occur when one organism makes the local environment more favourable for another, either directly or indirectly, and they can include tightly co-evolved, mutually obligate relationships, as well as facultative interactions. Many species modify their local environment and facilitate neighbouring species simply through their presence (Bruno *et al.* 2003). Experimental investigations from a wide variety of habitats have demonstrated the strong effects of facilitation on individual fitness, on population distribution and growth rates, on species composition and diversity, and even on landscape-scale community dynamics (Stachowicz 2001). The

application of facilitative interactions is perfect for the deep-sea ecosystems, as results that have been reported from these ecosystems suggest that many deep-sea species can benefit from the presence of others, leading to mutual enhancement of their performances (Danovaro *et al.* 2008b). Exponential relationships are also found when different biodiversity measures—including the richness of all of the higher fauna taxa—and independent measures of ecosystem functioning are used (Figure 9.4). Exponential relationships have been observed in all of the different deep-sea regions investigated: the subtropical Pacific Ocean, the temperate north-eastern Atlantic, the warm deep Mediterranean, and the cold deep Antarctic (Danovaro *et al.* 2008b). The systems investigated have shown different assemblage compositions, and environmental conditions—e.g. from very cold to warm, from very oligotrophic to meso-eutrophic, different salinities, different topographic settings. It is known that in all these systems, a greater biomass can be associated to a greater species richness. Greater biomass can occur if there is greater food input. So in all systems where there is a gradient in food input it can be expected to find a gradient in abundance—greater number of individuals in rich systems—and biomass. Higher abundances typically lead to more species, most of which are rare, but in deep-sea sediments species richness/expected number of species are largely independent from food supply as shown either from analysis of seasonal trends and from in-depth analysis of all food sources used as covariates (Danovaro *et al.* 2008b).

Similar data have been obtained in investigations of the relationships between deep-sea ecosystem efficiency (Naeem *et al.* 1994)—which reflects the ability of an ecosystem to exploit the available energy (food sources) and thereby to maximize the biomass and its production (Naeem *et al.* 1994; Loreau *et al.* 2001) and benthic biodiversity. All of the independent indicators of ecosystem efficiency—e.g. ratio of fauna biomass to organic C fluxes, ratio of prokaryote C production to organic C flux, ratio of total benthic meiofauna biomass to biopolymeric C content in the sediment—when related to normalized values species richness (ES 51) have been significantly and exponentially related to benthic biodiversity (Figure 9.5).

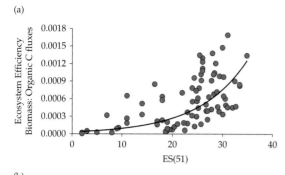

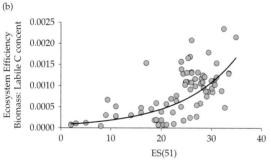

Figure 9.5   Relationships between deep-sea biodiversity (expected species number; ES(51)) and ecosystem efficiency, as: (a) ratio of fauna biomass to organic C flux; (b) ratio of fauna biomass to biopolymeric C concentration in the sediment. The exponential relationships shown used two indicators of ecosystem efficiency. The data are from the Mediterranean Sea ($R^2 = 0.55$). Depth range, 1078–3870 m.

Overall, these findings indicate that the exponential relationships between deep-sea biodiversity and ecosystem functioning and efficiency are consistent across a wide range of deep-sea ecosystems. Therefore they reflect interactions between organism life and deep-sea ecosystem processes that occur on a global scale (Danovaro *et al.* 2008b).

Recent studies have emphasized the importance of functional diversity traits that influence ecosystem functioning (Loreau *et al.* 2001; Tilman *et al.* 2001; Heemsbergen *et al.* 2004). Understanding how species interactions influence the relationships between biodiversity and ecosystem functioning or efficiency implies a thorough knowledge of the processes regulating deep-sea benthic food-webs and the ecological role of each species.

In deep-sea sediments, species number and diversity of functional traits are directly and positively related (Figure 9.6), so that changes in species rich-

ness apparently have a direct effect on functional diversity and related ecological processes. Taken together, these relationships suggest that more diverse deep-sea systems are characterized by higher rates of ecosystem processes than less diverse systems, as well as by an increased efficiency with which these processes are performed.

Although experimental approaches are the only tools to unequivocally demonstrate the effects of biodiversity loss on deep-sea-ecosystem functioning, one case study from the deep eastern Mediterranean identified a clear link between ecosystem functioning and functional diversity (Danovaro *et al.* submitted). After the extreme climate event known as the Eastern Mediterranean Transient, which caused a rapid change in water salinity and temperature (by ca. 0.4 °C), significant changes in species compositions were seen. In particular, ca 50% of the species that were present before this episodic event were replaced. These changes resulted in a decrease in the functional diversity (by ca 35%), which in turn was linked to a decrease in the benthic fauna biomass (ca. 40% reduction). Moreover, despite the increased input of organic nutrients associated with the Eastern Mediterranean Transient, a major decrease in the prokaryote biomass (by > 80%) was seen, along with a significant accumulation of organic C and N, which reflects a strong decrease in ecosystem functioning—e.g. nutrient regeneration rates (Danovaro *et al.* 2001). These data have promoted the hypothesis that there is a

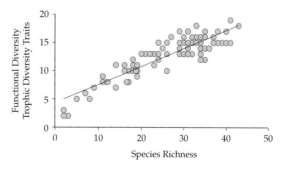

Figure 9.6   Relationship between biodiversity (species richness) and functional diversity (total number of trophic traits). The equation for the full dataset (solid line; log transformed) is $Y = 0.32\,X + 4.3$ ($R^2 = 0.80$, $p < 0.01$).

direct link between ecosystem functioning and deep-sea biodiversity, and they support the correlative finding that a reduction in biodiversity appears associated with an exponential decline in ecosystem processes.

The exponential relationships between biodiversity and ecosystem functioning that have been seen for deep-sea ecosystems are consistent with previous studies that have hypothesized the existence of mutually positive functional interactions—ecological facilitation (Cardinale *et al.* 2002). However, they also apparently expand the range and ecological relevance of these interactions, as the global distribution of these relationships at regional scales would suggest that ecological facilitation:

i) is dominant in deep-sea ecosystems;

ii) occurs in different deep-sea habitats, and with a wide range of longitude and latitude; and

iii) involves a huge number of species and assemblages with completely different compositions.

Moreover, the presence of exponential relationships has important theoretical and practical consequences. Indeed, the saturating relationships between biodiversity and ecosystem functioning—i.e. relationships reaching a threshold level of functioning—suggest that the loss of any species can lead to a significant decline in function, although as more disappear, there are fewer that can have a large impact. In contrast, the exponential relationship suggests that a declining diversity is associated with a loss of facilitation among species, which leads to an exponential decline in ecosystem functioning.

It is evident that the exponential relationships reported for deep-sea ecosystems cannot be indefinite, as the number of species cannot grow indefinitely, and nor can the ecosystem functioning grow exponentially, without end. However, it is worth noting that these exponential relationships are observed only at regional scales, and as the diversity and functional variables are measured synoptically and contextually, they refer to the in-situ interactions at the moment of sampling.

The number of species encountered in different samples—the alpha diversity—varies notably among different samples within the same regions,

reaching values higher than 100 species per sample. These data indicate that at values of alpha diversity that are even higher than 100 species, there are no signs of saturating conditions in the relationships between biodiversity and ecosystem functioning.

Deep-water hypoxic settings are predicted to progressively increase in deep-sea ecosystems as a result of global changes—e.g. ocean acidification—and increasing human pressure on marine ecosystems (Weaver *et al.* 2009), and they are ideal environments for testing the effects of a lowered biodiversity of deep-sea ecosystem functioning. Indeed, in the oxygen minimum zones, sedimentary C cycling is depressed at depths where oxygen depletion is most severe, and metazoan abundance and diversity are lowest. Investigations conducted in oxygen minimum zones have suggested that where biodiversity is lower, all of the biological and biogeochemical processes are depressed. Since enzymatic activities in deep-sea sediments are generally not depressed by oxygen concentrations, it can be hypothesized that these processes are decreased because of the reduced bioturbation and the range of functional activities carried out by benthic organisms. At slightly higher bottom-water oxygen concentrations within the oxygen minimum zones, high abundance, virtually monospecific populations of macrofauna metazoans (polychaetes) are active in the short-term uptake of organic matter, and the biogeochemical cycles are faster and more efficient (Levin *et al.* 2000). Finally, in the permanently anoxic and sulphidic deep Black Sea, and in the deep hypersaline anoxic basins of the Mediterranean Sea, where conditions are even more extreme and the sediments virtually lacking metazoan diversity, all processes—including prokaryotic C production and C cycling—are extremely low.

These data suggest that higher biodiversity can enhance the ability of deep-sea benthic systems to perform the key biological and biogeochemical processes that are crucial for their sustainable functioning. Moreover, data obtained from deep-sea ecosystems suggest that the mechanisms regulating the relationships between biodiversity and the functioning of natural ecosystems are different from those predicted by manipulative experiments in coastal systems—i.e. in most cases, null, positive or idiosyncratic (Emmerson *et al.* 2001; Bolam *et al.* 2002; O'Connor and Crowe 2005).

Such differences might reflect several factors, including the characteristics of deep-sea ecosystems, the number of species involved, and the nature of the relationships between structural and functional biodiversity. Notably, these results suggest that the effects of deep-sea benthic biodiversity on ecosystem functioning becomes more evident when biodiversity values are high. To date, the available information on the effects of changing species richness on ecosystem functioning has been based on descriptive studies or on investigations using a limited number of manipulated species (Loreau *et al.* 2001; Naeem *et al.* 1994; Heemsbergen *et al.* 2004; Emmerson *et al.* 2001; O'Connor and Crowe 2005), and typically one order of magnitude lower than in the present study. These findings provide valuable pointers to the mechanisms that might cause the patterns observed, and point to the need for new multi-species experiments that reflect the conditions occurring in systems characterized by high biodiversity levels.

9.4 Relationships between biodiversity and ecosystem functioning in different deep-sea ecosystems

Although exponential relationships have been reported for all deep-sea ecosystems investigated so far—with the exception of oxygen-depleted ecosystems—there are notable differences in the slopes—i.e. the power—of the exponential relationships among different regions and habitats (Figure 9.7). These differences indicate that regions with assemblages with similar biodiversity but different species composition can have significantly different levels of ecosystem functioning and efficiency. On the one hand, these findings suggest an important role for the functional traits and identity of the species involved in shaping the relations in different deep-sea ecosystems (O'Connor and Crowe 2005). On the other hand, theoretically, the different slopes allow the identification of different levels of vulnerability to biodiversity loss. A theoretical case is reported in Figure 9.7, where a continental slope and a deep basin are compared. Assuming an abrupt

species loss that is equivalent to 10% in both systems, the consequences in terms of the loss of ecosystem functions would be significantly higher along continental slopes (ca. 30%) than in the deep basin (ca. 15%). Therefore, it can be hypothesized that the coefficient (power) of the exponential relationships between biodiversity and ecosystem functioning in deep-sea ecosystems is a potential tool for estimating the vulnerability of deep-sea habitats to biodiversity loss. If this approach can be based on a solid dataset and confirmed by independent measures it will be a valuable tool for sustainable management of deep-sea ecosystems and habitats.

The analysis of the differences in assemblage composition that lead to different exponential coefficients appears to be related to differences in the specific functional traits of deep-sea species. For instance, in an ecosystem with the highest performance, the presence of a large fraction of rare species has been shown, with a large overall abundance, but also high biodiversity, of predators. Data reported from multitrophic deep-sea assemblages have promoted the hypothesis that rare species and/or species trophic groups can be supportive of mutualistic interactions (facilitation), as they are specifically associated with the higher efficiencies by which ecosystem functions are performed. If these apparent links are indicative of cause–effect processes, a high number of rare species would represent insurance for the functioning of an

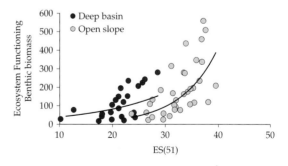

Figure 9.7   Relationships between biodiversity (expected species number, ES(51)) and ecosystem functioning (total fauna biomass; mg C g^{-1}) in deep basins and open slopes along continental margins. The equations for the full datasets (solid lines) are $Y = 19.3\ e^{0.07X}$ ($R^2 = 0.2$), and $Y = 1.18\ e^{0.15X}$ ($R^2 = 0.5$), respectively.

ecosystem, as this allows the maintenance of high levels of ecosystem functioning.

9.5 Conclusions and perspectives

Most of the relationships between biodiversity and ecosystem functioning have been established through controlled field experiments that have assembled model communities to measure the effects of changes in diversity on ecosystem processes (Loreau *et al.* 2001; Heemsbergen *et al.* 2004; Tilman *et al.* 2006). Meta-analyses of these studies have shown that species diversity generally has positive and saturating effects on ecosystem processes, and that these effects are remarkably consistent across trophic levels and different ecosystems (Cardinale *et al.* 2006; Balvanera *et al.* 2006). However, most of these studies have been carried out considering the biodiversity of a single trophic level or a single taxon, whilst multitrophic interactions are known to produce a richer variety of diversity-functioning relationships than monotonic changes predicted for single trophic levels (Duffy *et al.* 2007). Data collected from a large variety of deep-sea ecosystems have provided evidence that interactions among a large number of species (typically > 30–40) are needed to generate exponential relationships, and that no clear and significant patterns can be detected by considering or manipulating a limited number of species (typically < 5). A common problem with comparative studies is that correlation does not imply causation, and further experimental studies are needed to determine the mechanisms driving these relationships. However, to date, all of the rigorous manipulative experiments that have been performed in marine systems have been conducted in intertidal systems—characterized by extremely low biodiversity—or, to a lesser extent, in shallow subtidal systems; they have also always manipulated very few species. In the light of the data also reported from deep-sea systems, these approaches should be carefully reconsidered, as we risk the production of misleading or non-realistic results if we do not perform studies that can recreate the in-situ biodiversity levels and the trophic complexity.

Data from deep-sea ecosystems have potential implications far beyond these remote systems, not only for their relevance on a global scale, but also for their extremely reduced anthropogenic impact, which makes these systems possibly the only actually 'natural' systems remaining on the Earth. In these highly stable and uncontaminated systems, facilitative interactions prevail. At the same time, deep-sea ecosystems are highly vulnerable, sensitive to climate change (Danovaro *et al.* 2004), and susceptible to biodiversity losses (Gage and Tyler 1991; Grassle and Maciolek 1992; Levin *et al.* 2001). In modern oceans, deep-sea ecosystems are already being threatened by man through trawling, dumping, and oil, gas and mineral extraction, and other pollution sources (Glover and Smith 2003). Moreover, the impacts due to changes in the thermohaline circulation linked to global climate change (Bryden *et al.* 2005) are expected to be extremely severe (Danovaro *et al.* 2004).

Empirical and theoretical studies increasingly argue that biodiversity regulates ecosystem functioning (Cardinale *et al.* 2002; Stuart *et al.* 2003; Heemsbergen *et al.* 2004; Naeem *et al.* 1994; Worm *et al.* 2006). Therefore, data from deep-sea ecosystems strongly suggest that changes in deep-sea diversity can influence key ecosystem functions on a global scale.

If the mechanisms that have been widely demonstrated in a number of studies (Cardinale *et al.* 2002; Heemsbergen *et al.* 2004; Naeem *et al.* 2004; Worm *et al.* 2006) can be applied to the deep sea, then reductions in biodiversity, even by 25%, might be associated with reductions in marine ecosystem functions by as much as 50% or more. Given the profound involvement of deep-sea ecosystems in global biogeochemical processes, these data provide the scientific foundation for conserving deep-sea biodiversity so as to preserve deep-sea ecosystem functioning and promote sustainability of the exploitation of the goods and services provided by the largest ecosystems on Earth.

References

Balvanera, P., Pfisterer, A.B., Buchmann, N. *et al.* (2006) Quantifying the evidence for biodiversity effects on ecosystem functioning and services. *Ecol Lett*, **9**, 1146–56.

Bolam, S., Fernandes, T., Huxham, M. (2002) Diversity, biomass, and ecosystem processes in the marine benthos. *Ecol Monogr*, **72**, 599–615.

Bongers, T. and Ferris, H. (1999) Nematode community structure as a bioindicator in environmental monitoring. *Trends Ecol Evol*, **14**, 224–8.

Bruno, J.F., Stachowicz, J.J., Bertness, M.D. (2003) Inclusion of facilitation into ecological theory. *Trends Ecol Evol*, **18**, 119–25.

Bryden, H.L., Longworth, H.R., Cunningham, S.A. (2005) Slowing of the Atlantic meridional overturning circulation at 25 N. *Nature*, **438**, 655–7.

Buesseler, K.O., Lamborg, C.H., Boyd, P.W. *et al.* (2007) Revisiting carbon flux through the ocean's twilight zone. *Science*, **316**, 567–70

Cardinale, B.J., Palmer, M.A., Collins, S.L. (2002) Species diversity enhances ecosystem functioning through interspecific facilitation. *Nature*, **415**, 426–9.

Cardinale, B.J., Srivastava, D.S., Duffy, J.E. *et al.* (2006) Effects of biodiversity on the functioning of trophic groups and ecosystems. *Nature*, **443**, 989–92.

Danovaro, R. (2003) Organic inputs and ecosystem efficiency in the deep Mediterranean Sea. *Chem Ecol*, **19**, 391–8.

Danovaro, R. (2010) Methods for the study of deep-sea sediments, their functioning and biodiversity. Taylor and Francis Group, Boca Raton.

Danovaro, R., Dell'Anno, A., Fabiano, M. *et al.* (2001) Deep-sea ecosystem response to climate changes: the eastern Mediterranean case study. *Trends Ecol Evol*, **16**, 505–10.

Danovaro, R., Dell'Anno, A., Gambi, C. *et al.* Microscopic predators play a key role in deep-sea ecosystem functioning. (*submitted*).

Danovaro, R., Dell'Anno, A., Pusceddu, A. (2004) Biodiversity response to climate change in a warm deep sea. *Ecol Lett*, **7**, 821–8.

Danovaro, R., Gambi, C., Dell'Anno, A. *et al.* (2008b) Exponential decline of deep-sea ecosystem functioning linked to benthic biodiversity loss. *Curr Biol*, **18**, 1–8.

Danovaro, R., Gambi, C., Lampadariou, N. *et al.* (2008a) Deep-sea biodiversity in the Mediterranean Basin: testing for longitudinal, bathymetric and energetic gradients. *Ecography*, **31**, 231–44.

De Mesel, I., Derycke, S., Moens, T. *et al.* (2004) Top-down impact of bacterivorous nematodes on the bacterial community structure: a microcosm study. *Environ Microbiol* **6**, 733–44.

Dell'Anno, A. and Danovaro, R. (2005) Extracellular DNA plays a key role in deep-sea ecosystem functioning. *Science*, **309**, 2179.

Duffy, J.E., Cardinale, B.J., France, K.E. *et al.* (2007) The functional role of biodiversity in ecosystems: incorporating trophic complexity. *Ecol Lett*, **10**, 522–38.

Emmerson, M.C., Solan, M., Emes, C. *et al.* (2001) Consistent patterns and the idiosyncratic effects of biodiversity in marine ecosystems. *Nature*, **411**, 73–7.

Gage, J.D. and Tyler, P.A. (1991) *Deep-Sea Biology: A Natural History of Organisms at the Deep Sea Floor.* Cambridge University Press, Cambridge.

Giere, O. (2009) Meiobenthology. *The microscopic motile fauna of aquatic sediments.* Springer, Berlin Heidelberg.

Glover, A.G., Smith, C.R. (2003) The deep-sea floor ecosystem: current status and prospects of anthropogenic change by the year 2025. *Environm Conserv*, **30**, 219–41.

Glover, A.G., Smith, C.R., Paterson, G.L.J. *et al.* (2002) Polychaete species diversity in the central Pacific abyss: local and regional patterns and relationships with productivity. *Mar Ecol Prog Ser*, **240**, 157–70.

Gooday, A.J., Bett, B.J., Shires, R. *et al.* (1998) Deep-sea benthic foraminiferal diversity in the NE Atlantic and NW Arabian sea: a synthesis. *Deep-Sea Res II*, **45**, 165–201.

Grassle, J.F. and Maciolek, N.J. (1992) Deep-sea species richness: regional and local diversity estimates from quantitative bottom samples. *Am Nat*, **139**, 313–41.

Griffin, J.N., De la Haye, K., Hawkins, S.J. *et al.* (2008) Predator diversity and ecosystem functioning: density modifies the effect of resource partitioning. *Ecology*, **89**, 298–305.

Heemsbergen, D.A., Berg, M.P., Loreau, M. *et al.* (2004) Biodiversity effects on soil processes explained by interspecific functional dissimilarity. *Science*, **306**, 1019–20.

Heussner, S., Monaco, A., Carbonne, J. (1990) The PPS 3 timeseries sediment trap and the sample processing technique used during the ECOMARGE experiment. *Cont Shelf Res*, **10**, 943–58.

Higgins, R.P. and Thiel, H. (1988) *Introduction to the study of meiofauna.* Smithsonian Institution Press, Washington, DC.

Hooper, D.U., Chapin III, F.S., Ewel, J.J. *et al.* (2005) Effects of biodiversity on ecosystem functioning: a consensus of current knowledge. *Ecol Monogr*, **75**, 3–35.

Jensen, P. (1987) Feeding ecology of free-living aquatic nematodes. *Mar Eco. Progr Ser*, **35**, 187–96.

Lambshead, P.J.D. (2004) Marine nematode biodiversity. In: ZX Chen, SY Chen, DW Dickson eds. *Nematology: Advances and Perspectives Vol. 1: Nematode Morphology, Physiology and Ecology,* pp. 436–67. CABI Publishing, London.

Lambshead, P.J.D., Brown, C.J., Ferrero, T.J. *et al.* (2002) Latitudinal diversity patterns of deep-sea marine nematodes and organic fluxes: a test from the central equatorial Pacific. *Mar Ecol Progr Ser*, **236**, 129–35.

Levin, L. (2003) Oxygen minimum zone benthos: adaptation and community response to hypoxia. *Oceanogr Mar Biol Annu Rev*, **41**, 1–45.

Levin, L.A., Etter, R.J., Rex, M.A. *et al.* (2001) Environmental influences on regional deep sea species diversity. *Ann Rev Ecol Syst*, **32**, 51–93.

Levin, L.A., Gage, J.D., Martin, C. *et al.* (2000) Macrobenthic community structure within and beneath the oxygen minimum zone, NW Arabian Sea. *Deep-Sea Res II*, **47**, 189–226.

Lohrer, A.M., Thrush, S.F., Gibbs, M.M. (2004) Bioturbators enhance ecosystem function through complex biogeochemical interactions. *Nature*, **431**, 1092–5.

Loreau, M. (2008) Biodiversity and ecosystem functioning: the mystery of the deep sea. *Curr Biol*, **18**, 126–8.

Loreau, M., Naeem, S., Inchausti, P. *et al.* (2001) Biodiversity and ecosystem functioning: current knowledge and future challenges. *Science*, **294**, 804–8.

Martinez, P. and Schminke, H.K. (2005) DIVA-1 expedition to the deep sea of the Angola Basin in 2000 and DIVA-1 workshop 2003. *Org Divers Evol*, **5**, 1–2.

McIntyre, A. (ed). *Life in the World's Oceans: Diversity, Distribution and Abundance.* Wiley-Blackwell, Oxford.

Meysman, J.R.F., Middelburg, J.J., Heip, C.H.R. (2006) Bioturbation: a fresh look at Darwin's last idea. *Trends Ecol Evol*, **21**, 688–95.

Micheli, F., Benedetti-Cecchi, L., Gambaccini, S. *et al.* (2005) Cascading human impacts, marine protected areas, and the structure of Mediterranean reef assemblages. *Ecol Monogr*, **75**, 81–102.

Naeem, S., Thompson, L.J., Lawler, S.P. *et al.* (1994) Declining biodiversity can alter the performance of ecosystems. *Nature*, **368**, 734–6.

O'Connor, N.E. and Crowe, T.P. (2005) Biodiversity loss and ecosystem functioning: Distinguishing between number and identity of species. *Ecology*, **86**, 1783–96.

Petchey, O.L. and Gaston, K.J. (2006) Functional diversity: back to basics and looking forward. *Ecol Lett*, **9**, 741–58.

Pike, J., Bernhard, J.M., Moreton, S.G. *et al.* (2001) Microbioirrigation of marine sediments in dysoxic environments: implications for early sediment fabric formation and diagenetic processes. *Geology*, **29**, 923–6.

Pusceddu, A., Dell'Anno, A., Fabiano, M. *et al.* (2004) Quantity and biochemical composition of organic matter in marine sediments. *Biologia Marina Mediterranea*, **11**, 39–53.

Sardà, F., Calafat, A., Flexas, M.M. *et al.* (2004) An introduction to Mediterranean deep-sea biology. *Sci Mar*, **68**, 7–38.

Sih, A., Englund G., Wooster, D. (1998) Emergent impacts of multiple predators on prey. *Trends Ecol Evol*, **13**, 350–5.

Smith, C.R. and Demoupolos, A.W.J. (2003) Ecology of the Pacific Ocean floor. In PA Tyler ed. *Ecosystems of the World*, pp. 179–218. Elsevier, Amsterdam.

Snelgrove, P.V.R. (1999) Getting to the bottom of marine biodiversity: sedimentary habitats. *BioScience*, **49**, 129–38.

Snelgrove, P.V.R. and Smith, C.R. (2003) A riot of species in an environmental calm: the paradox of the species-rich deep-sea floor. *Oceanogr Mar Biol Annu Rev*, **40**, 311–42.

Stachowicz, J.J. (2001) Mutualism, facilitation, and the structure of ecological communities. *Bioscience*, **51**, 235–46.

Stuart, C.T., Rex, M.A., Etter, R.J. (2003) Large-scale spatial and temporal patterns of deep-sea benthic species diversity. In: *Ecosystems of the World, Volume 28: Ecosystems of the Deep Oceans*, Tyler, P.A. (ed), Elsevier, Amsterdam, 295–311.

Tilman, D., Reich, P.B., Knops, J. *et al.* (2001) Diversity and productivity in a long-term grassland experiment. *Science*, **294**, 843–5.

Tilman, D., Reich, P.B., Knops, J.M.H. (2006) Biodiversity and ecosystem stability in a decade-long grassland experiment. *Nature*, **441**, 629–32.

Tyler, P.A. (2003) The deep Atlantic Ocean. In: *Ecosystems of the World, Volume 28: Ecosystems of the Deep Oceans.* Elsevier, Amsterdam.

Weaver, P.E., Boetius, A., Danovaro, R. *et al.* (2009) The future of integrated deep-sea research in Europe: the Hermione project. *Oceanography* **22**(1), 170–83.

Wei, C-L., *et al.* (2010) Global Patterns and Predictions of Seafloor Biomass Using Random Forests. *PLoS One* 5 (12) e15323 http://dx.plos.org/10.1371/journal.pone.0015323.

Wieser, W. (1953) Die Beziehung zwischen Mundhöhlengestalt, Ernährungsweise und Vorkommen bei freilebenden marinen Nematoden. *Arkiv Zool*, 2–4, 439–84.

Witte, U., Wenzhofer, F., Sommer, S. *et al.* (2003) *In-situ* experimental evidence of the fate of a phytodetritus pulse at the abyssal sea floor. *Nature*, **424**, 763–5.

Worm, B., Barbier E.B., Beaumont, N. *et al.* (2006) Impacts of biodiversity loss on ocean ecosystem services. *Science*, **314**, 787–90.

Yool, A., Martin, A.P., Fernández, C *et al.,* (2007) The significance of nitrification for oceanic new production. *Nature*, **447**, 999–1002.

CHAPTER 10

Incorporating extinction risk and realistic biodiversity futures: implementation of trait-based extinction scenarios

Martin Solan, Finlay Scott, Nicholas K. Dulvy, Jasmin A. Godbold, and Ruth Parker

10.1 Introduction

Rates of species extinction have intensified over the last century (Barnosky *et al.* 2011), largely as a product of human activity (Sala *et al.* 2000; Halpern *et al.* 2008), climate forcing (Thomas *et al.* 2004), and their interactions (Brook *et al.* 2008), and it is presently anticipated that extinction rates will continue to increase well into the next century (Pereira *et al.* 2010). Concerns over the potentially important consequences that this may have for the stability and functioning of ecosystems has provided impetus for research, and an extensive body of literature now exists that explicitly focuses on the ecological and environmental consequences of biodiversity loss (reviewed in Balvanera *et al.* 2006; Cardinale *et al.* 2006; Schmid *et al.* 2009; Solan *et al.* 2009). Collectively, this knowledge base provides unambiguous evidence that, irrespective of the system and the cause of expiration, declining species richness tends to disrupt processes that maintain ecosystem integrity. This occurs because, as species are removed, their absence—or reduction in abundance and/or biomass—directly causes: (i) a reorganization of sampling/selection effects (Aarssen 1997; Loreau and Hector 2001); (ii) a reduction in the level of interspecific resource partitioning; and, (iii) a declining probability that non-additive interspecific interactions—facilitative or inhibitive——will take place. In addition, a range of functional effects may

occur in response to the indirect effects of species loss on specific surviving species through various mechanisms, including release from competition (Godbold *et al.* 2009) or predation (Burkepile and Hay 2007), breakdown of mutualistic associations (Harrison 2000; Hughes *et al.* 2009) or the cascading effects of co-extinction (Moir *et al.* 2009, Fowler 2010). Whilst it is likely that a subset of both direct and indirect mechanisms will simultaneously operate in natural systems subject to extinction forcing, their relative role in altering ecosystem properties is only just emerging; random assembly experiments where researchers have manipulated species diversity have, more often than not, predominantly attributed the functional consequences of species loss to a combination of the loss of the single most productive species and/or a reduction in species complementarity (Cardinale *et al.* 2006).

Accepting the mechanistic processes that may underpin the ecological consequences of biodiversity loss, a general difficulty of small-scale experiments is that it is not tractable to work with the full complement of species that represent a natural system, or to explore fully the range of likely—or anticipated——alternative extinction scenarios that such assemblages may encounter (Naeem 2008). A number of notable contributions have attempted to address these logistic difficulties by using information from naturally assembled communities to refine species selection and limit the number of

Marine Biodiversity and Ecosystem Functioning. First Edition. Edited by Martin Solan, Rebecca J. Aspden, and David M. Paterson.
© Oxford University Press 2012. Published 2012 by Oxford University Press.

experimental assemblage combinations (Zavaleta and Hulvey 2004; Srinivasan *et al.* 2007; Bracken *et al.* 2008; de Visser *et al.* 2011), but the majority of experiments examining how altered levels of biodiversity affect the functioning of ecosystems have incorporated communities that do not exist in natural systems, and they have generally assumed that species loss is random (Naeem 2008). These simplifying assumptions have received substantial criticism—summarized in Naeem *et al.* 2002, Solan *et al.* 2009—because, far from being a random process, extinction of species tends to follow an ordered sequence that reflects the relative distribution of extinction risk amongst species within an assemblage. Hence the ecological significance of extinction will be dependent on the sequential order of species loss and whether the extinction risk of each species correlates with the life-history traits that are important in mediating ecosystem functioning (Solan *et al.* 2004). If they do, on average the ecological outcome is likely to be exacerbated relative to that observed following a random extinction where extinction takes place irrespective of the distribution of species traits (Gross and Cardinale 2005). Hence there is a compelling need to determine the species and traits—effect and response traits (Hooper *et al.* 2002)—that are important for specific ecosystem functions, and assess the fate of species endangered by multiple drivers of change across a range of systems.

Identifying the threat status of species is difficult, partly because the necessary information—geographic distribution, temporal changes in population structure, changes to ecological function—is difficult to obtain or not available, but also because there is a lack of generally accepted criteria for identifying when species become impoverished or extinct (Nicholson *et al.* 2009). The International Union for Conservation of Nature (IUCN) Red List, a global index of the state of degeneration of biodiversity, is based on a set of transparent and quantitative criteria—largely based on geographical distribution and population status (IUCN 2001, Mace *et al.* 2008)—that assess the threat status of a species—i.e. least concern, near threatened, vulnerable, critically endangered, or endangered. These categorizations are useful in other contexts, but are

of limited value when parameterizing models with sequential species losses as the categorizations are very broad in coverage. Furthermore, there are difficulties in determining whether species are genuinely at risk at local scales (Dulvy *et al.* 2003). There are a variety of alternative ways, however, to estimate the relative extinction vulnerability of a species including, for example, the use of local knowledge and the grey literature (Castellanos-Galindo *et al.* 2011), the examination of the distribution of species along gradients of disturbance (Pearson and Rosenberg; 1978, Vitousek *et al.* 1994; Isbell *et al.* 2008; Hall-Spencer *et al.* 2008), and the collation of information on the physiological tolerance (Labrune *et al.* 2006; Srinivasan *et al.* 2007), or behavioural sensitivity (Liow *et al.* 2009) of species to specific agents of perturbation. In addition, a suite of metrics and protocols (reviewed in Schläpfer *et al.* 2005) are available that have been developed to determine objectively the conservation priority status of various groups of species, most notably species with high conservation status (fish, Dulvy *et al.* 2004; mangroves, Polidoro *et al.* 2010; seagrass, Short *et al.* 2011), but also for less charismatic but functionally important groups (benthic invertebrates, Freeman *et al.* 2010). These range from simple metrics based on specific traits, such as body size (Olden *et al.* 2007), to more sophisticated multivariate frameworks that encompass multiple traits (Bremner *et al.* 2003) and species–environment interactions (Branco *et al.* 2008; Graham *et al.* 2011). The resulting ranked inventories have the advantage that they provide continuous data on the relative vulnerability of species, allowing subtle differences in risk to be accounted for, and the arbitrary division of species into specific risk groupings of equal ranking to be avoided. Notwithstanding the need to test rigorously the sensitivity and appropriateness of such metrics, it is clear that most of these methodologies are readily transferrable to most marine species and habitats, thereby offering a means to generate credible sequences of extinction risk for specific populations and context.

For a predictive understanding of the ecological consequences of extinction, inventories of species vulnerabilities to extinction must be coupled with information on the magnitude, functional role, and extent to which species perform unique contributions—i.e. the degree of trait overlap within an assemblage (Naeem *et al.* 2002; Micheli and Halpern 2005; Zhang and Zhang 2007. Fortunately, there is a long tradition in ecology of documenting the functional effects of species (additive) and assemblages (non-additive) on a variety of ecosystem properties in the presence and absence of a wide range of abiotic stressors. From this repository of information, we know that the importance of individual species will be context-dependent, and even subtle changes in species behaviour in response to a range of abiotic (e.g. temperature, Beveridge *et al.* 2010; food availability, Nogaro *et al.* 2008; elevated CO_2, Bulling *et al.* 2010; habitat properties, Larsen *et al.* 2005, Godbold *et al.* 2011) and biotic (presence of predator, Maire *et al.* 2010) drivers of change can have dramatic effects on ecosystem properties. These changes in community dynamics and species interactions can frequently alter, or even reverse, species vulnerabilities to environmental forcing (Griffen and Drake 2008; Ball *et al.* 2008), and may be a product of multiple interacting components of an ecosystem (Edgar *et al.* 2010). Despite such complexities, however, species within an assemblage can be readily grouped into cohorts that share common traits based on the way they modify a given ecosystem process (Hooper *et al.* 2002), making it feasible to assemble inventories of species that are likely to respond to particular perturbations in a similar manner. Indeed, many readily discernable species traits, such as body size and trophic position, are known to form intimate linkages with many ecosystem properties (Bremner *et al.* 2006; Fisher *et al.* 2010; Laughlin 2011), such that *a priori* predictions can be made that the preferential loss or reduction of these traits within an assemblage may have disproportionate effects on ecosystem functioning (Hillebrand and Matthiessen 2009). For species where there are information gaps or incomplete knowledge about species traits, iterative hierarchal schemes for assigning functional group classifications can be adopted (see Figure 17.2 in Hooper *et al.* 2002; e.g. benthic macrofauna, Swift *et al.* 1993; Solan *et al.* 2004; fish, McIntyre *et al.* 2007) that allow functional groups to be assigned based on secondary information—i.e. possession of shared or similar traits and/or expert opinion.

Such techniques allow information on the role of individual species to be extended to entire assemblages and, when coupled with large datasets on extinction vulnerability, provide sufficient information to examine alternative ecologically realistic extinction scenarios (Coreau *et al.* 2009).

One way in which this can be achieved is through the use of trait-based extinction scenarios, where the sequence of species extinction can be ordered in accordance with specific extinction mechanisms, and the functional consequences of each scenario can be estimated. The consequences of possible biodiversity-environment futures for ecosystem functioning have been explored in this way across a range of freshwater, terrestrial, and marine habitats, and for a variety of ecosystem functions, including productivity (Smith and Knapp 2003; Schläpfer *et al.* 2005), resistance to invasive species (Zavaleta and Hulvey 2004), nutrient cycling (Solan *et al.* 2004, McIntyre *et al.* 2007), carbon storage (Bunker *et al.* 2005), and decomposition (Ball *et al.* 2008). The flexibility and adaptability of this approach is yet to be fully exploited, but marine ecologists are particularly well placed to take advantage of the technique as the data required to generate probabilistic numerical simulations is routinely collected as a matter of course. In this chapter, we provide the relevant model code for a range of extinction scenarios and demonstrate how this modelling framework can be applied at local and regional scales. Our objective is to encourage widespread application of modelled predictions and foster dialogue on how such methodologies can be expanded and improved.

10.2 How to implement non-random extinction scenarios

Generation of probabilistic numerical simulations that test how species loss may affect ecosystem functioning can be achieved using the R statistical and programming environment available from <http://cran.r-project.org/>. This process requires information on relevant species traits (abundance, A_i, and biomass, B_i), an indication of each species contribution to ecosystem functioning, and reference values pertaining to extinction risk for any extinction drivers of interest (Table 10.1). Following the study of

Solan et al. (2004) for marine benthic invertebrates, the latter can be achieved by characterizing the per capita contributions of each species using an appropriate trait-based index e.g. bioturbation potential, BP_i, that differentially scores each species with respect to their contribution to the relevant ecosystem process (in this case, bioturbation). These scores can then be collated at the population ($BP_p = A_i \times BP_i$) and community level ($BP_c = \sum BP_p$) and related to measured values of ecosystem functioning e.g. mixed depth of sediments, MD; chlorophyll a, Chl a; or organic carbon, Org-C. This relationship can then be used to predict the level of ecosystem functioning (i.e. MD, Chl a or Org-C) for novel communities following a scenario-based manipulation of the community i.e. recalculated levels of BP_c. Scenarios of extinction may be random = 1/n, i.e. extinction risk is a reciprocal function of the number of species in the assemblage and, therefore, is not linked to species traits, as in Table 10.1, or will match likely drivers of change, including extinctions ordered by body size (B_i), rarity (A_i), trophic position, or particular pressures, such as an anthropogenic activity or an element of environmental change. Extinction probabilities for the latter can be estimated, for example, by quantifying the proportional change in species abundance along a gradient of increasing pressure. Once the scenario is assembled, additional biotic conditions can be added. For example, as the functional consequences of extinction are known to depend on the response of surviving species, the simulation can include compensatory responses expressed as changes in population size in the surviving population following an extinction event. This can represent a range of assumed community responses, from no compensation i.e. loss of a species does not affect the population size of any surviving species following, for example, predation or competitive release to full numerical, or biomass, compensation i.e. the best-case outcome that assumes compensation is additive and substitutions of abundance, or biomass, maintain total community density. The following shows how to implement these basic functions in R; for convenience, we supply the implementation code as a Tinn-R file (*extinction_code.r*, available at http://ukcatalogue.oup.com/product/9780199642267.do).

To call your data (tab delimited text file required), copy and paste the following code into R:

```
rm(list = ls())
alldata ← read.table(file=file.
    choose(), header = T)
```

You can check that your data has imported correctly by entering:

Table 10.1 Example dataset for the generation of extinction scenarios detailing species identity (here, benthic invertebrates), species traits (abundance, A_i, and biomass, B_i), an indication of each species contribution to ecosystem functioning (bioturbation potential of the individual, BP_i, and at population level, BP_p), and reference values pertaining to extinction risk (*Extinction Probability*, here assumed to be random, 1/n).

Species or taxon	A_i	B_i	BP_i	BP_p	Extinction Probability
Capitellidae	8.5	0.00017	0.00044	0.00374	0.076923
Glycera lapidum	25.7	0.046681	0.23777	6.110689	0.076923
Gastrosaccus spinifer	2.9	0.117213	0.09627	0.279183	0.076923
Bathyporeia elegans	2.9	0.117213	0.57763	1.675127	0.076923
Atylus falcatus	8.5	0.117213	0.28882	2.45497	0.076923
Echinocyamus pusillus	2.9	0.311372	1.41271	4.096859	0.076923
Nemertea	2.9	0.333333	1.49926	4.347854	0.076923
Lanice conchilega	8.5	0.342163	0.25561	2.172685	0.076923
Eteone foliosa	5.8	0.342163	1.53366	8.895228	0.076923
Ophelia borealis	2.9	0.416667	1.21014	3.509406	0.076923
Harmothoe lunulata	2.9	0.5	2.1131	6.12799	0.076923
Abra alba	2.9	0.754535	0.97665	2.832285	0.076923
Nephtys cirrosa	25.7	2.5	6.52882	167.7907	0.076923

```
dim(alldata)
names(alldata)
```

A summary of the structure of the data confirming that you have the correct number of observations (x rows = x species) with their relevant traits in columns (y columns = y traits plus an additional column providing any extinction scenario information), followed by a list of your column names will be returned.

The sample dataset we have provided to accompany this chapter (*sampledata.csv*, available at http://ukcatalogue.oup.com/product/9780199642267.do) is for a subset of benthic invertebrates and has six columns: *Species*—the name of the species or taxon; A_i—the mean abundance of individuals within each species population; B_i—the mean biomass or other appropriate size metric of individuals within each species population; BP_i—the bioturbation potential of an individual; BPp—the bioturbation potential of the population; and *ExtinctProb*—the probability of extinction. In the example dataset, the extinction probabilities are equal indicating that in this scenario the extinctions are random. Alternatively, extinction probabilities in this column could be set to match specific traits, such as A_i, B_i or other information—e.g. vulnerability to a specific extinction driver—such that the most numerous, largest or most susceptible species are more likely to become extinct. To change the risk of extinction to relate to B_i, for example, use the following code:

```
alldata$ExtinctProb ← alldata$Bi
  / sum(alldata$Bi)
```

Before the simulation can be implemented, it is necessary to generate two extra columns in the data set to record the abundances and extinction probabilities during each simulation. Their values are set to *NA* to show that no data has yet been generated:

```
alldata$AiSim ← NA
alldata$EPSim ← NA
```

For convenience, a couple of variables are set that store the number of species in the community and the original total biomass of the community—this is needed in case biomass compensation is used:

```
nsp ←nrow(alldata)
OrigTotalBiomass ← sum(alldata$Bi
  * alldata$Ai)
```

Two parameters then need to be set to determine how many simulations will be run (for brevity, we use 100 here, but ≥ 1000 simulations is more appropriate for full datasets) and whether or not biomass compensation should be included (*TRUE* or *FALSE*):

```
nsims ← 100
BioCompFlag ← TRUE
```

Finally, an output data frame is set up to record the results of the simulation. This has the following columns to record the respective information:

Simulation—the simulation number,
Nsp—the number of species remaining in the community,
ExtinctSpecies—the species that will go extinct at the next iteration,
Measure—the community function that is being calculated; e.g. Org-C or Chl *a*,
Value—the value of that community function,

Note that the *ExtinctSpecies* column contains the species that will go extinct at the next iteration, i.e. the first row records the function of the full community and describes which species will be the next to go extinct. This is to save staggering the *ExtinctSpecies* column when it comes to the final species going extinct.

The dataset is now established to run the simulations. The simulations are run inside a large *for* loop which counts down the number of simulations:

```
for (sim_count in 1: nsims){
```

At the start of each simulation the abundances and extinction probabilities need to be reset—otherwise they will hold the results left over from any previous simulations:

```
alldata$AiSim ← alldata$Ai
alldata$EPSim ←
    alldata$ExtinctProb
```

We now set up another *for* loop—still inside the main simulation *for* loop—which counts down the number of species in the community:

```
for (sp_count in nsp:1){
```

Each time around the species loop, we record the value of the community functions, select a species for extinction, apply biomass compensation, if chosen, and then remove the species that has become extinct from the community.

Following Solan *et al.* (2004), we first calculate *BPc* (see section 10.3.2 below) using the current abundances and calculated bioturbation potentials of each species:

```
BPc ←sum(alldata$AiSim *
    alldata$BPi)
```

We then relate measured levels of ecosystem functioning to the value of *BPc* using an appropriate relationship, such as a linear regression. Here, we use calculated values for *Org-C* and *Chl a*—an equivalent equation will relate to your own data:

```
Chla ← 10^(-1.14372 +
    0.50808*log10(BPc))
Org C ← sin(0.108291
    - 0.017583*log10(BPc))*100
```

We then store these values in the output data frame:

```
output[output$Simulation ==
    sim_count and output$Nsp==sp_
    count and
    output$Measure=="Chla",
    "Value"] ← Chla
output[output$Simulation == sim_
    count and output$Nsp==sp_count
    & output$Measure=="Org C",
    "Value"] ← Org C
```

Now we select the species to remove based on the extinction probabilities. This is achieved by picking a random number from a uniform distribution, with a minimum of 0 and maximum of 1, and comparing it to the current extinction probabilities. The value of *Extinct* is the row number of the species that will go extinct:

```
Extinct ← which(cumsum(alldata$EP
    Sim)>=runif(1))[1]
```

If the extinction scenario included biomass compensation, then the biomass that will be lost during each species extinction is calculated. To implement biomass compensation, the abundances of the surviving species are increased by the same proportion as the lost biomass to the total biomass. This ensures that the total biomass is kept constant:

```
if(BioCompFlag==TRUE) {
BiomassLost
    ← alldata[Extinct,"AiSim"] *
    alldata[Extinct,"Bi"]
alldata$AiSim ← alldata$AiSim *
    (OrigTotalBiomass /
    (OrigTotalBiomass
    - BiomassLost))
}
```

The abundance and the extinction probability of the expired species is set to 0, thereby removing it from the community:

```
alldata[Extinct,c("EPSim",
    "AiSim")] ← 0
```

The remaining extinction probabilities need to be normalized so that they sum to 1:

```
alldata$EPSim ← alldata$EPSim /
    sum(alldata$EPSim)
```

Finally, we record the identification number—the row number in the original data—of the expired species in the output data frame:

```
output[output$Simulation ==
    sim_count & output$Nsp==sp_
```

```
count,"ExtinctSpecies"]
<- alldata
[Extinct,"Species"]
```

We are now at the end of the species loop so it returns to the beginning, removes another species, records the corresponding community function values, and so on, until all species have become extinct. When the final species is reached, the simulation ends and we return to the start of the simulation loop, ready for another simulation with a full community. After all the simulations have been run, the results in the *output* data frame can be explored.

10.3 Case study: implications of regional biodiversity loss on carbon cycling in the shelf sea sediments of the North Sea

Carbon components in shelf sediments are controlled by supply from the water column, incorporation into the sediments, and degradation or uptake processes which are either microbially driven (redox transformations) or governed by meio- and macrofaunal consumption. In the North Sea, the supply of carbon fractions to the sediment is driven by physicochemical processes which control phytoplankton growth, zooplankton grazing, and sinking or mixing of detritus into the seabed. These can vary across large spatial scales, or ecohydrodynamic regions, and are associated with distinct benthic communities. In turn, these communities can affect the rate of organic matter decomposition by controlling sediment redox via bioturbation processes (Ziebis *et al.* 1996; Arzayus and Canuel 2004; Michaud *et al.* 2006) or through the active uptake and incorporation of carbon fractions across the sediment–water interface (Lohrer *et al.* 2004; Boon and Duineveld 1998; Boon *et al.* 1999; Witbaard *et al.* 2000). Here, we explore how various scenarios of extinction—random, body size, and rarity at local and regional scales—for marine benthic invertebrates are likely to influence the overall community potential for sediment carbon coupling via bioturbation mechanisms and, hence, the sediment levels of comparatively labile (Chlorophyll a, Chl a) and refractory (total organic carbon, TOC) carbon components.

Chl *a* is a labile carbon fraction which is preferentially utilized by benthic communities—macro-, meio-, and micro-benthos—and is, therefore, comparatively short-lived and quickly turned over in sediments, providing a proxy for fast coupling and carbon incorporation processes, including bioturbation. TOC is an integrated measure of all carbon in the sediment, derived from any detrital source (e.g. plankton, pigments, fish faeces) and describes a more refractory (slower) and residual fraction that reflects longer-term turnover of sediment carbon.

10.3.1   Study sites and data collection

Data was obtained from the North Sea Benthos Survey for the southern and central North Sea conducted by the ICES Benthos Ecology Working Group (1986). This dataset includes 109 stations that contain information on benthic macrofaunal species abundance and biomass, as well as measures of benthic ecosystem processes: Chl *a*—extraction and fluorimetry, chloroplast pigment equivalents (CPE), $\mu g\ 5\ cm^{-3}$ sediment—and organic carbon—Org-C, wet oxidation of upper 5 cm sediment, %. These stations are located in an area of approximately 350 000 km^2, extending from 51.5°N to 58.0°N, and from 2.5°W to 8.0°E (Figure 10.1).

The benthic macrofaunal assemblages of the North Sea originate from two distinct regional species pools, which correspond to contrasting hydrogeographical characteristics of the southern and northern regions (Künitzer *et al.* 1992; Basford *et al.* 1993; Callaway *et al.* 2002). At each station, the macrofauna were sampled, either by box sampler (3 replicates) and van Veen grab (1 replicate) or, during inclement weather, only by van Veen grab (3 replicates) (ICES, 1986). Species were identified to the lowest possible taxon (87% to species, 8% to genus, 3% to family, 2% higher miscellaneous) from a total macrofaunal species pool of 557 taxa for the North Sea. For each station, the benthic community was characterized in terms of species composition (station species richness ranged from 13 to 106 species), abundance, biomass, and functional types. Mean weight was calculated by multiplying numerical density with average weight. The average weight for each species was derived from the best

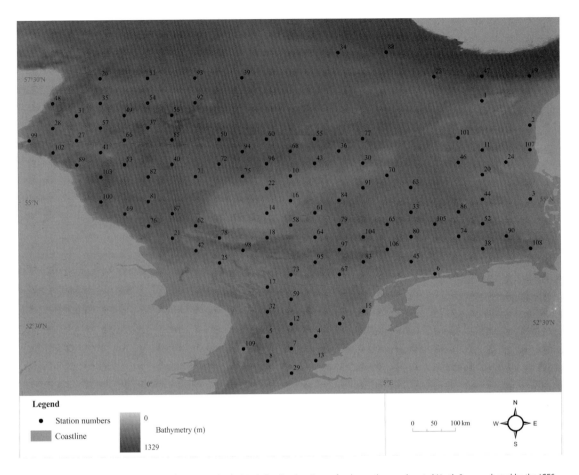

Figure 10.1 The location of the 109 sampling stations in the North Sea Benthos Survey for the southern and central North Sea, conducted by the ICES Benthos Ecology Working Group in 1986.

species-specific estimates obtained from multiple sources (Huys *et al.* 1992; Callaway *et al.* 2002; Mackinson and Daskalov 2007) including North Sea infauna and epifauna surveys.

10.3.2 Benthic bioturbation characterization

We used an index of benthic bioturbation to characterize the per capita effect of each macrofaunal species on sediment mixing (Solan *et al.* 2004). To ensure that species impacts conformed to a linear scale, our data required a modification of Equation 1 in Solan *et al.* (*loc. cit.*) to incorporate a $\log_{10}$ transformation:

$$BP_i = \log_{10}(B_i) \times M_i \times R_i \qquad \text{(Equation 10.1)}$$

This index accounts for three biological traits known to influence sediment bioturbation: (i) body size (B_i, mean wet biomass in grams), (ii) propensity to move through the sedimentary matrix (M_i, mobility), and (iii) the way in which species influence sediment particle exchange (R_i, reworking mode). M_i was scored on a categorical scale that reflects increasing activity of the species (1 = in a fixed tube, 2 = limited movement, sessile, but not in tube, 3 = slow movement through sediment, 4 = free movement via burrow system; Swift, 1993). R_i was also scored on a categorical scale to reflect increasing impacts on the sediment turnover (1 = epifauna that bioturbate at the sediment-water interface; 2 = surficial modifiers, whose activities are restricted to < 1–2 cm of the sediment profile; 3 = head-down/

head-up feeders that actively transport sediment to/from the sediment surface; 4 = biodiffusers whose activities result in a constant and random diffusive transport of particles over short distances; and 5 = regenerators that excavate holes, transferring sediment at depth to the surface). These functional associations are well-established classifications in the marine literature and explicitly recognize that individual species have differing impacts on ecosystem process (Meysman *et al.* 2003). To determine the population-level bioturbation potential, BP_p, for a given species, per capita effects (BP_i) were multiplied by mean species abundances (A_i, m^{-2}) at each station, and BP_p was summed across species in a station to estimate the community-level bioturbation potential, BP_c.

We used hierarchical clustering on sediment grain size, bed stress, bottom temperature, and depth to identify contrasting hydrogeographical regions within the North Sea. This identified a northern and southern sector and, for each region, the relationship between BP_c and ecosystem functioning (Chl *a* and Org-C) was fitted empirically using standard linear regression and validation procedures:

For Chl a (northern):
$$Log(Chla) = -1.14372 \\ + (0.50808 \times (log(BP_c))) \quad \text{(Equation 10.2)}$$

For Chl a (southern):
$$Log(Chla) = -1.1282 \\ + (0.5046 \times (log(BP_c))) \quad \text{(Equation 10.3)}$$

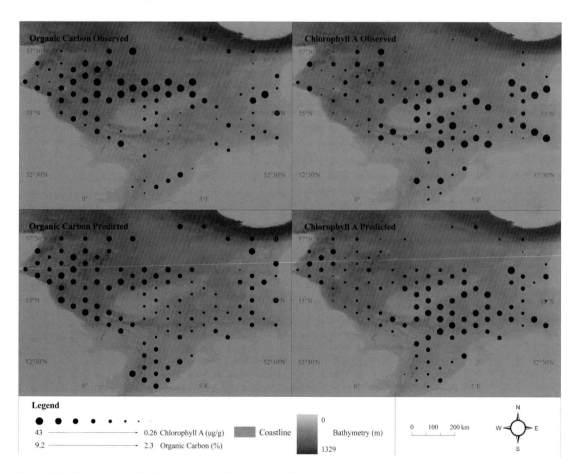

Figure 10.2   Observed and predicted levels of ecosystem functioning for sediment Org-C (left-hand panels) and Chl *a* (right-hand panels) for each station in the North Sea. Predictions were fitted empirically using standard linear regression (Equations 10.2–10.5). See also Plate 2.

For Org-C (northern):

$$Arc \sin (Org\text{-}C) = 0.121421$$
$$- (0.019874 \times (log(BP_c)))$$

(Equation 10.4)

For Org-C (southern):

$$Arc \sin (Org\text{-}C) = 0.108291$$
$$- (0.017583 \times (log(BP_c)))$$

(Equation 10.5)

The predictive capacity of these models was satisfactory (Figure 10.2). We compared the predictions of Equations 10.2–10.5 to alternative GLM and GAM models, but found that they provided no improvement.

10.3.3 Modelling

We modelled scenarios of extinction that match likely drivers of change in marine shelf systems (see below). For each scenario, we ran 2000 simulations each representing one trajectory of species loss from the full complement of species to 1 species using equations 2–5 to predict ecosystem functioning (Chl a or Org-C) of each community at each station (n = 109). In each case, the starting community consisted of the full complement of species that occurred at each station during the 1986 survey. Subsets of each station assemblage were then selected to represent communities of reduced richness with the probability of extinction being either random (1/n), or assumed to be proportional to specific traits: mean abundance within a station (rare species first, lowest to highest A_i); mean abundance across the regional species pool (rare species first, lowest to highest A_i); or body size (largest species first, largest to smallest B_i).

10.3.4 Estimating non-linear changes in ecosystem functioning

In order to generate maps of biodiversity futures based on the level of species richness at which the effect on ecosystem function changes abruptly, we used segmented regression (Muggeo 2003). This technique defines the point of change (= breakpoint) using the slope parameters and the break-points

where the linear relation changes. From these estimates we derived the level of ecosystem process and the percentage species loss at the break-point for each station.

10.4 Results and discussion

Overall, our models predict a loss of functioning (= reduction in Chl a and increased Org-C, Figures 10.3 and 10.4) with increasing levels of species loss, irrespective of extinction scenario and location. In each case, however, the rate of change, level of species richness at which the decline in functioning rapidly begins to accelerate, and the variation in the magnitude of ecosystem functioning all vary as a function of extinction scenario. This is because the traits that most influence bioturbation covary with the vulnerability of species to particular drivers of extinction (Figure 10.5), consistent with previous findings based on single locations elsewhere (Solan et al. 2004).

Comparisons of our model predictions across all 109 stations of the North Sea, however, reveal that these general patterns and relationships are underpinned by complex species–environment interactions (Godbold and Solan 2009). Indeed, the details of how macrofaunal extinctions affect carbon cycling in shelf sediments vary as a function of subtle inter-station differences in species composition (species richness, abundance, biomass, evenness) that relate to large-scale hydrodynamic phenomena, in particular the separation of mixed and stratified water masses (Callaway et al. 2002). This suggests, at least for the drivers of change examined here, that the loss of entire functional groups from regional seascapes is unlikely because hydrogeographical areas could form vital refugia for functionally important species. If true, these broader scale areas could form important foci for the restoration of degraded benthic environments, management of fisheries, and designation of marine protected areas (Noss 2010).

One of the most striking outcomes of our analyses is that several species, rather than one functionally dominant species, exert a strong influence on ecosystem functioning, providing direct support for

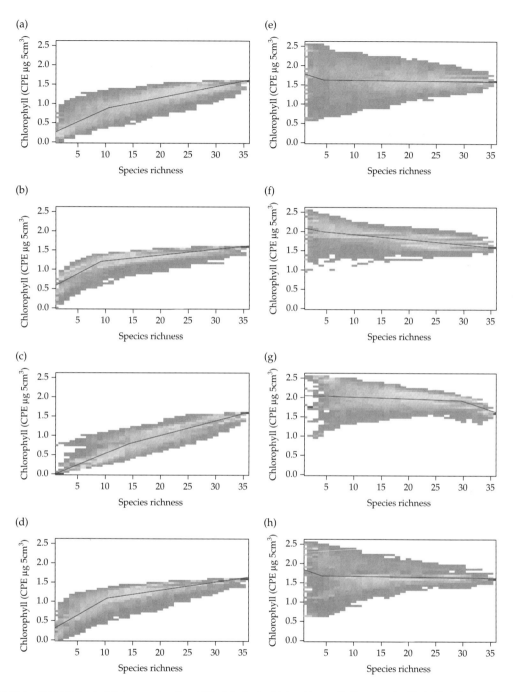

Figure 10.3 Representative predicted changes in sediment Chl *a* following benthic invertebrate extinctions at station 8 in the North Sea (predictions for the other 108 stations are available from the authors). Each panel within a station shows the results of 2000 simulations, each representing one trajectory of species loss from the full complement of species to 1 species. The probabilistic order of species extinction was either random (panels a and e), or assumed to be proportional to specific traits: mean abundance within a station (rare species first, lowest to highest, panels b and f); mean abundance across the regional species pool (rare species first, lowest to highest, panels c and g); or related to body size (largest to smallest, panels d and h). Simulations (a), (b), (c), and (d) are for a non-interactive model of community assembly assuming no compensation by surviving species. Simulations (e), (f), (g), and (h) are for an interactive model that assumes full compensation by the surviving community following extinction. Lines represent segmented linear regression with the break-point indicating the level of species richness at which the effect on ecosystem process changes abruptly. See also Plate 3.

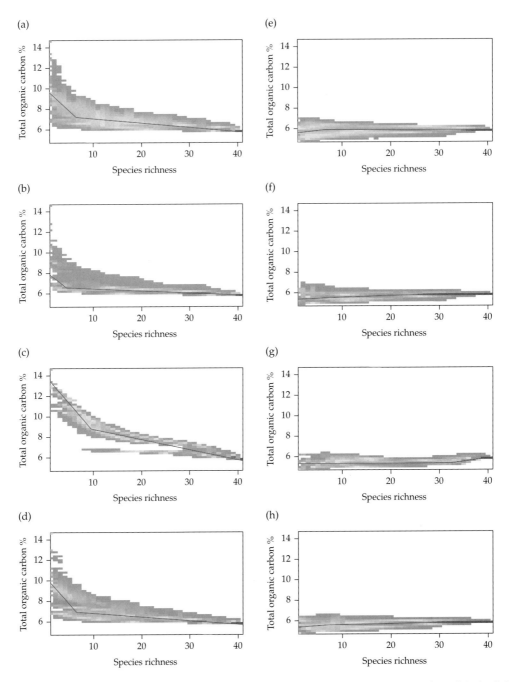

Figure 10.4 Representative predicted changes in sediment Org-C following benthic invertebrate extinctions at station 54 in the North Sea (predictions for the other 108 stations are available from the authors). Each panel follows the probabilistic order of species extinctions as listed in Figure 10.3. Lines represent segmented linear regression with the break-point indicating the level of species richness at which the effect on ecosystem process changes abruptly. See also Plate 4.

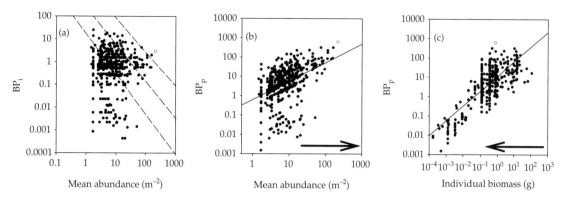

Figure 10.5   The relation between per capita bioturbation, BP_i, and mean species abundance (a) reveals that at the population level (diagonal dashed lines, each an order of magnitude difference in bioturbation), a substantial number of species contribute little to bioturbation (left of short-dashed line). Population level bioturbation, BP_p, is proportional to species abundance (b) and body size (c). Arrows indicate order of extinctions in the simulations.

the view that ecosystem functioning at habitat scales is supported by multiple species (Hector and Bagchi 2007). Certainly, in the North Sea, there are at least six species that are particularly functionally important—*Amphiura filiformis, Chamelea gallina, Echinocardium cordatum, Mysella bidentata, Nephtys cirrosa, Ophelia borealis* (Figure 10.6a)—and, in all cases, our simulations indicate that their removal leads to dramatic step-changes in functional loss. These species frequently co-occur, but the representation of each species at specific locations can radically differ (1–6 spp., Figure 10.6b), minimizing the chance of cumulative impacts. Perhaps more important, however, is the observation that the functional role of each of these species is not consistent in terms of either their absolute contribution to functioning or with respect to their functional ranking (Figure 10.6a). Whilst some of this variation undoubtedly relates to differences in population size and hydrogeographical setting (for review, see Teal *et al.* 2008), it is not clear why the most productive species may differ between assemblages which share a similar community structure and are in close proximity with one another. Nevertheless, it is clear that conclusions about the functional importance of species drawn from single locations, or under certain conditions, are unlikely to reflect the full contribution that a particular species may make to ecosystem functioning and, more impor-

tantly, such summaries are likely to misrepresent the levels of biodiversity required to maintain functions and services at broader scales.

When surviving communities were assumed to undergo full numeric compensation, our simulations reveal that the average level of ecosystem functioning (Chl *a* or Org-C) could either remain unchanged, increase, or decrease, relative to the no compensation scenario, indicating that a number of mechanisms may operate to effect post-extinction functioning depending on the composition of the surviving community. Presumably, given our earlier observations on the switchable importance of functionally important species, similar reshuffling processes will occur within the surviving community. However, the point at which far-reaching changes in ecosystem functioning begin to occur—i.e. break-points—appear to change, often radically, post-extinction. This means that the compensatory responses of surviving species may lead to short-term changes in functioning but, because the structure of the community lacks the most productive species, the likelihood and severity of further deleterious change will increase as extinction forcing continues.

It is important to put the findings of our simulations into the context of the North Sea. Our simple numerical models, parameterized using data from a standard benthic survey and monitoring programme, suggest that some fundamental

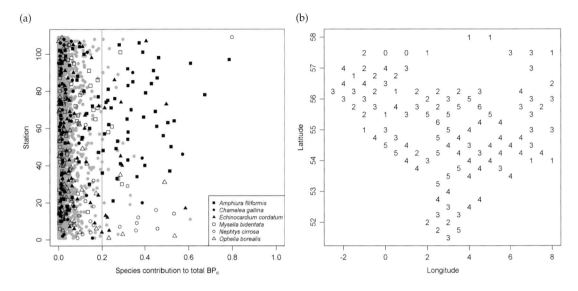

Figure 10.6 Examination of the contribution of species to total BP_c across the 109 stations of the North Sea reveals (a) a total of six species (indicated in the inset) frequently, but not always, individually contribute to > 20% (right of the vertical solid line) of $total_{BPc}$, and that (b) the number of these functionally important species varies between location tremendously.

changes to carbon cycling (Chl *a*, Figure 10.7; Org-C, Figure 10.8) can be expected given the wide spectrum of pressures that are present in marine shelf environments (Halpern *et al.* 2008). These effects will not be universally expressed and certain regions are more likely to benefit than others—i.e. winners and losers, compare circles in Figures 10.7 and 10.8—irrespective of whether full compensation occurs or otherwise. Moreover, the loss of species may not necessarily negatively impact levels of ecosystem functioning in the short term and at a local scale, but our simulations do suggest that a decline in benthic diversity will generally lead to the loss of function and a reduction in community resilience over the longer term and at larger scales. Changes to ecosystem structure are likely which, crucially, may fundamentally alter which species mediate important ecosystem processes and functions and, in turn, lead to abrupt changes in ecosystem performance. These wholesale transformations suggest dramatic reductions in carbon cycling and turnover that are likely to have significant effects on benthic metabolism (e.g. Tang and Kristensen 2007) and the food-web.

10.5 Conclusions and recommendations

We have demonstrated a framework that provides the opportunity to explore and predict how ecosystems may respond to a variety of extinction drivers and that can be applied to regional datasets. Like all models, the resulting probabilistic distributions will reflect the underpinning assumptions, and we reiterate here that these should be continually refined and reinforced with ecological theory and empirical investigation (Coreau *et al.* 2009). Whilst it is tempting to speculate on the implications of specific predictions from the model outcomes, it is important to emphasize that a more prudent approach is to use them to provide insights into processes and mechanisms that may be applicable in natural systems. It would be incorrect to use the absolute values obtained from the predictions to underpin specific biodiversity–environment forecasts. Nevertheless, there is a clear need to develop these techniques further, particularly with respect to other habitats, time periods, and more realistic scenario generation. For example, here we have considered each extinction driver independently and have assumed that partial extinctions—reductions

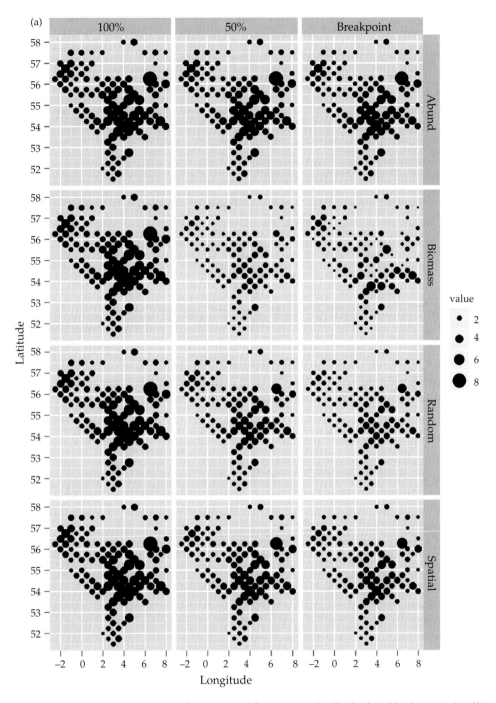

Figure 10.7   The present (100%) versus predicted levels of Chl *a* (µg 5 cm⁻³) for extinctions ordered by abundance (Abund, rarest species within station first), biomass (Biomass, largest first), random (1/n, equal probability), or regional abundance (Spatial, rarest species across region first), assuming 50% species loss (50%) or an extinction set at the level of species richness where the effect on ecosystem functioning changes abruptly (Breakpoint, depicted in Figure 10.3) for scenarios with (a) no compensation versus (b) full compensation at the 109 sampling stations in the North Sea (listed in Figure 10.1). Values for Chl *a* increase with circle diameter (indicated by the key).

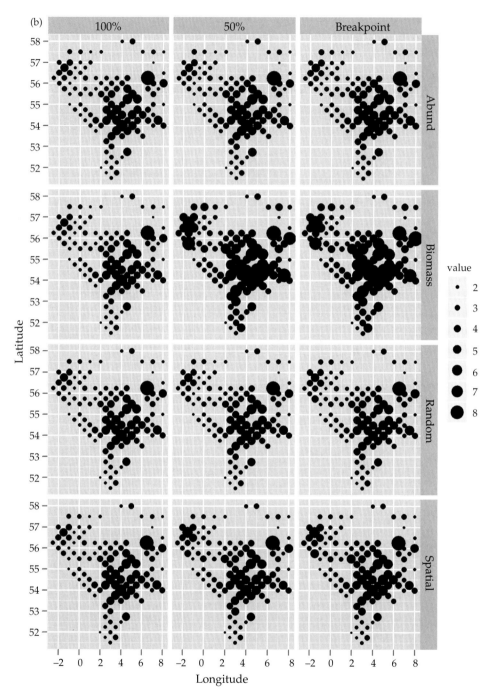

Figure 10.7 Continued

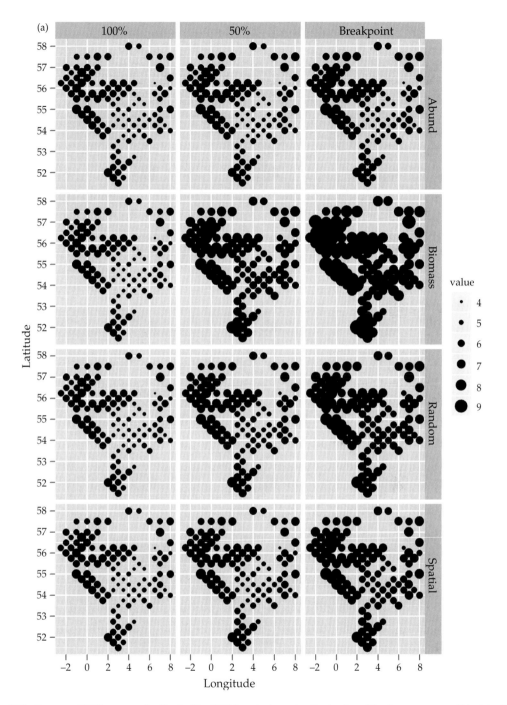

Figure 10.8   The present (100%) versus predicted levels of Org-C (%) for extinctions ordered by abundance (Abund, most numerous within station first), biomass (Biomass, largest first), random (1/n, equal probability), or regional abundance (Spatial, most numerous across region first), assuming 50% species loss (50%) or an extinction set at the level of species richness where the effect on ecosystem functioning changes abruptly (Breakpoint, depicted in Figure 3) for scenarios with (a) no compensation versus (b) full compensation at the 109 sampling stations in the North Sea (listed in Figure 10.1). Values for Org-C increase with circle diameter (indicated by the key).

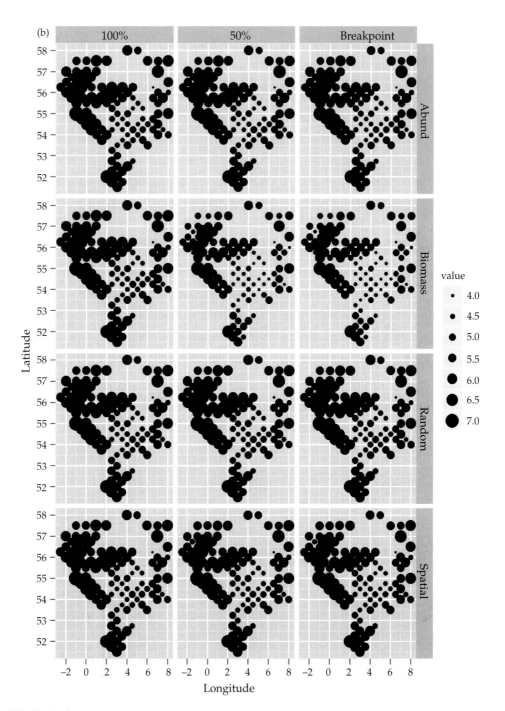

Figure 10.8 Continued

in biomass and/or abundance—extinction cascades or ecological rebounds and revolutions do not occur. Incorporation of multiple interacting drivers and community responses, including seasonal and multi-annual cycles, would aid the understanding of inherent versus environmentally forced variation in ecosystem performance. Similarly, there is scope to incorporate alternative community compensation scenarios, abrupt ecosystem shifts, or human intervention, including the effects of policy and management. In short, the work presented here, and that of others (e.g. Bunker *et al.* 2005; McIntyre *et al.* 2007), demonstrates the practical potential for understanding the likely ecological consequences of biodiversity–environment futures. Our hope is that others will use, adapt and expand this methodology beyond current capacity and across multiple systems.

References

Aarssen, L. W. (1997) High productivity in grassland ecosystems: effected by species diversity of productive species? *Oikos*, **80**, 183–4.

Arzayus, K. M. and Canuel, E. A. (2005) Organic matter degradation in sediments of the York River estuary: effects of biological vs. physical mixing. *Geochimica et Cosmochimica Acta*, 69, 455–64.

Ball, B. A., Hunter, M. D., Kominoski, J. S. *et al.* (2008) Consequences of non-random species loss for decomposition dynamics: experimental evidence for additive and non-additive effects. *Journal of Ecology*, **96**, 303–13.

Balnavera, P., Pfisterer, A. B., Buchmann, N. *et al.* (2006) Quantifying the evidence for biodiversity effects on ecosystem functioning and services, *Ecology Letters* 9, 1146–1156.

Barnosky, A. D., Matzke, N., Tomiya, S. *et al.* (2011) Has the Earth's sixth mass extinction already arrived? *Nature*, **471**, 51–7.

Basford, D. J., Eleftheriou, A., Davies, I. M. *et al.* (1993) The ICES North Sea benthos survey—the sedimentary environment. *ICES Journal of Marine Science*, **50**, 71–80.

Beveridge, O. S., Petchey, O. L., Humphries, S. (2010) Direct and indirect effects of temperature on the population dynamics and ecosystem functioning of aquatic microbial ecosystems. *Journal of Animal Ecology*, **79**, 1324–31.

Boon, A. R. and Duineveld, G. C. A. (1998) Chlorophyll a as a marker for bioturbation and carbon flux in southern and central North Sea sediments. *Marine Ecology Progress Series*, 162, 33–43.

Boon, A. R., Duineveld, G. C. A., Kok, A. (1999) Benthic organic matter supply and metabolism at depositional and non-depositional areas in the North Sea. *Estuarine, Coastal and Shelf Science*, 49, 747–61.

Bracken, M. E., Friberg, S. E., Gonzales-Dorantes, C.A. *et al.* (2008). Functional consequences of realistic biodiversity changes in a marine ecosystem. *Proceedings of the National Academy of Sciences USA*, **105**, 924–8.

Branco, P., Costa, J. L., de Almeida, P. R. (2008) Conservation Priority index for estuarine fish (COPIEF). *Estuarine, Coastal and Shelf Science*, **80**, 581–8.

Bremner, J., Rogers, S. I., Frid, C. L. J. (2003) Assessing functional diversity in marine benthic ecosystems: a comparison of approaches. *Marine Ecology Progress Series*, **254**, 11–25.

Bremner, J., Rogers, S. I., Frid, C. L. J. (2006) matching biological traits to environmental conditions in marine benthic ecosystems. *Journal of Marine Systems*, **60**, 302–16.

Brook, B. W., Sodhi, N. S., Bradshaw, C. J. A. (2008) Synergies among extinction drivers under global change. *Trends in Ecology and Evolution*, **23**, 453–60.

Bulling, M. T., Hicks, N., Murray, L. *et al.* (2010) Marine biodiversity-ecosystem functions under uncertain environmental futures. *Philosophical Transactions of the Royal Society* 365, 2107–2116.

Bunker, D. E., DeClerck, N., Bradford, J. C. *et al.* (2005). Species loss and aboveground carbon storage in a tropical rainforest. *Science*, **310**, 1029–31.

Burkepile, D. E. and Hay, M. E. (2007) Predator release of the gastropod *Cyphoma gibbosum* increases predation on gorgonian corals. *Oecologia*, **154**, 167–73.

Callaway, R. *et al.* (2002). Diversity and community structure of epibenthic invertebrates and fish in the North Sea. *ICES Journal of Marine Science*, **59**, 1199–1214.

Cardinale, B. J., Srivastava, D. S., Duffy, J. E. *et al.* (2006) Effects of biodiversity on the functioning of trophic groups and ecosystems. *Nature* 443, 989–992.

Callaway, R., Alsvag, J., de Boois, I. *et al.* (2002). Diversity and community structure of epibenthic invertebrates and fish in the North Sea. *ICES Journal of Marine Science*, 59: 1199–1214.

Castellanos-Galindo, G. A., Cantera, J. R., Espinosa, S. *et al.* (2011) Use of local ecological knowledge, scientist's observations and grey literature to assess marine species at risk in a tropical eastern Pacific estuary. *Aquatic Conservation: Marine and Freshwater Ecosystems*, 21, 37–48.

Coreau, A., Pinay, G., Thompson, J. D. *et al.* (2009) The rise of research on futures in ecology: rebalancing scenarios and predictions. *Ecology Letters* 12: 1277–1286.

De Visser, S., Freymann, B., Olff, H. (2011) The Serengeti food web: empirical quantification and analysis of toplogical changes under increasing human impact. *Journal of Animal Ecology*, **80**, 465–75.

Dulvy, N., Ellis, J., Goodwin, N. *et al.* (2004) Methods of assessing extinction risk in marine fishes. *Fish and Fisheries*, **5**, 255–76.

Dulvy, N., Sadovy, Y., Reynolds, J. (2003) Extinction vulnerability in marine populations. *Fish and Fisheries*, **4**, 25–64.

Edgar, G. J., Banks, S. A., Brandt, M. *et al.* (2010) El Nino, grazers and fisheries interact to greatly elevate extinction risk for Galapagos marine species. *Global Change Biology*, **16**, 2876–90.

Fisher, J. A. D., Frank, K. T., Leggett, W. C. (2010) Global variation in marine fish body size and its role in biodiversity-ecosystem functioning. *Marine Ecology Progress Series*, **405**, 1–13.

Fowler, M. S. (2010) Extinction cascades and the distribution of species interactions. *Oikos* 119: 864–873.

Freeman, D. J., Marshall, B. A., Ahyong, S. T. *et al.* (2010) Conservation status of New Zealand marine invertebrates, 2009. *New Zealand Journal of Marine and Freshwater Research*, **44**, 129–48.

Godbold, J. A., Bulling, M. T., Solan, M. (2011) Habitat structure mediates biodiversity effects on ecosystem properties. *Proceedings of the Royal Society of London B*, **278**, 2510–18.

Godbold, J. A., Rosenberg, R., Solan, M. (2009) Species-specific traits rather than resource partitioning mediate diversity effects on resource use. *PLoS One*, **4(10)**, e7423.doi:10.1371/journal.pone.0007423

Godbold, J. A. and Solan, M. (2009) Relative importance of biodiversity and the abiotic environment in mediating an ecosystem process. *Marine Ecology Progress Series*, **396**, 281–90.

Graham, N. A. J., Chabanet, P., Evans, R. D. *et al.* (2011) Extinction vulnerability of coral reef fishes. *Ecology Letters*, **14**, 341–8.

Griffen, B. D. and Drake, J. M. (2008) A review of extinction in experimental populations. *Journal of Animal Ecology*, **77**, 1274–87.

Gross, K., Cardinale, B. J. (2005) The functional consequences of random vs. ordered species extinctions. *Ecology Letters* 8: 409–418

Hall-Spencer, J. M., Rodolfo-Metalpa, R., Martin, S. *et al.* (2008) Volcanic carbon dioxide vents show ecosystem effects of ocean acidification. *Nature*, 454, 96–9.

Halpern, B. S. Walbridge, S., Selkoe, K. A. *et al.* (2008) A global map of human impact on marine ecosystems. *Science* 319: 948–952.

Harrison, R. D. (2000) Repercussions of El Nino: drought causes extinction and the breakdown of mutualism in Borneo. *Proceedings of the Royal Society B*, **267**, 911–15.

Hector, A. and Bagchi, R. (2007) Biodiversity and ecosystem functionality. *Nature*, **448**, 188–90.

Hillebrand, H., Matthiessen, B. (2009) Biodiversity in a complex world: consolidation and progress in functional biodiversity research. *Ecology Letters* 12: 1405–1419.

Hooper, D. U., Chapin, F. S., Ewel, J. J. *et al.* (2005) Effects of biodiversity on ecosystem functioning: A consensus of current knowledge. *Ecological Monographs* 75:3–35.

Hooper, D. U., Solan, M., Symstad, A. *et al.* (2002) Species diversity, functional diversity, and ecosystem functioning. In: *Biodiversity and ecosystem functioning. Synthesis and perspectives.* Loreau, M., Naeem, S., Inchausti, P. (eds) Oxford University Press, Oxford.

Hughes, A. R., Williams, S. L., Duarte, C.M. *et al.* (2009) Associations of concern: declining seagrasses and threatened dependent species. *Frontiers in Ecology and the Environment*, **7**, 242–6.

Huys, R., Herman, P. M. J., Heip, C. H. R. *et al.* (1992) The meiobenthos of the North Sea: density, biomass trends and distribution of copepod communities. *ICES Journal of Marine Science*, **49**, 23–44.

ICES (1986) Fifth report of the Benthos Ecology Working Groups, Ostende, 12–15 May 1986. ICES 1986/C:27.

Isbell, F. I., Iosure, D. A., Yurkonis, K. A. *et al.* (2008) Diversity-productivity relationships in two ecologically realistic rarity-extinction scenarios. *Oikos*, **117**, 996–1005.

IUCN. (2001) IUCN Red List categories and criteria: version 3.1. IUCN, Gland, Switzerland, and Cambridge, United Kingdom. Available at: <http://www.iucnredlist.org/technical-documents/categories-and-criteria> (accessed July, 2011).

Künitzer, A. *et al.* (1992) The benthic infauna of the North Sea: species distribution and assemblages. *ICES Journal of Marine Science*, **49**, 127–43.

Labrune, C., Amouroux, J.M., Sarda, R. *et al.* (2006) Characterisation of the ecological quality of the coastal Gulf of Lions (NW Mediterranean). A comparative approach based on three biotic indices. *Marine Pollution Bulletin*, 52, 34–47.

Larsen, T. H., Williams, N. M., Kremen, C. (2005) Extinction order and altered community structure rapidly disrupt ecosystem functioning. *Ecology Letters*, **8**, 538–47.

Laughlin, D. C. (2011) Nitrification is linked to dominant leaf traits rather than functional diversity. *Journal of Ecology* 99: 1091–1099. doi: 10.1111/j.1365-2745.2011.01856.x

Liow, L. H., Fortelius, M., Lintulaakso, K. *et al.* (2009) Lower extinction risk in sleep-or-hide mammals. *The American Naturalist*, 173, 264–72.

Lohrer, A. M., Thrush, S. F., Gibbs, M. M. (2004) Bioturbators enhance ecosystem function through complex biogeochemical intercations. *Nature*, **431**, 1092–95.

Loreau, M. and Hector, A. (2001) Partitioning selection and complementarity in biodiversity experiments. *Nature*, **412**, 72–6.

Mace, G. M., Collar, N. J., Gaston, K. J. *et al.* (2008) Quantification of extinction risk: IUCN's system for classifying threatened species. *Conservation Biology*, **22**, 1424–42.

Mackinson, S. and Daskalov, G. M. (2007) An ecosystem model of the North Sea. Cefas, Lowestoft, UK.

McIntyre, P. B., Jones, L. E., Flecker, A. S. *et al.* (2007) Fish extinctions altr nutrient recycling in tropical freshwaters. *Proceedings of the National Academy of Sciences, USA*, **104**, 4461–6.

Meysman, F. J. R., Boudreau, B. P. and Middelburg, J. J. (2003) Relations between local, nonlocal, discrete and continuous models of bioturbation. *Journal of Marine Research*, **61**, 391–410.

Michaud, E., Desrosiers, G., Mermillod-Blondin, F. *et al.* (2006) The functional group approach to bioturbation: II. The effects of the *Macoma balthica* community on fluxes of nutrients and dissolved organic carbon across the sediment-water interface. *Journal of Experimental Marine Biology and Ecology*, **337**, 178–89.

Micheli, F. and Halpern, B. S. (2005) Low functional redundancy in coastal marine assemblages. *Ecology Letters*, **8**, 391–400.

Moir, M.L., Vesk, P.A., Brennan, K.E.C. (2009) Current Constraints and Future Directions in Estimating Coextinction. Conservation Biology 24: 682–690.

Muggeo, V. M. R. (2003) estimating regression models with unknown break-points. *Statistics in Medicine*, **22**, 3055–71.

Naeem, S. (2008) Advancing realism in biodiversity research. *Trends in Ecology and Evolution*, **23**, 414–16.

Naeem, S., Loreau, M., Inchausti, P. (2002) Biodiversity and ecosystem functioning: the emergence of a synthetic ecological framework. In: *Biodiversity and ecosystem functioning. Synthesis and perspectives.* Loreau, M., Naeem, S., Inchausti, P. (eds) Oxford University Press, Oxford.

Nicholson, E., Keith, D. A., Wilcove, D. S. (2009) Assessing the threat status of ecological communities. *Conservation Biology*, **23**, 259–74.

Nogaro G., Charles F., de Mendonca, J. B., Mermillod-Blondin, F., Stora, G., Francois-Carcaillet, F. (2008) Food supply impacts sediment reworking by *Nereis diversicolor*. *Hydrobiologia*, **598**, 403–8.

Noss, R. F. (2010) Local priorities can be too parochial for biodiversity. *Nature*, **463**, 424.

Olden, J. D., Hogan, Z. S., Zanen, M. J. V. (2007) Small fish, big fish, red fish, blue fish: size based extinction risk of the world's freshwater and marine fishes. *Global Ecology and Biogeography*, **16**, 694–701.

Pearson, T. H., Rosenberg, R. (1978) Macrobenthic succession in relation to organic enrichment and pollution of the marine environment. *Oceanography and Marine Biology: an Annual Review*, 16: 229–311.

Pereira, H. M., Leadley, P. W., Proenca, V. *et al.* (2010) Scenarios for Global Biodiversity in the 21st Century. *Science*, **330**, 1496–501.

Polidoro, B. A., Carpenter, K. E., Collins, L., *et al.* (2010) The loss of species: mangrove extinction risk and geographic areas of global concern. *PLoS ONE* 5(4): e10095. doi:10.1371/journal.pone.0010095.

Sala, O. E., Chapin, F. S., Armesto, J. J. *et al.* (2000) Biodiversity – biodiversity scenarios for the year 2100. Science 287: 1770–1774.

Schläpfer, F., Pfisterer, B., Schmid, B. (2005) Non-random species extinction and plant production: implications for ecosystem functioning. *Journal of Applied Ecology*, **42**, 13–24.

Schmid, B., Balnavera, P., Cardinale, B. J. *et al.* (2009) Consequences of species loss for ecosystem functioning: meta-analyses of data from biodiversity experiments. In: *Biodiversity, ecosystem functioning and human wellbeing: an ecological and economic perspective.* Naeem, S., Bunker, D.E., Hector, A. *et al.* (eds), Oxford University Press, Oxford, 14–29.

Short, F. T., Polidoro, B., Livingstone, S. R. *et al.* (2011) Extinction risk assessment of the world's seagrass species. Biological Conservation 144, 1961–71.

Smith, M. D., Knapp, A. K. *et al.* (2003) Dominant species maintain ecosystem function with non-random species loss. *Ecology Letters* 6: 509–517.

Solan, M., Cardinale, B. J., Downing, A.L. *et al.* (2004) Extinction and ecosystem function in the marine benthos. *Science* **306**, 1177–80.

Solan, M., Godbold, J. A., Symstad, A. *et al.* (2009) Biodiversity–ecosystem function research and biodiversity futures: early bird catches the worm or a day late and a dollar short? In: *Biodiversity, ecosystem functioning and human wellbeing: an ecological and economic perspective.* Naeem, S., Bunker, D.E., Hector, A. *et al.* (eds), Oxford University Press, Oxford, 30–45.

Srinivasan, U. T., Dunne, J. A., Harte, J. *et al.* (2007) Response of complex food webs to realistic extinction sequences. *Ecology*, **88**, 671–82.

Swift, D. J. (1993) The macrobenthic infauna off Sellafield (North-eastern Irish Sea) with special reference to bioturbation. *Journal of the Marine Biological Association of the United Kingdom*, **73**, 143–62.

Tang, M. and Kristensen, E. (2007) Impact of microphytobenthos and macroinfauna on temporal variation of benthic metabolism in shallow coastal sediments. *Journal of Experimental Marine Biology and Ecology*, **349**, 99–112.

Teal, L. R., Bulling, M. T., Parker, E. R. *et al.* (2008) Global patterns of bioturbation intensity and mixed depth of marine soft sediments. *Aquatic Biology*, **2**, 207–18.

Thomas, C. D., Cameron, A., Green, R. E. *et al.* (2004) Extinction risk from climate change. *Nature*, **427**, 145–8.

Vitousek, P. M., Turner, D. R., Parton, W. J. *et al.* (1994) Litter decomposition on the Mauna Loa environmental matrix, Hawaii'i: patterns, mechanisms and models. *Ecology*, 75, 418–29

Witbaard, R., Dunieveld, G. C. A., Van der Weele, J. A. *et al.* (2000) The benthic response to the seasonal deposition of phytopigments at the Porcupine Abyssal Plain in the North East Atlantic. *Journal of Sea Research*, **43**, 15–31.

Zavaleta, E. S., Hulvey, K. B. (2004) Realistic species losses disproportionately reduce grassland resistance to biological invaders. *Science* 306: 1175–1177.

Zhang, Q-G. and Zhang, D-Y. (2007) Consequences of individual species loss in biodiversity experiments: an essentiality index. *Acta Oecologica*, **32**, 236–42.

Ziebis, W., Forster, S., Huettel, M. *et al.* (1996) Complex burrows of the mud shrimp *Callianassa truncata* and their geochemical impact in the sea bed. *Nature*, **382**, 619–22.

CHAPTER 11

Biodiversity and ecosystem functioning: an ecosystem-level approach

David Raffaelli and Alan M. Friedlander

11.1 The need to work at seascape scales

The great interest in how ecosystems can be managed so that they can continue to deliver ecosystem services (Millennium Ecosystem Assessment 2005) has recently found fresh articulation in a broad range of follow-up initiatives. These include the frequent reference to ecosystem services provisions within the top 40, 50, or 100 research questions in ecology identified by various national forums (Rudd 2011), the establishment of the UN Intergovernmental Panel on Biodiversity and Ecosystem Services (Larigauderie and Mooney 2010), and the reframing of the Convention on Biological Diversity's (CBD) biodiversity targets towards maintaining ecosystem services and the sustainable use of biodiversity (Conference of the Parties to the CBD, Nagoya, Japan, October 2010). Yet the contribution of biodiversity to ecosystem services is still far from clear in many specific instances (Balvanera *et al.* 2006). Despite the large literature that presently exists on biodiversity and ecosystem functioning research (Caliman *et al.* 2010), the majority of studies are small-scale and their extension to ecosystem services is not easy. In part this is because ecosystem services are usually underpinned by many ecological functions and hence sets of biodiversities, and, related to this, because many services are generated at the much larger landscape and seascape scales.

The need to understand biodiversity–ecosystem functioning at larger spatial and at longer temporal scales is especially relevant for marine ecosystems.

Their openness, interconnectedness, and complexity makes extrapolation from small-scale studies problematic. Marine ecosystem processes are inherently scale-dependent (Denman 1994; Palumbi *et al.* 2009), often with abrupt discontinuities between processes operating at smaller and larger scales (Holling 1992; Raffaelli and Frid 2010). A more complete understanding of biodiversity–ecosystem functioning at seascape scales will demand that we work at those larger scales.

Much of the marine research in the area of biodiversity and ecosystem functioning has been done at smaller scales which are more tractable for tighter experimental control and where high levels of replication can be achieved (Bulling *et al.* 2006). Replication and statistical power often allows the most powerful inferences and is needed to persuade reviewers and critics of the effects of biodiversity change on functioning (Raffaelli and Moller 2000). Moving to larger spatial and temporal scales implies a shift from using highly controlled and replicated approaches towards more correlative or model-based approaches (e.g. Emmerson and Huxham 2001; Raffaelli 2006). Such observational approaches are often seen to allow the weakest form of inference, since they are inherently inductive, but, as MacNeil (2008) points out, this does not have to be the case. If peers and critics are to be persuaded of biodiversity change effects inferred for observation studies, then those effects will have to be large or dramatic (Raffaelli and Moller 2000). In addition, working at larger scales also brings with it greater complexity, both in biodiversity and in functioning, as the dynamics of many trophic levels need to be accommodated as well as the physical, especially hydrographic, processes that come into play at these scales. Such complexity only serves to increase the risk that multiple interaction processes other than those inferred might be responsible for the observed effects, so that it becomes important to challenge continually interpretation with competing hypotheses as we are performing more 'do-able' experimental tests of falsifiable hypotheses on smaller scale processes (MacNeil 2008).

Here, we describe some approaches to understanding biodiversity–ecosystem functioning at these large scales which we hope will stimulate further work in this area. First, we consider what might be needed to build a credible evidence base of biodiversity change effects on functioning at seascape scales. Second, we explore the different aspects of biodiversity change, compositional and species richness, within contrasting bottom-up controlled (estuaries) and top-down controlled (coral reefs) systems, affected by two dominant impacts in marine systems, eutrophication and over-fishing.

11.2 Building a credible evidence base

For scientific evidence to be of use to those engaged with managing marine systems for their ecosystem services, that evidence needs to be persuasive and defendable. By far the most persuasive kind of evidence comes from demonstrations that a change in biodiversity results in an outcome—an effect—which is outwith the normal variation in the system. In highly controlled and replicated manipulative experiments where only one factor—biodiversity—is allowed to vary, the normal variation is captured by statistical variation between replicates in the controls which is then compared with the variation within the experimental treatments (MacNeil 2008). In this way, even quite small changes in ecosystem functioning can be convincingly shown to be due to biodiversity change. In contrast, it is hard to persuade others of the functional effects caused by biodiversity change when working at large scales, because trying to achieve any kind of replication is a non-trivial issue (Raffaelli and Moller 2000), and it is usually the case that factors other than biodiversity

will be different between natural and disturbed systems, so that interpretation of biodiversity effects per se may be confounded. However, if the effect is so large that it is outwith the widely accepted—but not measured statistically—range of variation, then the case for a biodiversity effect will often be persuasive, despite the lack of control and statistical replication. As important in this respect is to challenge the interpretation with alternative, competing hypotheses (MacNeil 2008). Examples of studies where the effects of altering biodiversity are demonstrably clear without replication or controls include the keystone consumer removal experiments carried out by Paine on predatory starfish (*Pisaster*) in the 1970s—the majority of which had no controls or replication—and by Lodge on limpets (*Patella*) in the late 1940s—both reviewed in Raffaelli and Moller 2000. In both cases, the effect of consumer removal was impressive: starfish removal leads to a rapid shift in mussels out of their normal zone to completely dominate the shore and causes many secondary extinctions. Limpet removal from a 10 m-wide strip produced a bright green algal band that was in stark contrast to the normal shore on either side. Nobody would question that those effects were caused by the loss of a keystone consumer. The persuasiveness of the evidence for keystone consumers is in the magnitude of the effect, not the rigour of the experimental design and herein lies the challenge for seascape scale studies of biodiversity–ecosystem functioning research: if sufficient (or any) replication cannot be achieved for statistical testing, and potentially confounding factors cannot be ruled out or controlled for, then managers are only likely to be convinced from observational studies of the role of biodiversity in maintaining ecosystem functioning if the effects demonstrated are large, and if no other competing hypotheses can sensibly account for the observations.

Below, we look at biodiversity change caused by eutrophication and fishing using two case studies. In presenting these studies, it should be realized that 'biodiversity' is a contested term. In terrestrial systems at least, different groups have different mental constructs of 'biodiversity', and most are not restricted to the 'species richness' construct beloved of experimentalists working in this area (Fischer and Young 2007). Indeed, when working with complex, multitrophic systems at seascape scales, looking at the effects of changes in the species richness of the entire system may not sensible. The impacts of biodiversity change may be explored better by looking at the effects of species loss or compositional change within key functional groups, such as top predators or primary producers. This is the approach adopted here.

11.3 Case study 1: The Ythan estuary, Scotland

The Ythan estuary, Aberdeenshire, Scotland, is one of the smallest estuaries in the UK, but probably one of the best documented and understood in the world with respect to its biodiversity dynamics. The River Ythan catchment is relatively small (ca 640 km^2 of low-elevation highly productive agricultural land) and the estuary itself is only about 8 km in length. At low tide, much of the estuary is mudflats, sand flats, and intertidal mussel (*Mytilus edulis*) beds, which together comprise ca 190 ha (Baird and Milne 1981). Since the late 1950s, research on the Ythan (Hall and Raffaelli 1991; Gorman and Raffaelli 1993; Raffaelli 2011) has allowed the system to be documented extensively with respect to the population numbers and biomasses of the primary producers, benthic invertebrates, and their consumers (predatory polychaetes, crustaceans, fish, and shorebirds). Water column food chains—phytoplankton, zooplankton, planktivorous fish and their piscivores—are only loosely associated with the predominant benthic compartment of the Ythan food-web (Raffaelli 2011) and are not included in this analysis. Over the 50-year period for which the tropho-dynamics of the Ythan have been studied, the river draining the agricultural hinterland, and hence the estuary itself, have experienced nitrogen enrichment due to changes in land-use driven by agricultural policies that promoted the growing of nitrogen-intensive crops such as wheat, barley, and oil seed rape, as well as intensive pig production (Raffaelli *et al.* 1999; Wiegand *et al.* 2010). These agricultural changes led to severe eutrophication symptoms in the estuary by the mid-1980s, manifested as a two to threefold increase in river nitrogen

(almost all nitrate) and increasingly heavy blooms of the opportunistic green macro-algae, mostly *Ulva intestinalis,* that form mats 1–3 kg wet weight.m^{-2} over the mudflats. The impact of the mats on the underlying invertebrates is dramatic, with up to tenfold losses of the amphipod *Corophium volutator.* This invertebrate is an important prey for all of the Ythan's higher consumers and there have been concomitant changes in numbers of shorebirds (Raffaelli, Hull, and Milne 1989; Raffaelli *et al.* 1999). Along with the elevated levels of nitrogen in the Ythan, these biodiversity changes were significantly dramatic to force designation of the Ythan catchment as a Nitrogen Vulnerable Zone under the European Nitrates Directive in 1998: the Ythan ecosystem had moved to an ecological state which was far beyond the scale of natural variation. Many replicated manipulated experiments of the smaller-scale processes—interactions between weed mats and sediment chemistry, weed mats and invertebrates, and weed mats and shorebirds (Raffaelli *et al.* 1999)—have confirmed the causal nature of these relationships.

The changes in biodiversity and ecosystem functioning as the system moved from relatively undisturbed to eutrophic conditions are captured here through comparison of two periods: the late 1960s/early 1970s, and the late 1980s/early 1990s, largely based on Baird and Milne (1981), Baird and Ulanowicz (1993), Buchan (1997), Goyal (2005), Raffaelli *et al.* (2004), and Raffaelli (2011). Aspects of ecological functioning were derived from mass-balance models for the two periods built and analysed using Ecopath (<http:www.ecopath.org>). Production/biomass (P/B) ratios required for Ecopath come from Baird and Milne (1981); Buchan (1997), and several web-based databases—e.g. fishbase, Chesapeake Bay laboratory. P/B values were the same for the two versions of the model, but the diet matrices differed to reflect the known trophic responses (e.g. Dayawansa 1995; Raffaelli *et al.* 1999) of the consumers of the main invertebrates *Corophium volutator, Nereis (Hediste) diversicolor, Hydrobia ulvae,* and *Macoma balthica,* all of which showed a marked biomass difference between the two periods. Ecotrophic efficiencies (EE) were initially set at 0.95 for invertebrates, and for shorebirds (mostly top predators) at 0.1, but the majority were re-estimated by Ecopath as slightly different values (all less than unity) in order to balance the models (Raffaelli 2011).

11.3.1 Biodiversity in the two periods

The biomasses estimated for the two periods are shown in Table 11.1. The biomass of macroalgae (*Ulva intestinalis* and *Chaetomorpha spp.*) increased by almost 150% between the two periods, as a result of an increase in riverine nitrogen (nitrate) by a similar amount, so that biomasses of > 1 kg wet weight.m^{-2} now cover ca 40% of estuary's mudflats in bloom years (Raffaelli *et al.* 1989,

1999). Biomass of the amphipod *Corophium* was reduced by ca 90% between the two periods (Table 11.1), due to the deteriorating sediment physico-chemical environment (highly reduced redox conditions) under the algal mats, and by the weed physical interfering with the amphipod's feeding behaviour (Raffaelli 1999). The other three major invertebrates on mudflats increased in biomass, *Nereis* by > 100%, *Hydrobia* by 65%, and *Macoma* by ca 16% (Table 11.1), reflecting their different feeding behaviours and sensitivities to the sediment environment, compared to *Corophium* (Raffaelli *et al.* 1999). No other invertebrates changed in abundance between the two periods (Table 11.1).

Table 11.1 Biomasses (t.km^{-2}) of biodiversity groups used in mass-balance analysis of the Ythan estuary for two periods, undisturbed and eutrophic, and the percentage change between the periods (from Raffaelli 2011).

	undisturbed	eutrophic	change
Cormorant	0.040	0.040	0.00
Shelduck	0.043	0.015	−65.12
Red Breasted Merganser	0.012	0.012	0.00
Oystercatcher	0.058	0.058	0.00
Dunlin	0.006	0.009	50.00
Bar-Tailed Godwit	0.001	0.004	300.00
Curlew	0.005	0.076	1420.00
Redshank	0.021	0.019	−9.52
Wigeon	0.078	0.078	0.00
Mute Swan	0.066	0.066	0.00
Flounder	20.00	20.00	0.00
Goby	0.70	0.70	0.00
Other Fish Species	5.07	5.07	0.00
Shore Crab	8.5	8.5	0.00
Mytilus edulis	185.0	185.0	0.00
Nereis diversicolor	82.44	179.00	117.13
Crangon crangon	7.38	7.38	0.00
Corophium volutator	22.40	2.27	−89.87
Gammarus spp	36.67	36.67	0.00
Littorina spp	16.67	16.67	0.00
Hydrobia ulvae	33.60	55.46	65.06
Macoma balthica	138.00	160.15	16.05
Meiofauna	34.17	34.17	0.00
Ulva and Chaetomorpha	308.30	762.00	147.16
Other macrophytes	320.01	320.01	0.00
Benthic microphytes	5.000	5.00	0.00
Total	1225.4	1799.6	46.9

The shorebirds, shelduck (*Tadorna tadorna*) and redshank (*Tringa totanus*), declined between the two periods by ca 65% and ca 10%, respectively, consistent with the effects of algal mats on their feeding behaviour, and their dependence on *Corophium* (Buxton and Young 1981; Tubbs and Tubbs 1983; Dayawnsa 1995). In contrast, Dunlin (*Calidris alpina*), Bar-tailed Godwit (*Limosa lapponica*). and Curlew (*Numenius arquata*) increased in biomass between the two periods by 50%, 300%, and 1420%, respectively (Table 11.1). It should be noted that other species of shorebird—e.g. oystercatcher *Haematopus ostralegus*—changed in abundance on the Ythan over the eutrophication period, peaking in the mid-1980s, but their numbers in the two focal periods are similar due to this non-monotonic trend (Raffaelli *et al.* 1999). The two herbivorous shorebirds in the system, wigeon (*Anas penelope*) and mute swan (*Cygnus olor*), did not change in abundance, probably because they also feed extensively off-estuary.

There is no evidence of significant population changes for epibenthic crustaceans and fishes between the two time periods (e.g. Healey 1971; Summers 1974, 1980; Raffaelli and Milne 1987; Jaquet and Raffaelli 1989; Raffaelli *et al.* 1989) and biomasses are therefore assumed the same for the two periods (Table 11.1).

Clearly, the biodiversity of the system changed dramatically between the two periods associated with eutrophication, enough to warrant the river catchment's protection under a European Directive. However, these changes did not involve losses or additions of any species—only their relative abundances changed. This is also reflected in the Information Index derived from the models, a measure of the system's complexity.

11.3.2   Ecological functioning in the two periods

The total biomass in the system increased by 47% between the two periods (Table 11.1) and this is reflected in several measures of ecological functioning (Figures 11.1–11.3). All measures of functioning increased between the two periods, except for the cycling indices. Cycling indices, such as

Finn's index and the Predator Cycling index (Figure 11.2), are thought to capture the functions of mineral and nutrient cycling, using the proxy of energy cycling in the system (Christensen, Walters, and Pauly 2005). The above increases are also reflected in network characteristics (Figure 11.3), including the Total System Throughput, a measure of the size of the system, (Kay, Graham, and Ulanowicz, 1989), which increased by 38%. System Overhead (calculated as the difference between the network properties Capacity and Ascendancy), is seen as the 'strength in reserve' within the ecosystem which can be used to cope with external perturbations (Christensen, Walters, and Pauly 2005) and as such provides a measure of resilience. This metric increased by 25% whilst another proxy for stability, Ascendancy, increased by 35% (Figure 11.3). However, the relative measure (Ascendancy/Capacity) was similar for the two periods (24.6% and 24.9%, respectively), despite the large-scale changes in nutrient loading, primary production, and invertebrate biomass.

In summary, the Ythan case study has shown that biodiversity change—here compositional rather than species loss—had profound effects on ecosystem functioning. Measures of primary and secondary production all increased markedly, proxies for rates of cycling declined, whilst measures of resilience (Overhead and Ascendancy) increased.

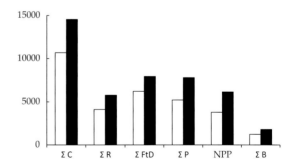

Figure 11.1   Measures of total ecological functioning from mass balance models, Ythan estuary (data from Raffaelli 2011). Open bars refer to the undisturbed and back bars to the eutrophic state. C = consumption; R = respiratory flows; FtD = flows to detritus; P = total production; NPP = net primary production; B = total biomass. All units = tons/km^2/yr, except B = tons/km^2

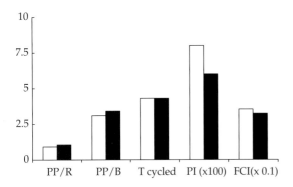

Figure 11.2   Relative measures of ecological functioning from mass balance models, Ythan estuary (data from Raffaelli 2011). Open bars refer to the undisturbed and back bars to the eutrophic state Key as for Figure 11.1, except T = throughput (tons/km^2/yr); PI = predatory index; FCI = Finn's cycling index. PI and FCI = % of throughput without detritus.

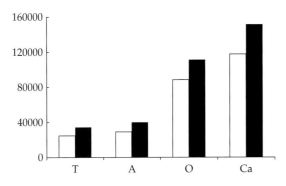

Figure 11.3   Network characteristics of mass balance models, Ythan estuary (data from Raffaelli 2011). T = total system throughput (tons/km^2/yr); A = ascendancy (flowbits); O = overhead (flowbits); Ca = capacity (flowbits).

Of course, the changes in primary production between the two periods are almost certainly entirely attributable to the direct, bottom-up effects of nutrient enrichment, rather than changes in other components of biodiversity. This increase in algae resulted in cascading changes on invertebrates and shorebird biodiversity up through the food-web, so that the relative abundance of species is dramatically different between the two periods (Table 11.1), and it is those biodiversity changes which are responsible for changes in secondary production and higher level-metrics in Figures 11.2 and 11.3. Consumer biomass in the undisturbed and eutrophic states were 48% and 40% of total

biomass, respectively, whilst secondary production remained at ca 15% of total production.

The Ythan is clearly functioning differently in its eutrophic state compared to its undisturbed state and there is little evidence that its resilience has been compromised on the basis of the network characteristics calculated here. But is the Ythan now functioning better—more productive—or worse—with fewer individuals of some species? This question highlights the difficulty in making general statements about the effects of biodiversity change on ecosystem functioning, as discussed later. It should be noted that the eutrophication process is well known to be non-monotonic and it is entirely possible that a threshold may be reached whereby the stimulatory effects on secondary production of nutrient enrichment and primary production are negated by a complete coverage of the mudflats by algal mats, at which stage the system would collapse (Raffaelli *et al.* 1999).

11.4   Case study 2: Hawaii and the northern Line Islands, central Pacific

Our understanding of what is natural in the marine environment is becoming increasingly compromised by the absence of locations that are not impacted by human activities and nowhere is this more acute than on coral reefs, where overexploitation and severe depletion have occurred on a global scale (Bellwood *et al.* 2004; Pandolfi *et al.* 2005). Although numerous factors have contributed to the decline in coral reefs worldwide, fishing has historically exerted the most direct and pervasive influence through the disproportionate removal of large-bodied top-level predators that exert strong control over the productivity of the entire ecosystem (Pauly *et al.* 1998; Jackson *et al.* 2001). Protected, remote locations are some of the few remaining examples of coral reefs without major anthropogenic influence and these 'ecological baselines' provide fundamental insights into natural patterns and processes in these high biodiversity ecosystems and allow for comparisons with more impacted human-dominated coral reefs (Knowlton and Jackson 2008; Sandin *et al.* 2008). This type of 'space-for-time' substitution is an alternative to long-term studies to assess the impact of human-induced changes where pre-impact records are sparse or non-existent.

11.4.1   Hawaii

The Hawaiian archipelago consists of the high, populated main Hawaiian Islands (MHI) to the southeast, and the low-lying north-western Hawaiian islands (NWHI) that stretch for more than 2000 km to the north-west of the MHI (Figure 11.4a). The MHI are heavily populated—1.2 million people—and highly urbanized while the NWHI have had limited human activity due to their remoteness. Recent designation of the NWHI as US Marine National Monuments makes it the largest single conservation area under US jurisdiction (362 074 km²). Intensive fishing pressure by

commercial, recreational, and subsistence fishers in the MHI has led to severe depletion, particularly around the more populated areas of the state, while the reefs of the NWHI are recognized as some of the most intact on earth (Friedlander and DeMartini 2002; Pandolfi *et al.* 2005; Williams *et al.* 2008).

11.4.2   Northern Line Islands

The northern Line Islands (NLI), located ca 1500 km to the south-south-west of the Hawaiian Archipelago, in the central Equatorial Pacific (Figure 11.4b), includes islands that range from relatively densely populated, where near-shore reef resources are heavily fished, to uninhabited atolls whose reefs were historically little fished and which are unfished at present (Charles and Sandin 2009).

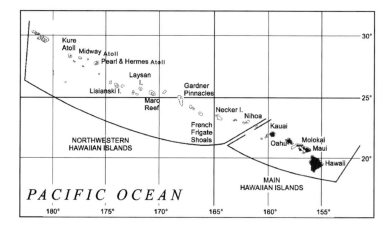

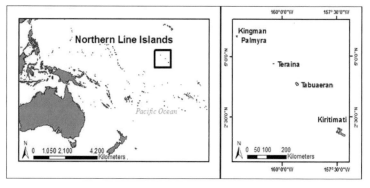

Figure 11.4   A. Hawaiian Archipelago with boundaries of the north-western Hawaiian Islands National Marine Monument. B. The northern Line islands. Palmyra and Kingman are US Marine National Monuments and the islands of Kiritimati, Tabuaeran, and Teraina belong to the country of Kiribati.

The two northernmost Line Islands—Kingman and Palmyra atolls—are US possessions that are currently protected as US Marine National Monuments (53 509 km²). The other two islands—Tabuaeran and Kiritimati atolls—belong to the Republic of Kiribati. In 2005, estimates of human population sizes were 5100 at Kiritimati and 2500 at Tabuaeran, although the actual numbers are likely to be much higher. Fisheries data indicate that subsistence fishing dominates reef activities at these atolls, with some commercial extraction of food fish, shark fins, and ornamentals (Sandin *et al.* 2008).

11.5   Effects of fishing on fish assemblage structure

11.5.1   Hawaii

The most conspicuous difference between fished and unfished areas in Hawaii is the strikingly high biomass densities and greater average body sizes of reef fishes in the NWHI compared to the MHI, particularly for large jacks, reef sharks, and other apex predators (Figure 11.5a, Friedlander and DeMartini 2002; DeMartini and Friedlander 2006). Biomass is considered a good proxy of ecosystem function as it represents metabolic requirements and therefore energy fluxes in the ecosystem (Cardinale *et al.* 2006; Brown *et al.* 2004). Overall fish standing stock in the NWHI was nearly three-fold greater than in the MHI with > 54% of the total fish biomass in the NWHI consisting of apex predators, whereas this trophic level accounted for less than 3% of the fish biomass in the MHI (Friedlander and DeMartini 2002). A 30–50% lower biomass of herbivores and lower-level carnivores in fished areas in the MHI compared to the NWHI illustrates the effects that intensive fishing also has on lower trophic levels and hence the entire reef fish assemblage in the MHI. The strongly inverted biomass pyramid—more predators than prey—found in the NWHI results from short-lived, lower trophic level fishes which serve as energetic shunts, converting food—e.g., plankton, algae—into fish biomass, which is quickly transferred up the food chain and stored in long-lived, higher trophic level fish through predation.

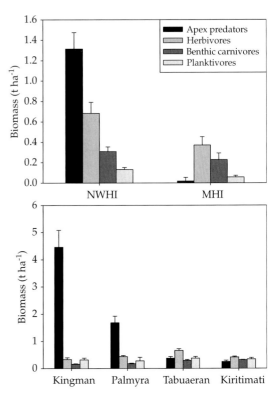

Figure 11.5   Trophic fish biomass (t ha⁻¹) A. Hawaiian Islands—NWHI = north-western Hawaiian Islands, MHI = main Hawaiian Islands (adapted from Friedlander and DeMartini 2002). B. Northern Line Islands (adapted from Sandin *et al.* 2008; DeMartini *et al.* 2010).

The effects of apex predation, primarily by giant trevally and sharks, are pervasive: they structure prey population sizes and age distributions, and strongly influence the reproductive and growth dynamics of harvestable fishes as well as smaller-bodied, lower-trophic-level fishes on shallow NWHI reefs (Friedlander and DeMartini 2002; Demartini and Friedlander 2006). Diversity based on trophic guilds is significantly higher (p < 0.001) in the NWHI compared with the MHI as a result of the near absence of apex predators in the MHI (Figure 11.6a, Friedlander unpublished data). The size distribution of prey reef fishes within the NWHI is strongly skewed toward smaller individuals within a species and smaller species overall with higher apex predator abundance (DeMartini *et al.* 2005). These changes in body size are likely to have important implications for the energetics of the entire ecosystem through higher turn-over rates and greater ecosystem efficiency (Hildrew *et al.* 2007). In contrast, the average size of prey is smaller on average in the MHI compared with the NWHI due to the high levels of fishing pressure on lower trophic levels in the MHI but maximum size within species is greater owing to the lack of natural predation on these smaller species (Figure 11.7a).

Apex predators can also influence the reproductive dynamics of prey taxa through size-selective predation. Protogynous (female-to-male sex-changing) labroid fishes (primarily parrotfishes), the adult sexes of which conspicuously differ in body coloration, are preferred prey of the giant trevally in Hawaii (Sudekum *et al.* 1991). The size at sex change in these labroids is inversely related to the density of giant trevally and other apex predators in Hawaii,

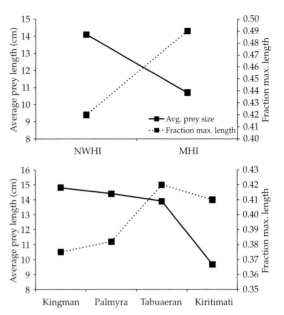

Figure 11.7 Fraction of maximum length within species and average length of prey species. A. Hawaiian Islands—NWHI = north-western Hawaiian Islands, MHI = main Hawaiian Islands. B. Northern Line Islands.

which strongly suggests higher mortality from predation at greater predator densities and plasticity in sex change that has implications for reproductive success (DeMartini *et al.* 2005).

The inverted trophic biomass pyramid observed in the NWHI may well be the natural state, but it contains the species that are most susceptible to, and so those rapidly removed by, human activities (Friedlander and DeMartini 2002). As a result of intensive fishing pressure in the MHI, these coral reefs are stressed and do not contain the full complement of species and interrelationships that would prevail in the absence of humans. Owing to the harvest of preferred food fish and desirable aquarium fish species, the only abundant fishes remaining on MHI coral reefs are often drab, small-bodied species of no ornamental and little food value.

11.5.2   Northern Line Islands

Virtually pristine and uninhabited Kingman Atoll in the northern Line Islands has an enormous fish biomass, and the fish assemblage is overwhelmingly

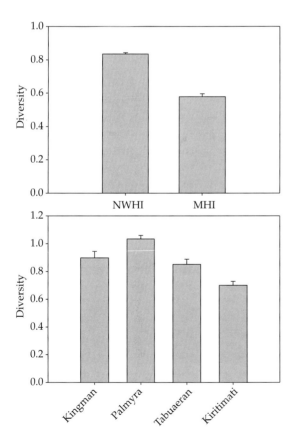

Figure 11.6 Shannon diversity based on trophic guilds for A. Hawaiian Islands (NWHI = north-western Hawaiian Islands, MHI = main Hawaiian Islands). B. Northern Line Islands. Values are means and SE.

dominated by large piscivores—mostly sharks up to > 2 m in length and snappers—(Demartini *et al.* 2008; Sandin *et al.* 2008). At the other end of the gradient, Kiritimati Atoll, which is the most populated island in the archipelago, biomass was five times lower than Kingman, and the assemblage was dominated by small-bodied species belonging to lower trophic levels, especially planktivores. Piscivores alone constituted 85% of the total biomass at Kingman and 65% at Palmyra, but contributed much less at Tabuaeran (21%) and Kiritimati (19%) most of which were snappers and groupers with a near total absence of sharks (Figure 11.6b). Accounts from eighteenth and nineteenth century explorers and whalers visiting Kiritimati specifically mention the great abundance of sharks and other large piscivores, but a recent and short-lived shark finning fishery has virtually extirpated all reef sharks from both Tabuaeran and Kiritimati (Sandin *et al.* 2008).

In contrast to predators, there was no clear difference in biomass of lower trophic levels among atolls in the northern Line Islands, although trophic diversity was lower at the fished atolls compared to the unfished ones (Figure 11.6b). Despite similarities in biomass of lower trophic levels, prey fishes showed a dramatic decline in abundance and body size from the most-exploited Kiritimati to the least-exploited Kingman. The intensive threat of predation appears to favour larger-bodied species within the prey assemblage of the Line Islands. The average size of prey is larger on average at unfished locations but within species sizes are smaller at these same locations (Figure 11.7b). Size and longevity of a top predator (*Lutjanus bohar*, red snapper) was lower at fished Kiritimati compared to unfished Palmyra (Ruttenberg *et al.* 2011). Demographic patterns also shifted dramatically for four of five fish species in lower trophic groups, opposite in direction to the top predator, including decreases in average size and longevity at unfished Palmyra relative to Kiritimati. The smaller size at sex change in parrotfishes at unfished Palmyra-Kingman compared to fished Tabuaeran -Kiritimati is consistent with the indirect response to increased piscivory by top predators in Hawaii (DeMartini *et al.* 2005). The size at maturation in parrotfishes was proportionally larger at fished locations and this has implications

for sex ratios, social dynamics, and reproductive success of this ecologically important family that is a major reef grazer and bio-eroder.

Despite relatively constant herbivore biomass among atolls, systematic shifts in the size-structure of herbivorous fishes (smallest at fished atolls) and the relative abundance of functional types (increase of territorial damselfish in the absence of predators) may have altered the patterns of herbivory at fished locations with implications for the entire benthic community. The release of algae from effective herbivory, along with increases in nutrient pollution, may have contributed to the higher macroalgal cover, lower coral cover, and increased microbial concentrations found at the atolls that experienced fishing (Sandin *et al.* 2008; Dinsdale *et al.* 2008; Madin *et al.* 2010a) showed that, on average, prey fishes on Palmyra's predator-rich reefs have smaller foraging excursions than do those on Kiritimati, but their overall aggregate feeding rates stayed the same. This led to more intensive grazing by herbivorous fishes in foraging areas with more predators compared with low-predator Kiritimati and Tabuaeran, where herbivores grazed more uniformly. Intensive grazing via behavioral responses provides space for corals to settle, and the higher coral cover at Kingman compared with Kirtimati may result from predators indirectly influencing the spatial heterogeneity and competitive balance among key benthic organisms, and in turn influence the spatial structure of the reef benthos (Madin *et al.* 2010b).

11.6 Implications for ecosystem function

The loss of apex predators impacts the entire ecosystem far beyond the absence of these megafauna alone. Large-bodied predators can influence reef productivity, both among prey fishes and more broadly across other functional groups. The lower growth rate of prey at fished locations suggests reduced fisheries productivity and population resilience. Exploitation of even a single keystone species, such as a top consumer, can destabilize ecosystems by decreasing redundancy and making them more susceptible to stressors such as climate change (Hughes *et al.* 2005). Top predators can regulate the structure of the entire community and have the potential to buffer some of the ecological effects that may occur from climate change (Sala 2006). Therefore intact ecosystems like the NWHI and the unfished northern Line Islands may be more resistant and resilient to stressors, including climate change, that can have a strong influence on overall biodiversity and ecosystem function.

Marine protected areas have been an effective management tool for conserving biodiversity in many locations (Claudet *et al.* 2008; Lester *et al.* 2009) but for the most part, their small size and limited habitat types do not allow for the entire fish assemblage to function in a natural manner compared to larger and relatively pristine areas such as the NWHI and northern Line Islands (Friedlander *et al.* 2007; Knowlton and Jackson 2008). Large intact ecosystems also allow us to observe global changes without the confounding influences of local impacts such as fishing and pollution (Knowlton and Jackson 208). From these two case studies, it is clear than the removal of top predator dramatically alters ecosystem function and biodiversity through both direct and indirect effects, and remote reference areas represent the few remaining baselines to understand natural ecosystem processes and help better protected human degraded reefs.

11.7 Conclusions

These two case studies clearly demonstrate the effects of biodiversity change on ecosystem functioning in seascape-scale systems. The effects are believable, not because of a statistically rigorous experimental design, but because the effect sizes are very large, and the altered biodiversity and ecological functioning are clearly outside the normal scales and patterns of variability or reference states. Given the weaker inferential power of observational studies, it is important to consider alternative explanations for the observed effects. A range of these have been considered over the history of the two case studies, often covariates of the long time scales over which the studies were done, and the most parsimonious and persuasive explanations for the observed changes in function remain the changes in biodiversity observed.

Where smaller changes in ecosystem functioning are observed, however ecologically significant these may be thought to be, it will be much harder to persuade others, including managers, that these are due to changes in biodiversity because potentially confounding factors cannot be easily dismissed. In other words, what may be socially unacceptable as a loss in functioning—e.g. a 10–20% change—may not be easy to demonstrate convincingly at these larger scales. Back-extrapolation of the observed—and convincing—large-scale effects in order to claim smaller effects of biodiversity change is not possible because the form of the biodiversity change-functioning relationship is unlikely to be linear. Abrupt changes in functioning may only occur for the greatest biodiversity changes because of redundancy in the system at higher biodiversities, so that it is only when the system is pushed beyond its resilience bounds that changes in functioning become apparent. Establishing how near a system is to those thresholds of rapid change, and thus how much biodiversity change is permissible, is an urgent area of research needed for the effective management of marine ecosystems. Naeem (Chapter 4) explores potential frameworks for assessing the location of the 'inflection point' in the typical saturation curves of biodiversity-ecosystem functioning relationships, which may represent critical levels of biodiversity for that system to continue to perform as normal. For both the Ythan and the Hawaii/NLI studies, an argument can be made for more of a qualitative regime shift occurring at critical changes of biodiversity, where the system continues to function but as a very different kind of system and in a very different way. This raises fundamental questions about management goals, societal choice, and what kind of system is desirable or acceptable, rather than whether the goal is to maximize a particular function and hence biodiversity set. Such choices will be profoundly influenced by stakeholder perceptions of what the system used to be like—the shifting baseline problem—and they may settle for the new system state.

The case studies also illustrate the need to embrace wider concepts of biodiversity and ecological functioning at these larger scales. The biodiversity and functioning metrics employed in small-scale experimental systems are not appropriate at the ecosystem scale, which makes translation from those experiments to larger scales problematic. Thus, species richness was not dramatically different between the impacted and reference states in either study. Rather, functional group diversity changed in the case of Hawaii/NLI as apex predators were removed, and there were only compositional changes in the case of the Ythan. Similarly, ecological functioning at the whole-system scale is better captured using system-level metrics, such as resilience and community production rather than specific processes such as nitrogen cycling or microbial film production.

In conclusion, we argue that high quality observational data can be used to explore biodiversity–ecosystem functioning at the seascape—and landscape—scale, despite their perceived weaker inferential power compared to highly controlled and replicated manipulative experiments (MacNeil 2008). In the present studies, the persuasiveness of biodiversity effects is the magnitude of the effect size and the robustness of our inferences to competing, alternative explanations. Nevertheless, where effect sizes are smaller and datasets are complex with potentially confounding variables that cannot be reasonably controlled for, it may still be possible to draw inferences which can be defended against alternative explanations using the powerful and complex statistical models and tools which are rapidly emerging in ecology, and where those working in the field of biodiversity and ecosystem functioning are playing a major role in their development and application (Zuur, Eno, and Smith 2007; Zuur et al. 2009).

Acknowledgements

We are grateful to the editors for asking us to explore this challenging area, to our colleagues past and present who have helped to make the two case studies so complete, and to anonymous reviewers for bringing to our attention literature and perspectives that have improved the chapter greatly.

References

Baird, D., and Milne, H. (1981) Energy flow in the Ythan estuary, Aberdeenshire, Scotland. *Est. Coast. Mar. Sci.* **13**: 217–32.

Baird, D. and Ulanowicz, R. E. (1993) Comparative study of the trophic structure, cycling and ecosystem properties of four tidal estuaries. *Marine Ecology Progress Series,* **99**, 221–37.

Balvanera, P., Pfisterer, A., Buchmann, N., He, J-S., Nakashizuka, T., Raffaelli, D. Schmid, B. (2006) Quantifying the evidence for biodiversity effects on ecosystem functioning and services. *Ecology Letters* **9**: 1–11

Bellwood, D. R., Hughes, T. P., Folke, C. et al. (2004) Confronting the coral reef crisis. *Nature* **429**:827–33.

Brown J. H., Gillooly J. F., Allen A. P. et al. (2004) Toward a metabolic theory of ecology. *Ecology* **85**, 1771–89

Buchan, R. G. (1997) A comparative study of two east coast Scottish estuaries, the Lower Forth and the Ythan, using the Ecopath trophic interaction model. MSc thesis, Napier University, Edinburgh.

Bulling, M. T., White, P. C. L. W., Raffaelli, D. G. et al. (2006) Using model systems to address the biodiversity-ecosystem functioning process. *Marine Ecology Progress Series* **311**: 295–309.

Buxton, N. E. and Young, C. M. (1981) The food of shelduck in North East Scotland. *Bird Study* **28**, 41–8.

Caliman, A., Pires, A. F., Esteves, F. A. et al. (2010) The prominence of and biases in biodiversity and ecosystem functioning research. *Biodiversity Conservation* **19**: 651–64.

Cardinale, B. J., Srivastava, D. S., Duffy, J. E. et al. (2006) Effects of biodiversity on the functioning of trophic groups and ecosystems. *Nature* **443**:989–92.

Chambers, M. and Milne, H. (1975a) The production of Macoma balthica (L.) in the Ythan estuary. *Est. Coast. Mar. Sci.* **3**: 443–55.

Chambers, M., and Milne, H. (1975b) Life cycle and production of Nereis diversicolor (O.F. Muller) in the Ythan estuary, Scotland. *Est. Coast. Mar. Sci.* **3**: 133–4.

Charles, C. and Sandin, S. A. (2009) Line Islands. In R.G. Gillespie and D.A. Clague (eds) The Encyclopedia of Islands, University of California Press, Berkeley, 553–8.

Christensen, V, Walters, C. J. and Pauly, D. (2005) Ecopath with Ecosim: a User's Guide, November 2005 Edition. Fisheries Centre, University of British Columbia, Vancouver, Canada.

Claudet, J., Osenberg, C. W., Benedetti-Cecchi, L. *et al.* (2008) Marine reserves: size and age do matter. *Ecology Letters* **11**:481–9.

Dayawansa, P. N. (1995) The distribution and foraging behaviour of wading birds on the Ythan estuary, Aberdeenshire, in relation to macroalgal mats, Ph.D. Thesis, University of Aberdeen.

DeMartini, E. E. and Friedlander, A. M. (2006) Predation, endemism, and related processes structuring shallow-water reef fish assemblages of the Northwestern Hawaiian Islands. *Atoll Research Bulletin* **543**:237–56.

DeMartini, E. E., Friedlander, A. M. and Holzwarth, S. (2005) Size at sex change in protogynous labroids, prey size distributions, and apex predator densities at NW Hawaiian atolls. *Marine Ecology Progress Ser*ies **297**:259–71.

DeMartini, E. E., Friedlander, A. M., Sandin, S. A. *et al.* (2008) Differences in fish-assemblage structure between fished and unfished atolls in the northern Line Islands, central Pacific. *Marine Ecology Progress Series* **365**: 199–215.

DeMartini, E. E., Anderson, T. W., Friedlander, A. M. (2010) Predator biomass, prey density, and species composition effects on group size in recruit coral reef fishes. *Mar. Biol.* **158**:2437–2447.

Denman K. L. (1994) Scale determining biological-physical interactions in oceanic food webs. In Giller, P.S., Hildrew, A.G. and Raffaelli, D.G. (eds), *Aquatic ecology. Scale, pattern and process. 34th Symposium of the British Ecological Society.* Blackwell Science, Oxford, 377–403.

Dinsdale, E. A., Pantos, O., Smriga, S. *et al.* (2008) Microbial ecology of four coral atolls in the Northern Line islands. *PLoS One* 3(2)e 1584.

Emmerson, M. and Huxham, M. (2001) How can marine ecology contribute to the biodiversity-ecosystem functioning debate? In: M. Loreau, S. Naeem, P. Inchausti (eds), *Biodiversity and Ecosystem Functio*ning, Oxford University Press, Oxford, 139–50.

Fischer, A. and Young, J.C. (2007) Understanding mental constructs of biodiversity: Implications for biodiversity management and conservation. *Biological Conservation* **136** (2). 271–82.

Friedlander, A. M., Brown, E. K. and Monaco, M. E. (2007) Coupling ecology and GIS to evaluate efficacy of marine protected areas in Hawaii. *Ecological Applications* **17**:715–30.

Friedlander, A. M., and DeMartini, E. E. (2002) Contrasts in density, size, and biomass of reef fishes between the northwestern and the main Hawaiian Islands: the effects of fishing down apex predators. *Marine Ecology Progress Series* **230**:253–64.

Gorman, M. L. and Raffaelli, D. G. (1993) Classic Sites—The Ythan Estuary. *The Biologist* **40**: 10–13.

Goss-Custard, J. (1966) The feeding ecology of the red-shank Tringa totanus (L.) in winter on the Ythan esturary, Aberdeenshire. PhD thesis, University of Aberdeen.

Goyal, A. R. (2005) The application of mass-balance models for evaluating the health of estuarine ecosystems. MSc thesis, University of York.

Greenstreet, S. (1986) Use of the intertidal habitat of the Ythan estuary, Aberdeenshire by the redshank Tringa totanus (L), Ph.D. thesis, University of Aberdeen.

Hall, S. J. and Raffaelli, D. G. (1991) Static patterns in food webs: lessons from a large web. *J. Anim. Ecol.* **63**: 823–42.

Healey, M.C. (1971) The distribution and abundance of sand gobies Gobius minutus in the Ythan estuary. Journal of Zoology (London), **163**, 117–29.

Hepplestone, P. B. (1968) An ecological study of the oyster-catcher (Haematopus ostralegus) in coastal and inland habitats of north-east Scotland. Ph.D Thesis, University of Aberdeen.

Hildrew, A., Raffaelli, D., Edmonds-Brown, R. (2007) *Body size: the structure and function of aquatic ecosystems.* Cambridge University Press. Cambridge.

Holling, C. S. (1992) Cross-scale morphology, geometry and dynamics of ecosystems. *Ecological Monographs* **62**(4):447–502.

Hughes, T. P., Bellwood, D. R., Folke, C. *et al.* (2005) New paradigms for supporting the resilience of marine ecosystems. *Trends in Ecology and Evolution* **20**: 380–6.

Hull, S. C. (1987) Macro-algal mats and species abundances: a field experiment, *Estuarine, Coastal and Shelf Science* **25**, 519–32.

Jackson, J. B. C., *et al.* (2001) Historical overfishing and the recent collapse of coastal ecosystems. *Science* **293**:629–38.

Jaquet, N. and Raffaelli, D. (1989) The ecological importance of sand gobies in an estuarine system. *J. Exp. Mar. Biol. Ecol.* **128**: 147–56.

Joffe, M. (1978) Factors affecting the numbers and distribution of waders on the Ythan estuary. Ph.D Thesis. University of Aberdeen.

Kay, J. J., Graham, L., and Ulanowicz, R. E. (1989) A Detailed Guide to Network Analysis. In: *Network Analysis in Marine Ecology: Methods and Applications.* Wulff, F., Field, J. and Mann, K. (eds), Berlin, Springer-Verlag.

Knowlton, N. and Jackson, J. B. C. (2008) Shifting baselines, local impacts, and global change on coral reefs. *PLoS Biology* **6**: e54, 6.

Larigauderie, A. and Mooney, H. A. (2010) The Inter-governmental science-policy Platform on Biodiversity and Ecosystem Services: moving a step closer to an IPCC-like mechanism for biodiversity. *Current Opinion in Environmental Sustainability (COSUST)* 2(1): 9–14.

Lester, S. E., Halpern, B. S., Grorud-Colvert, K. *et al.* (2009) Biological effects within no-take marine reserves: A global synthesis. *Marine Ecology Progress Series* **384**:33–46.

MacNeil, M. A. (2008) Making empirical progress in observational ecology. *Environmental Conservation* **35**: 193–6.

Madin, E. M. P., Gaines, S. D. and Warner, R. R. (2010) Field evidence for pervasive indirect effects of fishing on prey foraging behavior. *Ecology* **91**:3563–71.

Millennium Ecosystem Assessment. (2005) *Ecosystems and Human Well-being.* Synthesis Island Press, Washington DC.

Milne, H. and Dunnet, G. (1972) Standing crop, productivity and trophic relations of the fauna of the Ythan estuary. In *The Estuarine Environment*, R. Green and R. S. K. Barnes (eds), 56–106.

Palumbi, S. R., Sandifer, P. A., Allan, J. D. *et al.* (2009) Managing for ocean biodiversity to sustain marine ecosystem services. *Frontiers in Ecology and the Environment* **7**: 204–11.

Pandolfi, J., Jackson, J. B. C., Baron, N. *et al.* (2005) Are US coral reefs on the slippery slope to slime? *Science* **307**:1725–6.

Pauly, D., Christensen, V., Dalsgaaard, J. *et al.* (1998) Fishing down marine food webs. *Science* **279**:860–3.

Raffaelli, D. (1999) Nutrient enrichment and trophic organisation in an estuarine food web. *Acta Oecologia* **20**: 449–61.

Raffaelli, D. (2000) Interactions between macro-algal mats and invertebrates in the Ythan estuary, Aberdeenshire, Scotland. *Helgoland Meersunter* **54**: 71–9.

Raffaelli, D. (2006) Biodiversity and ecosystem functioning: issues of scale and trophic complexity. *Marine Ecology Progress Series* **311**: 285–94.

Raffaelli, D. (2011) Contemporary concepts and models on biodiversity and ecosystem function. In: *Treatise on Estuaries* (in press).

Raffaelli, D., Connacher, A., McLachlan, H. *et al.* (1989) The role of crustacean epibenthic predators in an estuarine food web. *Est. Coast. Shelf.Sci.* **28**: 149–60.

Raffaelli, D. and Frid C. J. (2010) The evolution of ecosystem ecology. In: *Ecosystem ecology: a new synthesis.* Raffaelli, D., Frid, C. J. (eds), Cambridge University Press, Cambridge, UK.

Raffaelli, D., and Hall, S. (1992) Compartments and predation in an estuarine food web. *J. Anim. Ecol.* **61**: 551–60.

Raffaelli, D., Hull, S. and Milne, H. (1989) Long-term changes in nutrients, weed mats and shorebirds in an estuarine system. *Cal. Biol. Mar.* **30**: 259–70.

Raffaelli, D. and H. Milne (1987) An experimental investigation of the effects of shorebird and flatfish predation on estuarine invertebrates. *Estuarine, Coastal and Shelf Science* **24**: 1–13.

Raffaelli, D. and H. Moller (2000) Manipulative experiments in animal ecology—do they promise more than they can deliver? *Adv. Ecol. Res.* **30**: 299–330.

Raffaelli, D., White, P. C. L., Renwick, A. *et al.* (2004) The Health of Ecosystems: the Ythan estuary case study. In *Handbook of Indicators for Assessment of Ecosystem Health.* Jørgensen, S. E., Costanza, R. and Xu, F-L. (eds), CRC Press, Florida, FLA, 379–394.

Raffaelli, D. G., Balls, P., Way, S. *et al.* (1999) Major changes in the ecology of the Ythan estuary, Aberdeenshire: how important are physical factors? *Aquatic Conservation* **9**: 219–36.

Rudd, M. A. (2011) Conservation science and its research impacts. *Conservation Biology* (in press).

Ruttenberg, B. I., Hamilton, S. L., Walsh, S. M. *et al.* (2011) Predator-Induced Demographic Shifts in Coral Reef Fish Assemblages. PLoS ONE 6(6):e21062.doi:10.1371/journal.pone.0021062.

Sala, E. (2006) Top predators provide insurance against climate change. *Trends in Ecology and Evolution* **21**:479–80.

Sandin, S. A., Smith, J. E. DeMartini, E. E. *et al.* (2008) Baselines and degradation of coral reefs in the northern Line Islands. *PLoS ONE* 3(2): e1548.

Snow, D. W. and Perrins, C. M. (1998) The Birds of the Western Palearctic Concise Edition. Oxford, Oxford University Press.

Sudekum, A. E., Parrish, J. D., Radtke, R. L. *et al.* (1991) Life history and ecology of large jacks in undisturbed, shallow oceanic communities. *Fishery Bulletin* **89**: 493–513.

Summers, R. W. (1974) The feeding ecology of the flounder Platichthys flesus in the Ythan estuary, Aberdeenshire. Ph.D Thesis, University of Aberdeen.

Summers, R. W. (1980) The diet and feeding behaviour of the flounder Platichthys flesus (L.) in the Ythan Estuary, Aberdeenshire, Scotland. *Estuarine and Coastal Shelf Science* **11**, 217–32.

Tubbs, C. R. and Tubbs, J. M. (1983) 'Macroalgal mats in Langstone Harbour, Hampshire, England. *Marine Pollution Bulletin* **14**, 148–9.

Ulanowicz, R. E. (1986) *Growth and Development: Ecosystem Phenomenology.* Springer-Verlag, New York.

Ulanowicz, R. E. and Kay, J. J. (1991) A package for the analysis of ecosystem flow networks. *Environmental Software* **6**, 131–42.

Wiegand, J., Raffaelli, D., Smart, J. C. R. *et al.* (2010) Assessment of temporal trends in ecosystem health using an holistic indicator. *Journal of Environmental Management* **91**: 1446–55.

Williams, I. D., Walsh, W. J., Schroeder, R. E. *et al.* (2008) Assessing the relative importance of fishing impacts on Hawaiian coral reef fish assemblages along regional-scale human population gradients. *Environmental Conservation* **35**:261–72.

Zuur, A. F., Eno, E. N. and Smith, G. M. (2007) *Analysing Ecological Data.* Springer, New York.

Zuur, A. F., Eno, E. N., Walker, N. J. *et al.* (2009) *Mixed Effects Models and Extensions in Ecology with R.* Springer, New York.

CHAPTER 12

Multitrophic biodiversity and the responses of marine ecosystems to global change

J. Emmett Duffy, John J. Stachowicz, and John F. Bruno

12.1 Introduction

The world is rapidly changing, both physically and biologically. On a global level, the major drivers of biodiversity change include habitat loss and conversion, overexploitation, pollution, and species invasion (Purvis *et al.* 2000; Sala and Knowlton 2006). Climate warming and ocean acidification due to fossil fuel combustion are now also well-recognized drivers of change in both the abiotic environment and in the biological composition of marine communities (Hoegh-Guldberg and Bruno 2010; Doney *et al.* 2011). The impacts of these drivers are felt in even the most remote regions of the central Pacific and Southern Ocean (Halpern *et al.* 2008) (Figure 12.1). Many of these stressors will likely affect the relationships between biodiversity and ecosystem functioning (BEF) both directly and indirectly by altering habitat structure, individual physiology, and interactions among species. In this chapter, we focus primarily on whether and how global change stressors are likely to change biodiversity—which we consider to include the number, composition, and trait distribution of species in communities—and how those changes might cascade to affect ecosystem processes indirectly. More specifically, we focus on how changing composition and diversity in multilevel food webs affects ecological and biogeochemical processes, with emphasis on benthic and pelagic marine ecosystems. We refer to these as multitrophic systems to distinguish them from the simpler competitive assemblages of species (typically plants) at a single trophic level

that have been the subject of most previous BEF research.

Heightened awareness that species worldwide are declining or disappearing, and that these losses may compromise human well-being, stimulated a concerted research effort in the 1990s to understand how changing biodiversity might affect functioning of ecosystems (BEF) (e.g. Schulze and Mooney 1994). Research on this topic has had two partially overlapping but distinct objectives: (1) understanding the fundamental patterns and mechanisms that mediate ecosystem processes in communities of multiple interacting species, and (2) understanding how functioning of complex natural ecosystems is likely to change as species are lost due to human impacts. While these basic and applied objectives are related, they entail somewhat different approaches and different sets of corollary questions, and conflating the two motivations has sometimes led to confusion.

Bringing academic BEF research to bear on goals of applied conservation and management—addressing the second of these two objectives—involves the empirical questions of how biodiversity is actually changing—and how it is expected to change—in response to human stressors, and how that change translates to affect functioning of complex natural ecosystems. This is turn involves understanding what types of species are most affected by particular stressors, how communities adjust to decline or loss of those species, and how the altered community structure translates to affect ecosystem-level processes (Lavorel and Garnier

Marine Biodiversity and Ecosystem Functioning. First Edition. Edited by Martin Solan, Rebecca J. Aspden, and David M. Paterson.
© Oxford University Press 2012. Published 2012 by Oxford University Press.

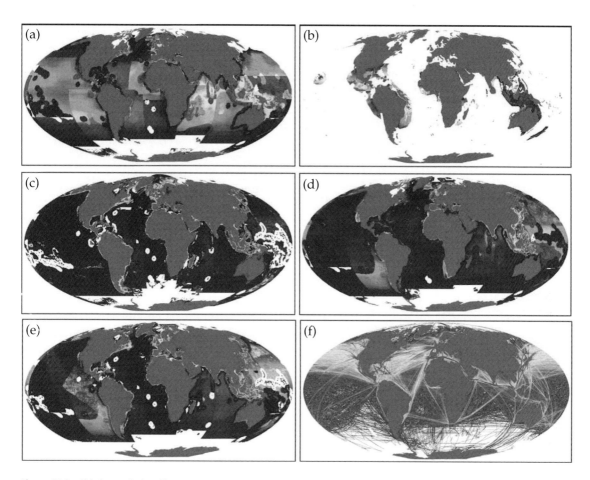

Figure 12.1   Global maps of selected human impacts on marine ecosystems. (a) pelagic fishing resulting in low bycatch (collateral damage to non-target species); (b) pelagic, high-bycatch fishing (taking substantial numbers of non-target species); (c) demersal (bottom) fishing that modifies habitat (such as reefs and seagrass beds); (d) demersal non-habitat modifying, low-bycatch fishing; (e) demersal non-habitat-modifying, high-bycatch fishing; (f) shipping. Blue and red denote low and high impacts, respectively, while white denotes no data. From Halpern *et al.* (2008)). See also Plate 5.

2002; Naeem and Wright 2003; Duffy *et al.* 2009). Answering these questions will benefit from integrating the primarily academic theme with the more applied theme of BEF research.

In this chapter, we explore how changing biodiversity influences the functioning of multitrophic ocean and estuarine ecosystems, particularly in the context of ongoing global environmental change. This involves four components. First, we sketch a brief summary of what is known about how diversity is changing in the sea. Second, to address how such changes might affect ecosystem functioning, we review experimental and observational research on links between biodiversity and ecosystem functioning in multitrophic systems. Third, we discuss several methodological and philosophical issues at the heart of transferring BEF academic research to applied conservation problems. Finally, we attempt to draw these lines of work together by asking how the conclusions emerging from BEF research might help us understand and predict consequences of global environmental change, notably including climate warming, ocean acidification, widespread invasions, and sustained intense harvesting of marine organisms. We close with thoughts on future research aimed at closing the gaps.

12.2 How and why biodiversity is changing in oceans and estuaries

Despite the plethora of different stressors affecting ecosystems worldwide, and regional variation in their importance, available evidence nevertheless suggests that human impacts tend to produce characteristic patterns in the structure of marine communities, which generally shift under intense human influence toward lower average trophic level, smaller body size, faster growth rates, and 'weedier' phenotypes. This conclusion comes from several lines of evidence, as follows.

Effects of fishing on biodiversity. Historically, fishing has represented the strongest and most consistent human impact on marine communities. In general, harvesting of marine animals targets, and most strongly affects, large animals at high trophic levels (Pauly *et al.* 1998; Jennings *et al.* 1999a; Jackson *et al.* 2001), although there are prominent exceptions (Essington *et al.* 2006). Decades of this size-selective fishing have resulted in declines in both abundance and body size of marine predators (Baum *et al.* 2003; Myers and Worm 2003; Hsieh *et al.* 2006; Myers *et al.* 2007). Quantitative analyses of such trends at the community level show that, from the North Sea to tropical coral reefs, sustained fishing has also shifted demersal fish communities to smaller, faster-maturing species that feed low in the food-web at the expense of larger, slower-growing predators (Jennings *et al.* 1999a; Jennings *et al.* 1999b). Using data from global tallies of marine extinctions (Dulvy *et al.* 2003), Byrnes *et al.* (2007) confirmed that 70% of documented global and local extinctions were of predators and higher-level omnivores. Thus, trophic skew appears to be a common result of fishing in marine systems.

Climate change, invasions, and trophic interactions. Marine communities have also been strongly altered by exotic invasions, which have accelerated in recent decades on a global scale (Ruiz *et al.* 2000). While many invasions are mediated by globalization of human commerce and associated shipping, climate warming is an increasingly important driver, most notably causing poleward shifts in geographic ranges of many marine species (Murawski 1993; Perry *et al.* 2005). Bioclimate envelope models predict that mean invasion intensity—the number of invasions relative to current species richness—may reach a global average 55%, with invasion concentrated at high latitudes (Cheung *et al.* 2009). These changes are already underway: in the North Sea, for example, fish species richness increased by roughly 50% between 1986 and 2005 in concert with rising temperatures, with a general trend toward greater representation of small-bodied species of southerly origin (Hiddink and ter Hofstede 2008). Quantitative analysis of combined extinction and invasion data from four well-studied temperate areas revealed that most invasions involved filter-feeding and scavenging invertebrates at lower trophic levels, exacerbating the trophic skew caused by predator extinctions (Byrnes *et al.* 2007).

An important question from the perspective of multitrophic BEF relationships is whether large-scale invasion patterns might influence trophic levels differently, thus changing average strength of trophic interactions. Evidence from morphology of predators and prey organisms, chemical defenses, and experiments indicate that, on average, consumer pressure tends to be stronger at low latitudes (Vermeij 1978; Bertness *et al.* 1981; Bolser and Hay 1996; Pennings and Silliman 2005; Ruzicka and Gleason 2008), although consumer impacts on prey may paradoxically be stronger at high latitudes, possibly because of lower diversity there (Frank *et al.* 2006). Thus, one possible outcome of poleward range extensions of lower-latitude species might be a strengthening of top-down control, even if consumer and prey species shift poleward at equal rates. One possibly relevant example involves a sharp temperature drop during the Eocene that eliminated shell-breaking sharks, rays, bony fishes, and decapod crustaceans from the Antarctic nearshore food-web; following these extinctions, dense assemblages of suspension-feeding echinoderm prey rapidly developed in the enemy-free space (Aronson *et al.* 2009). Aronson *et al.* (2009) speculate that current warming in polar regions will most likely allow shell-breaking crabs and fishes to return to the Antarctic Peninsula. If so, it remains an open question how these well-endowed predators might interact with the concurrent invasions of lower-level prey species expected from patterns documented elsewhere (Byrnes *et al.* 2007).

Pollution effects on biodiversity. Pollution from numerous sources has direct and indirect effects on species, but it is less clear whether such effects might differ systematically among trophic levels. For example, eutrophication-induced algal blooms can produce extensive hypoxia in bottom waters (Diaz and Rosenberg 2008). To the extent that large mobile animals are more sensitive to low oxygen it is possible that such events may cause disproportionate declines among predators. On the other hand, these animals can escape severe anoxic events by migrating to more favourable areas, whereas more sedentary animals—often at lower trophic levels—may be killed. Thus, the landscape-scale effects of such pollution on diversity and trophic interactions are unclear. Similarly, some forms of chemical contaminants increase in tissue concentrations of species at higher trophic levels. This bioaccumulation could disproportionately affect top predators, as has been seen for effects of DDT and several other pesticides. In general, most forms of pollution are likely to decrease community diversity, though the extent to which that decrease has a trophic level bias may depend on the type of pollution and severity of the event.

The question of how biodiversity influences ecosystem functioning has been addressed mainly by assembling experimental communities that differed in species richness, and comparing how processes such as productivity and resource use varied among these assemblages (Naeem *et al.* 1994; Kinzig *et al.* 2001; Loreau *et al.* 2001; Loreau *et al.* 2002; Hooper *et al.* 2005). Such experiments are inherently complex, and so a great deal of attention in the early years focused on details of appropriately designing, analysing, and interpreting such experiments (Huston 1997; Allison 1999; Loreau and Hector 2001). A primary focus was on rigorously separating effects of changing species richness per se from effects of changes in the identity of species present. Another major effort sought to tease out the mechanisms responsible for richness effects on functioning, which were divided into two major classes (Stachowicz *et al.* 2007): (1) *sampling effects,*

meaning the greater statistical probability of including a species with a dominant effect in an assemblage as species richness increases—i.e. as more species from the pool are 'sampled'—and (2) complementarity, the greater performance of a species in mixture than expected from its performance in monoculture, caused by interactions such as resource partitioning or facilitation among species. The intense focus on experimental design was critical in resolving criticisms of early BEF research, and in identifying the mechanisms mediating biodiversity effects (Schmid *et al.* 2002). But it also resulted in the main body of BEF research diverging from the original motivation of addressing conservation-related questions (Schwartz *et al.* 2000; Srivastava 2002; Duffy 2003; Srivastava and Vellend 2005).

One way in which early BEF research diverged from conservation-related concerns was in the focus on single trophic levels. Trophic interactions are a central feature of real-world ecological complexity, particularly in ocean and aquatic ecosystems (Duffy 2002; Raffaelli *et al.* 2002). Although several pioneering BEF experiments were conducted in systems with multiple trophic levels (Naeem *et al.* 1994; McGrady-Steed *et al.* 1997; Naeem and Li 1998), the goal of identifying mechanisms linking biodiversity to ecosystem functioning understandably concentrated on the simpler interactions within trophic levels, and the field was soon dominated by studies of interactions among plants in terrestrial grasslands (but see Mulder *et al.* 1999). In marine systems, however, incorporation of food-web interactions into BEF is especially important for two reasons. First, trophic interactions tend to be stronger in aquatic ecosystems than on land (Cyr and Pace 1993; Shurin *et al.* 2006; Cebrian *et al.* 2009) such that understanding consequences of marine biodiversity loss inherently requires a food-web context. Second, from a more specifically applied perspective, intense and widespread fishing tends to change systematically the trophic structure of marine communities, typically reducing abundance and diversity of top predators (Pauly *et al.* 1998; Byrnes *et al.* 2007; Duffy 2010). Thus, a characteristic aspect of diversity change in human-influenced marine systems is trophic skew (Duffy 2003)—

compaction in the average length of food chains. Consequently, a dominant pattern in erosion of marine biodiversity is changing trophic structure, and altered top-down control should have significant effects on both structure and functioning of marine ecosystems.

What if anything can 15 years of BEF research tell us about the likely consequences of ongoing changes in species and trait distribution for marine ecosystems? The most general conclusions from experimental BEF research, as shown by recent meta-analyses of diversity manipulations within trophic levels, is that declining diversity tends on average to reduce both productivity and resource use by an assemblage (Balvanera *et al.* 2006; Cardinale *et al.* 2006; Cardinale *et al.* 2011). While some of the patterns and mechanisms by which diversity acts within trophic levels translate simply to multitrophic level situations, interactions involving mobile heterotrophs are generally more varied and consideration of multiple trophic levels introduces a broad range of new complications (Thebault and Loreau 2003, 2005; Duffy *et al.* 2007; Bruno and Cardinale 2008). For example, a substantial body of experiments have explored Multiple Predator Effects (Sih *et al.* 1998) in a variety of communities including benthic marine systems (e.g. Martin *et al.* 1989; Crowder *et al.* 1997; Hixon and Carr 1997), and results have identified many mechanisms by which changing predator diversity can cascade to affect lower levels (Figure 12.2). Increasing predator diversity can either reduce abundance and impacts of their herbivore prey, via diet complementarity or facilitation among predators, or increase herbivory via intraguild interference or omnivory (arrows 3 and 7 in Figure 12.2). These processes will have positive and negative effects on plant biomass, respectively.

Over the last several years, a growing number of studies, particularly in marine systems, have addressed multitrophic aspects of biodiversity. This work has been synthesized in several reviews (Duffy *et al.* 2007; Stachowicz *et al.* 2007; Bruno and Cardinale 2008; Cardinale *et al.* 2009) and quantitative meta-analyses (Table 12.1). Nevertheless, the number of BEF experiments conducted with multiple trophic levels is considerably less than

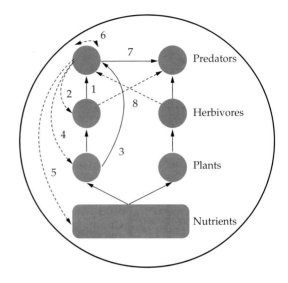

Figure 12.2 Complexity of potential effects of predators on communities and ecosystems. Solid and dashed lines show consumptive and nonconsumptive effects, respectively. (1) Direct consumption. (2) 'trait-mediated', nonlethal interactions that influence behavior or demographics of the affected species. (3) Omnivory, i.e. feeding on a non-adjacent trophic level. (4) trophic cascades, in which a consumer indirectly affects abundance or processes at more than one level below it. (5) nutrient recycling, by which a consumer returns inorganic nutrients to non-adjacent levels. (6) intraspecific interference, which may influence feeding rates. (7) intraguild predation, i.e. feeding of consumers on other individuals at the same trophic level (cannibalism is a special case). (8) competition. After Leroux and Loreau (2009), used with permission from John Wiley and Sons, and the British Ecological Society.

those for single trophic levels, limiting the power of formal meta-analyses. Moreover, many come from a small number of investigators and study systems, potentially raising issues of non-independence, so caution is in order in interpreting several of the results. We summarize the broad patterns emerging from this work in this section.

Biodiversity generally enhances production and resource use consistently across trophic levels. This conclusion now appears robust based on several meta-analyses. Cardinale *et al.* (2006) analysed data from 111 experimental manipulations of diversity spanning a range of trophic groups and habitats, and found that, on average, declining species richness decreased standing stocks of, and resource use by, the focal trophic group. Importantly, they found no significant difference in these patterns across trophic groups—including plants, herbivores decompos-

Table 12.1 A summary of recent quantitative meta-analyses that have analyzed studies exploring links between biodiversity and ecosystem functioning in multi-trophic systems.

Study	Questions	Organism and systems	Methods	Findings
Cardinale et al. 2006	Does diversity increase standing stock and resource use at a focal trophic level?	All organisms and systems	111 experiments manipulating species richness of organisms and measuring their abundance and resource depletion	Increasing species richness increased standing stock and resource use by focal trophic group, on average.
Worm et al. 2006	Does prey diversity increase fitness of consumers?	Marine primary producers	14 experiments manipulating species richness of marine primary producers and measuring performance of consumers	Diverse mixtures outperformed average monocultures for a variety of metrics, but data were insufficient to assess mechanism.
Schmid et al. 2009	Does changing diversity at lower trophic levels consistently affect functioning at higher trophic levels?	All organisms and systems	395 experimental and observational studies in which ecosystem responses were measured at higher trophic levels in response to changing diversity at a focal level	Effects of diversity were positive within trophic levels and two levels higher, but negative at the next level above the focal level manipulated.
Hillebrand and Cardinale 2004	Does prey (algal) diversity reduce impacts of grazers?	Aquatic algae (periphyton) and grazers	172 consumer-removal experiments; comparison of consumer effect size on algae across prey richness gradient.	Prey (algal) species richness was negatively related to herbivore impact, on average.
Edwards et al. 2010	Does prey (algal and sessile invertebrate) diversity reduce impacts of consumers on marine hard substrata?	Marine consumers of algae and sessile invertebrates on hard substrata	57 consumer removal experiments compared across prey richness gradient on marine hard substrata	Prey species richness was negatively related to consumer impact, on average, and diversity had a stronger effect than habitat or prey type.
Cardinale et al. 2009	Are plant diversity effects altered by ptresence of herbivores?	Plants and herbivores, all systems	143 experiments manipulating plant richness and measuring biomass production	Presence of herbivores had no consistent effect on plant diversity-productivity relationships.
Srivastava et al. 2009	Do changes in consumer diversity have similar effects as changes in resource diversity on detritus consumption?	Aquatic and terrestrial consumers of detritus	52 experiments that manipulated diversity of either detritus (dead plant matter) or detritivores (bacteria, fungi, macro-invertebrates)	Diversity had stronger top-down than bottom-up effects on decomposition.
Jiang and Pu (2009)	Does diversity reduce temporal variability in single-and multitrophic assemblages?	All organisms and systems	25 multitrophic, 27 single-trophic experimental and observational studies that measured temporal variability in community and/or population characteristics	Diversity reduced temporal variability in both population and community dynamics in multi-trophic but had no significant effect in single-trophic assemblages.

ers, and carnivores—nor between aquatic and ter-restrial experiments. Thus, the effects of diversity within trophic levels on basic processes of resource use and growth appear very general and funda-mental in ecological systems. This conclusion is supported by a more recent analysis based on an updated dataset (Cardinale *et al.* 2011), and by a similar meta-analysis (Balvanera *et al.* 2006) that included a broader range of observational studies and experimental designs. The mechanisms gen-erating these positive relationships between diver-sity and production are less well understood and probably involve both sampling effects and com-plementarity. Although meta-analysis found that production of diverse assemblages rarely exceeded that of the most productive single species (Cardinale *et al.* 2006), as shown in in Figure 12.3a, a careful reanalysis of grassland experiments con-cluded that this general pattern often masks an interaction between species dominance—sam-pling—and complementarity—niche partitioning and/or facilitation—among species (Cardinale *et al.* 2007).

Increasing prey diversity strengthens bottom-up control by enhancing consumer nutrition and fitness. Is the greater production in diverse assemblages discussed above transmitted up the food chain? Several marine experiments have found that mixed algal diets enhanced herbivore growth, biomass accumulation, and/or reproductive output com-

pared to average algal monocultures in nearly all cases examined for taxa as diverse as protozoa, crustaceans, and sea urchins (Worm *et al.* 2006); this result was supported by a recent comprehensive meta-analysis of experiments from marine, fresh-water, and terrestrial systems (Lefcheck *et al.* sub-mitted). An intriguingly similar pattern was found in a field survey across 80 lakes in Connecticut, USA (Olson *et al.* 2007), where prey species diversity was strongly positively correlated with growth rates of largemouth bass (*Micropterus salmoides*), a result that could not be explained by total fish abundance, predator abundance, or productivity. And in an ear-lier survey of North American lakes, total fish bio-mass was positively related to fish diversity whereas population biomass of individual species was lower in diverse than in single-species assemblages (Carlander 1955). The mechanisms underlying these effects are unclear: diverse diets could be better because of the provision of complementary nutri-ents, dilution of defensive chemicals, or both (Bernays *et al.* 1994; DeMott 1998). In some cases, diversity effects emerge most strongly when con-sidering integrated measures of fitness that include growth, survival, and fecundity, as different prey species may maximize each component, leading to highest consumer fitness only in mixed diets (Cruz-Rivera and Hay 2000; Dam and Lopes 2003), but meta-analysis did not support such synergistic effects as a general trend (Lefcheck *et al.* submitted)

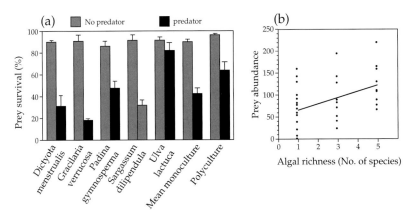

Figure 12.3 Effect of macroalgal identity and richness on carnivore-prey interactions. *Left*: Results of a mesocosm experiment show that amphipod survival in the presence of predators was higher, on average, in algal polycultures than monocultures. *Right*: Prey abundance increased with macroalgal richness in a field experiment. From Moran *et al.* (2010), used by permission of Inter-Research.

This notion that increasing the number of responses measured enhances the strength or likelihood of finding diversity effects has analogies in studies within trophic levels in which different species control different ecosystem processes; this has alternately been called multivariate dominance (Duffy *et al.* 2003), multivariate complementarity (Bracken and Stachowicz 2006), and multifunctionality (Hector and Bagchi 2007; Isbell *et al.* 2011). Intriguingly, there is also limited evidence that trophic transfer can be enhanced by interaction of both prey and consumer diversity: Gamfeldt *et al.* (2005) found in a microbial system that total consumer biomass was maximized when both prey and consumer diversity were highest.

Another mechanism by which plant—or sessile invertebrate—diversity may influence trophic transfer is via provision of habitat. It is well documented in a variety of systems that animal diversity is correlated with the diversity of plants that generate physical habitat structure (Murdoch *et al.* 1972; Southwood *et al.* 1979; Lawton *et al.* 1998a; Siemann 1998; Haddad *et al.* 2009). Recent experiments in a temperate subtidal system explored how basal species may affect upper levels through provision of habitat that mediates carnivore–herbivore interactions (Figure 12.3). Mesocosm results showed that increasing macroalgal richness strongly reduced the consumption of amphipods by a carnivorous fish, and this interpretation was consistent with findings from the field experiment that prey density was positively related to prey diversity (Moran *et al.* 2010). The enhancement of herbivore abundance in more diverse algal assemblages may derive from greater habitat complexity (Price *et al.* 1980), which can influence foraging efficiency of predators (Heck and Wetstone 1977; Diehl 1992; Bruno and Bertness 2001). Indeed, in the short term—and probably also the long term—such refuge effects on trophic transfer may well outweigh the purely nutritional effects of prey diversity mediated by diet mixing.

Increasing prey diversity tends to reduce the strength of top-down control. The last section reviewed evidence that prey diversity can enhance performance of upper trophic levels. The greater variety of traits in diverse prey assemblages may also reduce the effect of consumers on aggregate prey biomass due to (1) increased probability of including unpalatable species at high diversity; (2) higher productivity in diverse systems potentially compensating for losses due to consumption; and/or (3) unpalatable prey species providing associational refuges, i.e. microsites safe from consumers, that increase the persistence of palatable prey at high diversity (Duffy *et al.* 2007). Few experiments have explicitly tested effects of prey diversity on consumer control in marine systems, and although several have found that prey diversity can dampen ecosystem responses to top-down control, a recent meta-analysis of experiments manipulating prey diversity found little support for such effects on average (Cardinale *et al.* 2011).

Much of the evidence in support of this hypothesis has come from comparative studies across natural gradients in diversity of prey—broadly defined to include primary producers. Two studies in particular exploited the long tradition of consumer removal experiments that have been performed in a variety of habitats and under a range of ambient prey diversities (Figure 12.4). Hillebrand and Cardinale (2004) examined how impacts of aquatic grazers on periphyton (algal) biomass varied with algal diversity by amassing 87 field and lab experiments that manipulated grazer presence, while also measuring ambient algal diversity. They found that the strength of herbivore impacts on algae declined substantially with increasing diversity of the algal assemblage. Similarly, Edwards *et al.* (2010) examined effects of 47 consumer exclusion experiments on sessile invertebrate and seaweed prey on marine hard substrata. They found that richness of the prey assemblage was a strong and significant predictor of the overall impact of consumers on prey abundance across experiments, whereas habitat and prey type—algae versus invertebrate—had no significant influence on consumer impact. Total abundance of the prey assemblage was positively related to prey richness in general, consistent with high productivity in diverse assemblages, but this relationship became stronger when consumers were present, suggesting that top-down control of aggregate prey abundance is stronger in species-poor than species-rich prey communities, consistent with theory. In particular, studies with

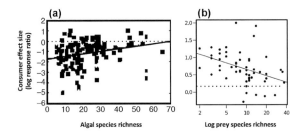

Figure 12.4 Prey diversity reduces top-down impact of consumers. Figures show change in consumer effect size (log ratio of treatment with vs without consumers) as a function of diversity of algae (left, Hillebrand and Cardinale 2004) or of sessile marine algae and invertebrates (right, Edwards *et al.* 2010) from meta-analyses. Note that consumer impacts are expressed as negative values in left panel and positive values in right panel, following the original publications, but in both cases consumer impacts are stronger (log response ratio farther from zero) at low prey diversity. Used with permission from John Wiley and Sons.

high prey diversity commonly found that, in the presence of consumers, susceptible prey species were replaced by consumer-resistant species, leading to large changes in species composition but no change in aggregate prey abundance, consistent with the first hypothesis above. In contrast, depauperate communities often lacked resistant prey species, leading to large changes in both composition and total prey abundance. It seems likely that several of the three mechanisms mentioned above might be at work in producing these patterns.

Finally, a global analysis of fisheries catch data has explored the question of how fish diversity affects vulnerability to the ultimate top predator, humans. Worm *et al.* (2006) analyzed relationships between species richness and fishery production in 64 'Large Marine Ecosystem' regions varying naturally in diversity—FAO, <http:www.fishbase.org>. Regions with naturally low diversity showed lower fishery catches, more frequent 'collapses'—strong reductions in fishery yield—and lower resilience—degree of recovery after overfishing—than naturally species-rich regions. The greater resilience of more diverse regions may result because fishers can switch more readily among target species when there are many species available—high richness—potentially providing overfished taxa with a chance to recover. This mechanism is consistent with theory, small-scale experiments, meta-analyses on the effects of prey richness on predator control of aggregate prey abundance (e.g. Hillebrand and Cardinale 2004; Edwards *et al.* 2010) (Figure 12.4), and with

the negative relationship Worm *et al.* (2006) found between fished taxa richness and variation in catch from year to year. The correlative approach employed in their comparison did not allow assignment of causation or mechanisms, and in fact these results must be interpreted cautiously because such fishery data are well-known to be affected by changes in gear use and markets, among other influences (Longhurst 2007). Nevertheless, such analyses do allow exploration of larger scale processes than is possible in small controlled experiments, and of potential connections between diversity and ecosystem services—fish production—with clear value to humans. Other cross-system comparative analyses support the idea that at high diversity compensatory population dynamics among predator species contributes to a greater stability of the predator community in the face of intense harvesting, and that this diversity is both a partial cause and a consequence of high primary production and biomass (Frank *et al.* 2006). Of course, such spatial comparisons are generally confounded by strong correlations of species richness with latitude and temperature, which can influence the strength of top-down control through effects on demographic rates (Frank *et al.* 2006; O'Connor *et al.* 2009), as well as with large-scale geographic trends in organismal traits relevant to productivity and resilience, such as body size (Fisher *et al.* 2010). Disentangling these influences remains a serious challenge of inferring mechanism from patterns. Finally, among the most extreme manifestations of altered stability are regime shifts to alternate semi-stable states (Scheffer *et al.* 2001). There is intriguing, if limited, evidence that biodiversity may partially mediate such shifts. For example, rapid transitions between ecosystem states are best documented in low diversity aquatic systems, such as temperate lakes (Scheffer and Jeppesen 2007), the Black Sea (Daskalov *et al.* 2007), and the North Atlantic (Frank *et al.* 2007). In particular, pelagic 'wasp-waist' ecosystems, in which the middle trophic level is dominated by one or a few species, appear especially prone to shifts between alternate community states, and this vulnerability has been linked to the low resilience of the intermediate trophic level dominated by one or a few species (Bakun 2006).

Diversity effects at one trophic level do not appear to be altered in any consistent way by presence of a higher level. Relatively few studies have addressed explicitly how interaction with adjacent trophic levels might alter diversity effects, and the results to date appear idiosyncratic. For example, the strength of diversity effects at a focal trophic level can be increased by the presence of consumers feeding on species at the focal level (Duffy *et al.* 2005; Hughes *et al.* 2010), reduced by such consumers (Mulder *et al.* 1999; Bruno *et al.* 2008), unaffected by consumers (Fox 2004; Douglass *et al.* 2008), or diversity effects can even change in sign in the presence of a consumer (Hattenschwiler *et al.* 2005; Wojdak 2005). To supplement the scarcity of such direct tests, the question of how higher consumers affect BEF relationships has also been approached indirectly. Cardinale *et al.* (2009) collated data on experiments that manipulated primary producer richness, and divided them into those that allowed herbivore access to experimental plots, versus those without herbivore access. They then compared the influence of plant richness on plant biomass in the two types of studies. Although plant species richness consistently increased plant biomass production, there was no evidence that herbivores altered the average magnitude of these plant diversity effects in either aquatic or terrestrial ecosystems. Thus, although data remain sparse, there is little support for the hypothesis that higher trophic levels consistently change the effects of diversity within prey communities in a consistent direction. This may however result from averaging the various opposing effects mentioned above, highlighting the need for more detailed study of context.

Predator diversity effects on the strength of trophic cascades are mixed. A topic of long-standing interest is whether species-poor systems show stronger top-down control generally, and trophic cascades specifically (Strong 1992; Schmitz *et al.* 2000; Duffy 2002). Experiments show that more diverse predator assemblages can either enhance positive indirect effects on plants (arrow 4 in Figure 12.2), where predators have complementary feeding preferences or modes that enhance the total risk of predation experienced by prey (herbivores), or alternatively, increased predator diversity can reduce indirect effects on plants when predators interfere with or eat one another (arrow 7 in Figure 12.2) (Sih *et al.* 1998; Casula *et al.* 2006), thus releasing herbivores from suppression. Experiments have confirmed that increasing diversity of species that are strict consumers of herbivores (i.e. no intraguild predation or omnivory, arrows 3 and 7 in Figure 2) can indirectly increase plant biomass in salt marsh (Finke and Denno 2005) and subtidal algal ecosystems (Bruno and O'Connor 2005; Byrnes *et al.* 2006). In most cases, this effect occurred because the presence of multiple predator species decreased herbivore activity or per capita feeding rate rather than by reducing herbivore density. Positive field correlations between predator diversity and plant biomass (Byrnes *et al.* 2006) suggest that these mechanisms also likely operate in natural systems. In contrast, experiments in a subtidal macroalgal system (Bruno and O'Connor 2005) and salt marsh (Finke and Denno 2005) found that when omnivores and intraguild predators were included in the most diverse predator communities, high predator diversity led to lower, not higher, plant biomass due to omnivory and intraguild predation. Such complex trophic interactions are a hallmark of even very simple natural ecosystems (Figure 12.2), and can reverse expected diversity effects based on niche partitioning and facilitation. The disparity between these results of studies in which omnivores were present or absent seems at odds with the consistency of patterns across trophic levels found in meta-analyses (Cardinale *et al.* 2006), and suggests that the average effects found in meta-analysis mask interesting and important details of how diversity influences functioning in multi-trophic communities.

Top-down effects of diversity are stronger than bottom-up effects in detritus-based systems. As summarized above, the generally positive effects of diversity within an assemblage on consumption of its own resources can enhance top-down control. But some data also show that diversity at a given trophic level tends to reduce its vulnerability to its own higher predators, which dampens top-down control. What is the relative importance of these opposing effects of diversity? Sufficient experimental data to answer this question rigorously are only available from detritus-based systems. A recent meta-analysis based on 90 observations from 28 studies in a range of habitats (Srivastava *et al.* 2009) asked: How do changes in consumer diversity compare with changes in resource diversity in influencing detritus consumption? The authors found that consumer diversity had stronger effects on the interaction than did resource diversity: on average, declining detritivore diversity reduced both rates of detritus consumption and standing stocks of detritivores. In contrast, detrital diversity—reflecting diversity of source plant species—had no effect on resource depletion rate and equivocal effects on detritivore stocks. The contrast between these results and the generally positive effects of plant diversity on bottom-up processes summarized above suggests either that effects of plant diversity on consumption change after plant death, or that the dynamic responses of living plants, are important in mediating effects of diversity.

Biodiversity stabilizes community and population dynamics more strongly in multi- than single-trophic level systems. A central question in BEF research has been whether and how diversity affects community stability, including several aspects of stochastic variation in composition and abundance, as well as resistance or sensitivity to disturbances (McCann 2000; Ives and Carpenter 2007). A recent meta-analysis of experiments across a range of habitats tested the link between species richness and one component of stability, temporal variabil-

ity in ecosystem-level properties (Jiang and Pu 2009). The authors found that, on average, increasing diversity indeed enhanced community-level temporal stability, but had no consistent effect on stability of the component populations. Intriguingly, the effects of diversity on stability differed strongly between experiments with single versus multiple trophic levels: diversity stabilized both population and community dynamics in multitrophic systems but there was no consistent effect of diversity in single-trophic assemblages. These patterns were broadly equivalent across experimental and observational studies as well as across terrestrial and aquatic studies, and suggest that effects of diversity on temporal stability may be generally enhanced by trophic complexity. What mechanisms might promote such stabilizing effects of diversity in food-webs? Several experiments offer intriguing clues. First, interactions within diverse natural assemblages may foster stability at the population level by imposing density-dependence in demographic rates, as shown for reef fishes (Hixon and Carr 1997; Carr *et al.* 2002). Second, theory (McCann *et al.* 1998) suggests that weak trophic interactions stabilize community dynamics as weak interactors dampen the destabilizing potential of strong interactors. O'Gorman and Emmerson (2009) tested this hypothesis by removing the strongest and weakest interactors from experimental food-webs on a rocky seashore. Extinction of strong interactors produced a strong trophic cascade and reduced temporal stability of community and ecosystem properties. Interestingly, loss of weak interactors also reduced temporal and spatial stability of community and ecosystem characteristics, as predicted by theory. Thus, even apparently insignificant species may play important roles in maintaining stability of natural ecosystems (Lyons *et al.* 2005).

12.3 Lessons learned: different designs for different questions

As mentioned at the beginning of this chapter, BEF studies have addressed two overlapping but distinct goals: mechanistic understanding of ecological interactions, and less commonly, applied questions about how extinction might affect ecosystem processes. Tension between these two threads, among other issues, has contributed to a history of debate about experimental design and inference in BEF research (Huston 1997; e.g. Lawton *et al.* 1998b). Key issues include the relative importance of dominance by particular species—sampling effects—versus niche partitioning and facilitation—complementarity—among species in explaining positive relationships between species richness and aggregate ecosystem properties such as total plant biomass (Huston 1997; Tilman 1997; Loreau and Hector 2001; Cardinale *et al.* 2007); relative value of additive versus replacement series designs (Griffin *et al.* 2008; Byrnes and Stachowicz 2009b); experimental duration (Cardinale *et al.* 2007; Stachowicz *et al.* 2008b); spatial scale and environmental heterogeneity (Fridley *et al.* 2007); and even confusion over subtle but important conceptions of what 'biodiversity' really means. In retrospect, many disagreements—including some that the co-authors of this chapter have had with each other!—have resulted in part from insufficient recognition that several distinct questions have been asked under the BDEF umbrella and that no single approach can address all questions simultaneously. They have also highlighted the central importance of experimental design for making inferences about how changing biodiversity will affect functioning of real-world ecosystems (Duffy 2009). For these reasons, we now turn to several issues of experimental design important to understanding the effects of changing biodiversity in a multitrophic context.

Testing specific extinction scenarios vs. general effects of diversity per se. An example of specific relevance to global change research involves the relative merits of experiments aimed at simulating realistic patterns of biodiversity loss (Lyons and Schwartz 2001; Zavaleta and Hulvey 2004; O'Connor and Crowe 2005; Bracken *et al.* 2008) versus those designed specifically to distinguish effects of species composition and richness. Experiments simulating realistic extinction scenarios reject the randomized assembly design typical of most BEF experiments on the grounds that they do not accurately mimic real trajectories of biodiversity loss or

future ecosystem functioning. This is a valid point, but it should not overlook the fact that the primary purpose of most BEF experiments to date has been not to predict the effects of specific extinction scenarios but to develop a general and rigorous understanding of the effects of changing species richness—usually—on ecosystem functioning, i.e. to separate the effects of the number of species from the long-recognized importance of effects of species composition or identity. To avoid confounding richness with composition, designs have included replicated monoculture treatments and a polyculture that includes all experimental taxa (Huston 1997; reviewed by Schmid *et al.* 2002). In contrast, studies focused on extinction are less concerned about separating richness from composition effects, and more interested in how loss of vulnerable species with certain traits—e.g. body size, high trophic level—might affect ecosystem processes (Duffy *et al.* 2009).

The contrast between the two objectives of BEF research raises the fundamental issue of what we mean by biodiversity. Studies focused on particular extinction scenarios necessarily involve covariance in richness and composition—indeed changes in species richness become inherently confounded with changes in species composition. Thus, in classical BEF experiments 'biodiversity' is typically equated with richness, whereas in studies of non-random extinction scenarios, changing 'biodiversity' is the result of simultaneous—and thus confounded—changes in both richness and composition (Bracken *et al.* 2008) or the presence of strong interactors (e.g. Solan *et al.* 2004). Neither of these definitions or approaches is inherently right or wrong; however, explicit recognition of the fundamentally different aims of the two types of studies would go some way towards reconciling apparently diverging conclusions. For example, at least two studies have shown that random changes in species richness had little or no effect on ecosystem processes, whereas diversity changes that mimic observed orders of species loss along gradients found substantial changes in exotic species invasion to a grassland (Zavaleta and Hulvey 2004), or nitrogen uptake by seaweeds (Bracken *et al.* 2008). While studies using random versus ordered extinction scenarios appear to reach contrasting conclusions, the discrepancy results primarily from manipulating different components of biodiversity, one focusing on species richness per se and another on specific combinations of species that covary in richness and composition.

Additive versus Replacement series designs. How are consumer density and diversity related in nature, and thus how should they covary in experiments? Studies motivated by understanding effects of multiple predators on total prey consumption—multiple predator effects literature—typically employ an additive design, whereas the BEF literature typically uses a replacement design. Each of these types of experiments confounds diversity and density in a different way: additive experiments confound total density with diversity, whereas replacement designs confound gain of interspecific interactions as diversity increases with the loss of intraspecific interactions as total predator density is held constant. Separately, these designs thus each detect only a subset of the changes in food-web interactions caused by predator loss (Byrnes and Stachowicz 2009b). This has led several researchers to employ a combination additive-replacement design to better assess the effect of diversity on the strength of top-down control (Douglass *et al.* 2008; Griffin *et al.* 2008; Byrnes and Stachowicz 2009b). However, such designs can rapidly become unwieldy and logistics quickly limit the maximum richness of predators that can be used. Thus, if the question of interest is to understand the consequences of changing predator diversity in a particular system, then the most appropriate design will mimic observed (or predicted) changes in both predator density and diversity in that system. This requires rigorous data on how diversity scales with total predator density in the field.

Experimental duration. A key question in adapting BEF research to broader scales is how diversity interacts with the environmental heterogeneity characteristic of most real ecosystems to affect functioning. Several experiments with primary producers have examined how diversity affects change with experimental duration, generally suggesting that longer experiments tend to show stronger diversity effects (Cardinale *et al.* 2007; Stachowicz

et al. 2008b; but see Fox 2004). While this pattern is consistent with theory (Pacala and Tilman 2002), it is important to point out that for fast growing species, or highly disturbed or seasonal communities, shorter experiments may be more realistic than longer ones. Perhaps more importantly, these studies have highlighted the difference between experiments that measure two distinct phenomena: (1) effects mediated only by individuals present at the establishment of the experiment, versus (2) effects involving population-level changes that extend over multiple generations. Individual processes might include vegetative growth, resource consumption, etc., and effects on such processes are often manifested over fairly short time scales, much less than one generation. In contrast, population processes include recruitment, realized reproduction, compensatory dynamics, etc., and generally manifest only in experiments run for multiple generations. Diversity effects that increase with experimental duration may imply that population-level processes are critical to diversity effects, as suggested by theory (Pacala and Tilman 2002; Cardinale *et al.* 2004); nevertheless, studies of varying duration are instructive in teasing apart the relative influence of individual- versus population-level mechanisms (Stachowicz *et al.* 2008a). Furthermore, optimizing experimental duration becomes more challenging in multitrophic BEF studies, in part because generation times of consumers can be long relative to producers in aquatic systems, and in part because there are many more mechanisms by which consumer diversity can affect function (Figure 12.2) and many of these may occur rapidly with very strong effects. Predation can produce large effects on prey density quickly, and on prey behavior virtually instantaneously. Logistical challenges and artifacts of long term caging experiments can also be severe. For example, in closed systems, prey can become unnaturally depleted if experiments run too long.

Environmental and resource heterogeneity. A number of the mechanisms that potentially connect diversity to function in both single and multitrophic systems involve differential resource use or response to the environment. Such effects are fundamental to expression of complementarity. Greater heterogeneity within an experimental unit provides a greater range of niches/resources, and thus a greater likelihood that complementary effects of species will be detected (Tilman 1999). This prediction is consistent with the stronger correlations between species richness and ecosystem processes observed across gradients of habitat heterogeneity in the field (Tylianakis *et al.* 2008). Since experiments may vary widely in the degree of heterogeneity present, this needs to be accounted for in interpreting experimental results (Stachowicz *et al.* 2008a). From a multitrophic perspective, for example, the effects of predator diversity on ecosystem functioning can depend on the diversity of resource (prey) species present (Gamfeldt *et al.* 2005). Heterogeneity of resources or environmental conditions is a prerequisite for complementarity, but the mere presence of heterogeneity by no means implies that complementarity will be a dominant force.

12.4 Biodiversity and ecosystem functioning in the Anthropocene

BEF research has shown clearly, as have many other lines of ecological and oceanographic research, that there are many interacting mechanisms by which species richness, identity, and composition can affect ecosystem-level properties and processes. But BEF is moving—or attempting to move—from a basic field of science to a more applied field designed to address specific questions. Applying insights from the primarily academic tradition of BEF to understand how complex, real-world ecosystems will respond to particular environmental stressors is a daunting challenge and has only begun. As the previous section showed, pursuing this goal will require alternative designs and approaches to test the projected scenarios of extinction, depletion, and invasion that are of interest in applied ecology and management. At the same time, we must realize that the insights emerging from designs aimed at specific inferences will necessarily be less general, and will apply primarily to particular sites, species, and sets of conditions. Thus, there is an inherent trade-off: more realistic diversity gradients may tell us a

great deal about likely effects of diversity change in a particular system, but relatively little about the function of changing diversity per se.

We are nevertheless optimistic that research linking biodiversity to ecosystem functioning can valuably inform our approach to dealing with real-world challenges of global change (see also Duffy 2009; Palumbi *et al.* 2009). Although important details will inevitably vary across systems and taxa, previous research appears to be converging on several fairly robust generalizations, outlined above, regarding both how diversity is changing and how it is likely to affect marine ecosystem structure and processes. Notably, the research summarized in section 12.2 indicates that decline in density and diversity is typically most pronounced among higher consumers, whereas local diversity tends to be increasing at the level of detritivores and suspension feeders. These patterns clearly have implications for the functioning of marine food-webs and ecosystems. Below we consider a few of several possible hypotheses for how these changes may affect ecosystem functioning based on our current knowledge of BEF relationships outlined in section 12.3.

1. *Intense harvesting shifts marine communities toward smaller average body sizes and lower average trophic levels, altering top-down control.* Disproportionate impacts on large-bodied, slow-growing, and predatory species will tend, on average, to reduce food-chain length and shift communities toward dominance by small-bodied, fast-maturing, and ecologically generalized omnivores, detritivores, and suspension-feeders. The altered top-down control expected as a result of this trophic skew likely will be central to understanding effects of global change on marine ecosystems. However, the specific consequences of top predator decline will depend on the effective length of food chains (Wootton and Power 1993; Stibor *et al.* 2004) and various other factors (Bruno and Cardinale 2008, see also Figure 12.2), and thus may either increase or decrease top-down control on particular species or assemblages.

2. *Invasions by suspension feeders may increase local to regional scale primary consumer richness, potentially*

altering standing stocks of phytoplankton. The introduction of exotic suspension-feeders has sometimes dramatically reduced phytoplankton biomass in both estuarine (Alpine and Cloern 1992) and lake ecosystems (Nicholls and Hopkins 1993). Although an experiment that increased species richness of marine suspension-feeders did not affect community filtration rate in the short term, and there was no clear differences in filtration capacity between invaders and natives, differences in seasonal phenology of novel versus resident species may increase the consistency of filtration over longer periods (Byrnes and Stachowicz 2009a). The potential implications of changing suspension feeder diversity are broad since they can alter the stock of food at the base of the web. Whether changing diversity of suspension-feeders might make them more resistant to consumers, as meta-analyses suggest for prey generally (section 12.2 Figure 12.4), remains to be tested.

3. *Invasions will increase richness of decomposers and detritivores, leading to faster processing of detrital matter and nutrient cycling.* The growing diversity of species at low trophic levels in many marine systems (Byrnes *et al.* 2007), coupled with a general trend toward greater decomposition rate as detritivore richness increases (Srivastava *et al.* 2009), suggests that rates of organic matter turnover and nutrient recycling might increase in detritus-based marine ecosystems. These processes have important rippling affects through ecosystems, and could have profound consequences for functioning.

4. *Reduced diversity of primary producers, as a result of habitat degradation and/or dominance by exotic invaders, will reduce habitat and food quality for herbivores, decreasing trophic transfer and fish productivity.* Benthic habitats worldwide are being altered or degraded by dredging and trawling (Thrush and Dayton 2002), loss of macrophyte vegetation due to eutrophication (Duarte 2002), and invasions of aggressive habitat-forming species such as the alga *Caulerpa* sp and ascidian *Didemnum* (Ruiz *et al.* 1999; Airoldi and Beck 2007; Byrnes *et al.* 2007). In many cases, these processes convert diverse benthic communities to near monocultures or depauperate assemblages of basal species. Robust associations between

habitat diversity and associated animal diversity, and between prey—including algae—diversity and consumer fitness (Lefchek *et al.* submitted) suggest that these alterations are likely to reduce the diversity and productivity of animals in affected habitats, including both invertebrate primary consumers and the fishes that feed on them, with consequences for fisheries.

Finally, although we have striven to be as concrete and specific as possible in our predictions, it must be emphasized that human economies, ecosystems, and the multivariate stressors that affect them are all inherently complex and their interactions are often non-linear. Thus we can be certain of ecological surprises (Doak *et al.* 2008). The evolving distributions of species mediated by climate change and human commerce are creating 'novel' or 'emerging' ecosystems populated by species with little or no shared evolutionary history and without close analogues in natural communities (Hobbs *et al.* 2006; Williams and Jackson 2007). To the extent that our current understanding of BEF—and ecology generally—is based on existing communities, the prospect of these novel ecosystems emphasizes the need for testing the robustness of our conclusions and models under new conditions. Positive surprises are also possible. For example, human population growth is declining more rapidly than predicted in several parts of the world. Marine protected areas and other measures have successfully changed the trajectory of ecosystem decline in several systems (Halpern and Warner 2002; Edgar *et al.* 2009), and it is conceivable that more widespread implementation of sustainable fishing practices might turn the tide of marine degradation on a larger scale (Worm *et al.* 2009). Nevertheless, from a practical, management-oriented standpoint, the certainty of future uncertainty argues for a precautionary approach to managing our interaction with the biosphere. In this context, the practical lesson from research linking biodiversity to ecosystem functioning may be a very simple and general one: maintaining viable populations of as many species, genes, and landscape elements as possible may be our best insurance against ecosystem failure in the face of change.

References

Airoldi, L. and Beck, M. W. (2007) Loss, status and trends for coastal marine habitats of Europe. In: *Oceanography and Marine Biology, Vol. 45*, 345–405.

Allison, G. W. (1999) The implications of experimental design for biodiversity manipulations. *American Naturalist*, **153**, 26–45.

Alpine, A. E. and Cloern, J. E. (1992) Trophic interactions and direct physical effects control phytoplankton biomass and production in an estuary. *Limnology and Oceanography*, **37**, 946–55.

Aronson, R. B., Moody, R. M., Ivany, L. C. *et al.* (2009) Climate Change and Trophic Response of the Antarctic Bottom Fauna. *PLoS ONE*, 4, e4385.

Bakun, A. (2006) Wasp-waist populations and marine ecosystem dynamics: Navigating the 'predator pit' topographies. *Progress in Oceanography*, **68**, 271–88.

Balvanera, P., Pfisterer, A. B., Buchmann, N. *et al.* (2006) Quantifying the evidence for biodiversity effects on ecosystem functioning and services. *Ecology Letters*, **9**, 1146–56.

Baum, J. K., Myers, R. A., Kehler, D. G. *et al.* (2003) Collapse and conservation of shark populations in the Northwest Atlantic. *Science*, **299**, 389–92.

Bernays, E. A., Bright, K. L., Gonzalez, N. *et al.* (1994) Dietary Mixing in a Generalist Herbivore: Tests of Two Hypotheses. *Ecology*, **75**, 1997–2006.

Bertness, M. D., Garrity, S. D. and Levings, S. C. (1981) Predation Pressure and Gastropod Foraging: A Tropical-Temperate Comparison. *Evolution*, **35**, 995–1007.

Bolser, R. C. and Hay, M. E. (1996) Are tropical plants better defended? Palatability and defenses of temperate vs tropical seaweeds. *Ecology*, **77**, 2269–86.

Bracken, M. E. S., Friberg, S. E., Gonzalez-Dorantes, C. A. *et al.* (2008) Functional consequences of realistic biodiversity changes in a marine ecosystem. *Proceedings of the National Academy of Sciences of the United States of America*, 105, 924–8.

Bracken, M. E. S. and Stachowicz, J. J. (2006) Seaweed diversity enhances nitrogen uptake via complementary use of nitrate and ammonium. *Ecology*, **87**, 2397–403.

Bruno, J. F. and Bertness, M. D. (2001) Habitat modification and facilitation in benthic marine communities. In: *Marine Community Ecology* (eds. Bertness, M. D., Gaines, S. D. and Hay, M. E.). Sinauer Sunderland, MA, 201–18.

Bruno, J. F., Boyer, K. E., Duffy, J. E. *et al.* (2008) Relative and interactive effects of plant and grazer richness in a benthic marine community. *Ecology*, **89**, 2518–28.

Bruno, J. F. and Cardinale, B. J. (2008) Cascading effects of predator richness. *Frontiers in Ecology and the Environment*, **6**, 539–46.

Bruno, J. F. and O'Connor, M. I. (2005) Cascading effects of predator diversity and omnivory in a marine food web. *Ecology Letters*, **8**, 1048–56.

Byrnes, J. and Stachowicz, J. J. (2009a) Short and long term consequences of increases in exotic species richness on water filtration by marine invertebrates. *Ecology Letters*, **12**, 830–41.

Byrnes, J., Stachowicz, J. J., Hultgren, K. M., *et al.* (2006) Predator diversity strengthens trophic cascades in kelp forests by modifying herbivore behaviour. *Ecology Letters*, **9**, 61–71.

Byrnes, J. E., Reynolds, P. L. and Stachowicz, J. J. (2007) Invasions and Extinctions Reshape Coastal Marine Food Webs. *PLoS ONE*, 2, e295.

Byrnes, J. E. and Stachowicz, J. J. (2009b) The consequences of consumer diversity loss: different answers from different experimental designs. *Ecology*, **90**, 2879–88.

Cardinale, B. J., Duffy, J. E., Srivastava, D. S. *et al.* (2009) Towards a Food-web perspective on biodiversity and ecosystem functioning. In: *Biodiversity, ecosystem functioning, and human wellbeing: An ecological and economic perspective.* Naeem, S., Bunker, D.E., Hector, A, *et al.* (eds), Oxford University Press, New York, 105–20.

Cardinale, B. J., Ives, A. R. and Inchausti, P. (2004) Effects of species diversity on the primary productivity of ecosystems: extending our spatial and temporal scales of inference. *Oikos*, **104**, 437–50

Cardinale, B. J., Matulich, K. L., Hooper, D. U. *et al.* (2011) The functional role of producer diversity in ecosystems. *American Journal of Botany*, **98**, 572–92.

Cardinale, B. J., Srivastava, D. S., Duffy, J. E. *et al.* (2006) Effects of biodiversity on the functioning of trophic groups and ecosystems. *Nature*, **443**, 989–92.

Cardinale, B. J., Wright, J. P., Cadotte, M. W. *et al.* (2007) Impacts of plant diversity on biomass production increase through time because of species complementarity. *Proceedings of the National Academy of Sciences*, **104**, 18123–8.

Carlander, K. D. (1955) The standing crop of fish in lakes. *Journal of the Fisheries Research Board of Canada*, **12**, 543–70.

Carr, M. H., Anderson, T. W. and Hixon, M. A. (2002) Biodiversity, population regulation, and the stability of coral-reef fish communities. *Proceedings of the National Academy of Sciences of the United States of America*, **99**, 11241–5.

Casula, P., Wilby, A. and Thomas, M. B. (2006) Understanding biodiversity effects on prey in multi-enemy systems. *Ecology Letters*, **9**, 995–1004.

Cebrian, J., Shurin, J. B., Borer, E. T. *et al.* (2009) Producer Nutritional Quality Controls Ecosystem Trophic Structure. *PLoS ONE*, 4, e4929.

Cheung, W. W. L., Lam, V. W. Y., Sarmiento, J. L. *et al.* (2009) Projecting global marine biodiversity impacts under climate change scenarios. *Fish and Fisheries*, **10**, 235–51.

Crowder, L. B., Squires, D. D. and Rice, J. A. (1997) Nonadditive effects of terrestrial and aquatic predators on juvenile estuarine fish. *Ecology*, **78**, 1796–804.

Cruz-Rivera, E. and Hay, M. E. (2000) The effects of diet mixing on consumer fitness: macroalgae, epiphytes, and animal matter as food for marine amphipods. *Oecologia*, **123**, 252–64.

Cyr, H. and Pace, M. L. (1993) Magnitude and Patterns of Herbivory in Aquatic and Terrestrial Ecosystems. *Nature*, **361**, 148–50.

Dam, H. G. and Lopes, R. M. (2003) Omnivory in the calanoid copepod Temora longicornis: feeding, egg production and egg, hatching rates. *Journal of Experimental Marine Biology and Ecology*, **292**, 119–37.

Daskalov, G. M., Grishin, A. N., Rodionov, S. *et al.* (2007) Trophic cascades triggered by overfishing reveal possible mechanisms of ecosystem regime shifts. *Proceedings of the National Academy of Sciences of the United States of America*, **104**, 10518–23.

DeMott, W. R. (1998) Utilization of a cyanobacterium and a phosphorus-deficient green alga as complementary resources by daphnids. *Ecology*, **79**, 2463–81.

Diaz, R. J. and Rosenberg, R. (2008) Spreading dead zones and consequences for marine ecosystems. *Science*, **321**, 926–9.

Diehl, S. (1992) Fish predation and community structure: the role of omnivory and habitat complexity. *Ecology*, **73**, 1646–61.

Doak, D. F., Estes, J. A., Halpern, B. S. *et al.* (2008) Understanding and predicting ecological dynamics: Are major surprises inevitable? *Ecology*, **89**, 952–61.

Doney, S. C., Ruckelshaus, M., Duffy, J.E. *et al.* (2010) Climate Change Impacts on Marine Ecosystems. *Annual Review of Marine Science*, **13**: 194–201.

Douglass, J. G., Duffy, J. E. and Bruno, J. F. (2008) Herbivore and predator diversity interactively affect ecosystem properties in an experimental marine community. *Ecology Letters*, **11**, 598–608.

Duarte, C. M. (2002) The future of seagrass meadows. *Environmental Conservation*, **29**, 192–206.

Duffy, J. E. (2002) Biodiversity and ecosystem function: the consumer connection. *Oikos*, **99**, 201–19.

Duffy, J. E. (2003) Biodiversity loss, trophic skew and ecosystem functioning. *Ecology Letters*, **6**, 680–7.

Duffy, J. E. (2009) Why biodiversity is important to functioning of real-world ecosystems. *Frontiers in Ecology and the Environment*, **7**, 437–44.

Duffy, J. E. (2010) Sea changes: structure and functioning of emerging marine communities. In: *Community ecology. Processes, models and applications.* Verhoef, H. A. and Morin, P. J. (eds), Oxford University Press Oxford, Oxford, 95–114.

Duffy, J. E., Cardinale, B. J., France, K. E. *et al.* (2007) The functional role of biodiversity in ecosystems: incorporating trophic complexity. *Ecology Letters*, **10**, 522–38.

Duffy, J. E., Richardson, J. P. and Canuel, E. A. (2003) Grazer diversity effects on ecosystem functioning in seagrass beds. *Ecology Letters*, **6**, 637–45.

Duffy, J. E., Richardson, J. P. and France, K. E. (2005) Ecosystem consequences of diversity depend on food chain length in estuarine vegetation. *Ecology Letters*, **8**, 301–9.

Duffy, J. E., Srivastava, D. S., McLaren, J. *et al.* (2009) Forecasting decline in ecosystem services under realistic scenarios of extinction. In: *Biodiversity, ecosystem functioning, and human wellbeing: An ecological and economic perspective.* Naeem, S., Bunker, D.E., Hector, A., *et al.* (eds), Oxford University Press New York, 60–77.

Dulvy, N. K., Sadovy, Y. and Reynolds, J. D. (2003) Extinction vulnerability in marine populations. *Fish and Fisheries*, **4**, 25–64.

Edgar, G. J., Barrett, N. S. and Stuart-Smith, R. D. (2009) Exploited reefs protected from fishing transform over decades into conservation features otherwise absent from seascapes. *Ecological Applications*, **19**, 1967–74.

Edwards, K. F., Aquilino, K. M., Best, R. J. *et al.* (2010) Prey diversity is associated with weaker consumer effects in a meta-analysis of benthic marine experiments. *Ecology Letters*, **13**, 194–201.

Essington, T. E., Beaudreau, A. H. and Wiedenmann, J. (2006) Fishing through marine food webs. *Proceedings of the National Academy of Sciences of the United States of America*, **103**, 3171–5.

Finke, D. L. and Denno, R. F. (2005) Predator diversity and the functioning of ecosystems: the role of intraguild predation in dampening trophic cascades. *Ecology Letters*, **8**, 1299–306.

Fisher, J. A. D., Frank, K. T. and Leggett, W. C. (2010) Global variation in marine fish body size and its role in biodiversity-ecosystem functioning. *Marine Ecology-Progress Series*, **405**, 1–13.

Fox, J. W. (2004) Effects of algal and herbivore diversity on the partitioning of biomass within and among trophic levels. *Ecology*, **85**, 549–59.

Frank, K. T., Petrie, B. and Shackell, N. L. (2007) The ups and downs of trophic control in continental shelf ecosystems. *Trends in Ecology and Evolution*, **22**, 236–42.

Frank, K. T., Petrie, B., Shackell, N. L. *et al.* (2006) Reconciling differences in trophic control in mid-latitude marine ecosystems. *Ecology Letters*, **9**, 1096–105.

Fridley, J. D., Stachowicz, J. J., Naeem, S. *et al.* (2007) The invasion paradox: Reconciling pattern and process in species invasions. *Ecology*, **88**, 3–17.

Gamfeldt, L., Hillebrand, H. and Jonsson, P. R. (2005) Species richness changes across two trophic levels simultaneously affect prey and consumer biomass. *Ecology Letters*, **8**, 696–703.

Griffin, J. N., De la Haye, K. L., Hawkins, S. J. *et al.* (2008) Predator diversity and ecosystem functioning: Density modifies the effect of resource partitioning. *Ecology*, **89**, 298–305.

Haddad, N. M., Crutsinger, G. M., Gross, K. *et al.* (2009) Plant species loss decreases arthropod diversity and shifts trophic structure. *Ecology Letters*, **12**, 1029–39.

Halpern, B. S., Walbridge, S., Selkoe, K. A. *et al.* (2008) A global map of human impact on marine ecosystems. *Science*, **319**, 948–52.

Halpern, B. S. and Warner, R. R. (2002) Marine reserves have rapid and lasting effects. *Ecology Letters*, **5**, 361–6.

Hattenschwiler, S., Tiunov, A. V. and Scheu, S. (2005) Biodiversity and litter decomposition in terrestrial ecosystems. *Annual Review of Ecology, Evolution, and Systematics*, **36**, 191–218.

Heck, K. L. J. and Wetstone, G. S. (1977) Habitat complexity and invertebrate species richness and abundance in tropical seagrass meadows. *Journal of Biogeography*, **4**, 135–42.

Hector, A. and Bagchi R. (2007) Biodiversity and ecosystem multifunctionality. *Nature*, **448**, 188–U6.

Hiddink, J. G. and ter Hofstede, R. (2008) Climate induced increases in species richness of marine fishes. *Global Change Biology*, **14**, 453–60.

Hillebrand, H. and Cardinale B. J. (2004) Consumer effects decline with prey diversity. *Ecology Letters*, **7**, 192–201.

Hixon, M. A. and Carr, M. H. (1997) Synergistic Predation, Density Dependence, and Population Regulation in Marine Fish. *Science*, **277**, 946–9.

Hobbs, R. J., Arico S., Aronson, J. *et al.* (2006) Novel ecosystems: theoretical and management aspects of the new ecological world order. *Global Ecology and Biogeography*, **15**, 1–7.

Hoegh-Guldberg, O. and Bruno, J. F. (2010) The Impact of Climate Change on the World's Marine Ecosystems. *Science*, **328**, 1523–8.

Hooper, D. U., Chapin, F. S., Ewel, J. J. *et al.* (2005) Effects of biodiversity on ecosystem functioning: A consensus of current knowledge. *Ecological Monographs*, **75**, 3–35.

Hsieh, C. H., Reiss, C. S., Hunter, J. R. *et al.* (2006) Fishing elevates variability in the abundance of exploited species. *Nature*, **443**, 859–62.

Hughes, A. R., Best, R. J. and Stachowicz, J. J. (2010) Genotypic diversity and grazer identity interactively influence seagrass and grazer biomass. *Marine Ecology Progress Series*, **403**, 43–51.

Huston, M. A. (1997) Hidden treatments in ecological experiments: Re-evaluating the ecosystem function of biodiversity. *Oecologia*, **110**, 449–60.

Isbell, F., Calcagno, V., Hector, A. *et al.* (2011) High plant diversity is needed to maintain ecosystem services. *Nature*, **477**: 199–202.

Ives, A. R. and Carpenter, S. R. (2007) Stability and diversity of ecosystems. *Science*, **317**, 58–62.

Jackson, J. B. C., Kirby, M. X., Berger, W. H. *et al.* (2001) Historical overfishing and the recent collapse of coastal ecosystems. *Science*, **293**, 629–38.

Jennings, S., Greenstreet, S. P. R. and Reynolds, J. D. (1999a) Structural change in an exploited fish community: a consequence of differential fishing effects on species with contrasting life histories. *Journal of Animal Ecology*, **68**, 617–27.

Jennings, S., Reynolds, J. D. and Polunin, N. V. C. (1999b) Predicting the vulnerability of tropical reef fishes to exploitation with phylogenies and life histories. *Conservation Biology*, **13**, 1466–75.

Jiang, L. and Pu, Z. C. (2009) Different Effects of Species Diversity on Temporal Stability in Single-Trophic and Multitrophic Communities. *American Naturalist*, **174**, 651–9.

Kinzig, A. P., Pacala, S. W. and Tilman, D. (eds) (2001) *The functional consequences of biodiversity: empirical progress and theoretical extensions*. Princeton University Press, Princeton, NJ.

Lavorel, S. and Garnier, E. (2002) Predicting changes in community composition and ecosystem functioning from plant traits: revisiting the Holy Grail. *Functional Ecology*, **16**, 545–56.

Lawton, J. H., Bignell, D. E., Bolton, B. *et al.* (1998a) Biodiversity inventories, indicator taxa and effects of habitat modification in tropical forests. *Nature*, **391**, 72–6.

Lawton, J. H., Naeem, S., Thompson, L. J. *et al.* (1998b) Biodiversity and ecosystem function: getting the Ecotron experiment in its correct context. *Functional Ecology*, **12**, 848–52.

Leroux, S. J. and Loreau, M. (2009) Disentangling multiple predator effects in biodiversity and ecosystem functioning research. *Journal of Animal Ecology*, **78**, 695–8.

Longhurst, A. (2007) Doubt and certainty in fishery science: Are we really headed for a global collapse of stocks? *Fisheries Research*, **86**, 1–5.

Loreau, M. and Hector, A. (2001) Partitioning selection and complementarity in biodiversity experiments. *Nature*, **412**, 72–6.

Loreau, M., Naeem, S. and Inchausti, P. (eds) (2002) *Biodiversity and ecosystem functioning: synthesis and perspectives*. Oxford University Press, New York.

Loreau, M., Naeem, S., Inchausti, P. *et al.* (2005) Rare species and ecosystem functioning. *Conservation Biology*, **19**, 1019–24.

Lyons, K. G. and Schwartz, M. W. (2001) Rare species loss alters ecosystem function—invasion resistance. *Ecology Letters*, **4**, 358–65.

Martin, T. H., Wright, R. A. and Crowder, L. B. (1989) Nonadditive impact of blue crabs and spot on their prey assemblages. *Ecology*, **70**, 1935–42.

McCann, K. S. (2000) The diversity-stability debate. *Nature*, **405**, 228–33.

McCann, K. S., Hastings, A. and Huxel, G.R. (1998) Weak trophic interactions and the balance of nature. *Nature*, **395**, 794–8.

McGrady-Steed J., Harris P. M. and Morin P. J. (1997) Biodiversity regulates ecosystem predictability. *Nature*, **390**, 162–5.

Moran, E. R., Reynolds, P. L., Ladwig, L. M. *et al.* (2010) Predation intensity is negatively related to plant species richness in a benthic marine community. *Marine Ecology Progress Series*, **400**: 277–282.

Mulder, C. P. H., Koricheva, J., Huss-Danell, K. *et al.* (1999) Insects affect relationships between plant species richness and ecosystem processes. *Ecology Letters*, **2**, 237–46.

Murawski, S. A. (1993) Climate Change and Marine Fish Distributions: Forecasting from Historical Analogy. *Transactions of the American Fisheries Society*, **12**, 647–58.

Murdoch, W. W., Evans, F. C. and Peterson, C. H. (1972) Diversity and pattern in plants and insects. *Ecology*, **53**, 819–29.

Myers, R. A., Baum, J. K., Shepherd, T. D. *et al.* (2007) Cascading effects of the loss of apex predatory sharks from a coastal ocean. *Science*, **315**, 1846–50.

Myers, R. A. and Worm, B. (2003) Rapid worldwide depletion of predatory fish communities. *Nature*, **423**, 280–3.

Naeem, S. and Li, S. B. (1998) Consumer species richness and autotrophic biomass. *Ecology*, **79**, 2603–15.

Naeem, S., Thompson, L. J., Lawler, S. P. *et al.* (1994) Declining Biodiversity Can Alter the Performance of Ecosystems. *Nature*, **368**, 734–7.

Naeem, S. and Wright, J. P. (2003) Disentangling biodiversity effects on ecosystem functioning: deriving solutions to a seemingly insurmountable problem. *Ecology Letters*, **6**, 567–79.

Nicholls, K. H. and Hopkins, G. J. (1993) Recent changes in Lake Erie (north shore) phytoplankton—cumulative impacts of phosphorus loading reductions and the zebra mussel introduction. *Journal of Great Lakes Research*, **19**, 637–47.

O'Connor, M. I., Piehler, M. F., Leech, D. M. *et al.* (2009) Warming and Resource Availability Shift Food Web Structure and Metabolism. *Plos Biology*, **7**(8): e1000178:doi:10.1371/journal.pbio.1000178.

O'Connor, N. E. and Crowe, T. P. (2005) Biodiversity loss and ecosystem functioning: Distinguishing between number and identity of species. *Ecology*, **86**, 1783–96.

O'Gorman, E. J. and Emmerson, M. C. (2009) Perturbations to trophic interactions and the stability of complex food webs. *Proceedings of the National Academy of Sciences of the United States of America*, **106**, 13393–8.

Olson, M. H., Jacobs, R. P. and O'Donnell, E. B. (2007) Species diversity enhances predator growth rates. *Research Letters in Ecology* 94587, 1–5.

Pacala, S. and Tilman, D. (2002)The transition from sampling to complementarity. In: *The functional consequences of biodiversity. Empirical progress and theoretical extensions* (ed. Kinzig, A. P., Pacala, S. W. and Tilman, D.). Princeton University Press Princeton, 151–66.

Palumbi, S. R., Sandifer, P. A., Allan, J. D. *et al.* (2009) Managing for ocean biodiversity to sustain marine ecosystem services. *Frontiers in Ecology and the Environment*, **7**, 204–11.

Pauly, D., Christensen, V., Dalsgaard, J. *et al.* (1998) Fishing down marine food webs. *Science*, **279**, 860–3.

Pennings, S.C. and Silliman, B.R. (2005) Linking biogeography and community ecology: Latitudinal variation in plant-herbivore interaction strength. *Ecology*, **86**, 2310–19.

Perry, A. L., Low, P. J., Ellis, J. R. *et al.* (2005) Climate change and distribution shifts in marine fishes. *Science*, **308**, 1912–15.

Price, P. W., Bouton, C. E., Gross, P. *et al.* (1980) Interactions among three trophic levels: influence of plants on interactions between insect herbivores and natural enemies. *Annual Review of Ecology and Systematics*, **11**, 41–65.

Purvis, A., Jones, K. E. and Mace, G. M. (2000) P Extinction. *Bioessays*, **22**, 1123–33.

Raffaelli, D., van der Putten, W. H., Persson, L. *et al.* (2002) Multi-trophic dynamics and ecosystem processes. In: *Biodiversity and ecosystem functioning. Synthesis and perspectives*. Loreau, M., Naeem, S. and Inchausti, P. (eds), Oxford University Press, Oxford, 147–54.

Ruiz, G. M., Fofonoff, P., Hines, A. H. *et al.* (1999) Non-indigenous species as stressors in estuarine and marine communities: Assessing invasion impacts and interactions. *Limnology and Oceanography*, **44**, 950–72.

Ruiz, G. M., Fofonoff, P. W., Carlton, J. T. *et al.* (2000) Invasion of coastal marine communities in North America: Apparent patterns, processes, and biases. *Annu Rev Ecol Syst*, **31**, 481–531.

Ruzicka, R. and Gleason, D. F. (2008) Latitudinal variation in spongivorous fishes and the effectiveness of sponge chemical defenses. *Oecologia*, **154**, 785–94.

Sala, E. and Knowlton, N. (2006) Global marine biodiversity trends. *Annual Review of Environment and Resources*, **31**, 93–122.

Scheffer, M., Carpenter, S., Foley, J. A. *et al.* (2001) Catastrophic shifts in ecosystems. *Nature*, **413**, 591–6.

Scheffer, M. and Jeppesen, E. (2007) Regime shifts in shallow lakes. *Ecosystems*, **10**, 1–3.

Schmid, B., Balvanera, P., Cardinale, B. J. *et al.* (2009) *Consequences of Species Loss for Ecosystem Functioning: Meta-analyses of data from biodiversity experiments*. In: *Biodiversity and ecosystem functioning. Synthesis and perspectives* (eds. Loreau, M, Naeem, S and Inchausti, P). Oxford University Press, Oxford, 60–67.

Schmid, B., Hector, A., Huston, M. A. *et al.* (2002) The design and analysis of biodiversity experiments. In: *Biodiversity and ecosystem functioning. Synthesis and perspectives*.Loreau, M., Naeem, S. and Inchausti, P. (eds), Oxford University Press, Oxford, 61–75.

Schmitz, O. J., Hamback, P. A. and Beckerman, A. P. (2000) Trophic cascades in terrestrial systems: A review of the effects of carnivore removals on plants. *American Naturalist*, **155**, 141–53.

Schulze, E. D. and Mooney, H. A. (eds) (1994) *Biodiversity and ecosystem function*. Springer-Verlag, Berlin.

Schwartz, M. W., Brigham, C. A., Hoeksema *et al.* (2000) Linking biodiversity to ecosystem function: implications for conservation ecology. *Oecologia*, **122**, 297–305.

Shurin, J. B., Gruner, D. S. and Hillebrand, H. (2006) All wet or dried up? Real differences between aquatic and terrestrial food webs. *Proceedings of the Royal Society B-Biological Sciences*, **273**, 1–9.

Siemann, E. (1998) Experimental tests of effects of plant productivity and diversity on grassland arthropod diversity. *Ecology*, **79**, 2057–70.

Sih, A., Englund, G. and Wooster, D. (1998) Emergent impacts of multiple predators on prey. *Trends in Ecology and Evolution*, **13**, 350–5.

Solan, M., Cardinale, B. J., Downing, A. L. *et al.* (2004) Extinction and ecosystem function in the marine benthos. *Science*, **306**, 1177–80.

Southwood, T. R. E., Brown, V. K. and Reader, P. M. (1979) The relationship of plant and insect diversities in succession. *Biological Journal of the Linnean Society*, **84**, 143–60.

Srivastava, D. S. (2002) The role of conservation in expanding biodiversity research. *Oikos*, **98**, 351–60.

Srivastava, D. S., Cardinale, B. J., Downing, A. L. *et al.* (2009) Diversity has stronger top-down than bottom-up effects on decomposition. *Ecology*, **90**, 1073–83.

Srivastava, D. S. and Vellend M. (2005) Biodiversity-ecosystem function research: Is it relevant to conservation? *Annual Review of Ecology Evolution and Systematics*, **36**, 267–94.

Stachowicz, J. J., Best, R. J., Bracken, M. E. S. *et al.* (2008a) Complementarity in marine biodiversity manipulations: Reconciling divergent evidence from field and mesocosm experiments. *Proceedings of the National Academy of Sciences of the United States of America*, **105**, 18842–7.

Stachowicz, J. J., Bruno, J. F. and Duffy, J. E. (2007) Understanding the effects of marine biodiversity on communities and ecosystems. *Annual Review of Ecology Evolution and Systematics*, **38**, 739–66.

Stachowicz, J. J., Graham, M., Bracken, M. E. S. *et al.* (2008b) Diversity enhances cover and stability of seaweed assemblages: the role of heterogeneity and time. *Ecology*, **89**, 3008–19.

Stibor, H., Vadstein, O., Diehl, S. *et al.* (2004) Copepods act as a switch between alternative trophic cascades in marine pelagic food webs. *Ecology Letters*, **7**, 321–8.

Strong, D. R. (1992) Are trophic cascades all wet—differentiation and donor-control in speciose ecosystems. *Ecology*, **73**, 747–54.

Thebault, E. and Loreau, M. (2003) Food-web constraints on biodiversity-ecosystem functioning relationships. *Proceedings of the National Academy of Sciences of the United States of America*, **100**, 14949–54.

Thebault, E. and Loreau, M. (2005) Trophic interactions and the relationship between species diversity and ecosystem stability. *American Naturalist*, **166**, E95–E114.

Thrush, S. F. and Dayton, P. K. (2002) Disturbance to marine benthic habitats by trawling and dredging: Implications for marine biodiversity. *Annu Rev Ecol Syst*, **33**, 449–73.

Tilman, D. (1997) Distinguishing between the effects of species diversity and species composition. *Oikos*, **80**, 185–185.

Tilman, D. (1999) The ecological consequences of changes in biodiversity: A search for general principles. *Ecology*, **80**, 1455–74.

Tylianakis, J. M., Rand, T. A., Kahmen, A. *et al.* (2008) Resource Heterogeneity Moderates the Biodiversity-Function Relationship in Real World Ecosystems. *Plos Biology*, **6**, e122.

Vermeij, G. J. (1978) *Biogeography and Adaptation*. Harvard University Press, Cambridge, MA.

Williams, J. W. and Jackson, S. T. (2007) Novel climates, no-analog communities, and ecological surprises. *Frontiers in Ecology and the Environment*, **5**, 475–82.

Wojdak, J. M. (2005) Relative strength of top-down, bottom-up, and consumer species richness effects on pond ecosystems. *Ecological Monographs*, **75**, 489–504.

Wootton, J. T. and Power, M. E. (1993) Productivity, Consumers, and the Structure of a River Food-Chain. *Proceedings of the National Academy of Sciences of the United States of America*, **90**, 1384–7.

Worm, B., Barbier, E. B., Beaumont, N. *et al.* (2006) Impacts of biodiversity loss on ocean ecosystem services. *Science*, **314**, 787–90.

Worm, B., Hilborn, R., Baum, J. K. *et al.* (2009) Rebuilding Global Fisheries. *Science*, **325**, 578–85.

Zavaleta, E. S. and Hulvey, K. B. *et al.* (2004) Realistic species losses disproportionately reduce grassland resistance to biological invaders. *Science*, **306**, 1175–7.

CHAPTER 13

Reality check: issues of scale and abstraction in biodiversity research, and potential solutions

Tasman P. Crowe, Matthew E. S. Bracken, and Nessa E. O'Connor

13.1 Introduction

Over the past few decades, there has been extensive research into the relationship between biodiversity and the functioning of ecosystems—BEF research. Much of it has been done in terrestrial systems, and many of the seminal papers were terrestrially focused (e.g. Walker, 1992; Naeem *et al.* 1994; Tilman *et al.* 1996; Hector *et al.* 1999). This dominance and the tendency for terrestrial ecologists to ignore aquatic literature have both been documented (e.g. Raffaelli *et al.* 2005; Menge *et al.* 2009). However, BEF research in marine systems has proliferated in recent years (Stachowicz *et al.* 2007). There is a clear need to bridge the gap between terrestrial and marine ecologists, and to foster greater exchange of concepts, approaches and technology (Menge *et al.* 2009). Marine systems have a number of advantages for BEF research and are now yielding some influential findings (e.g. Stachowicz *et al.* 1999; Duffy *et al.* 2003; Solan *et al.* 2004; Bracken *et al.* 2008). Based on evidence to date, there appear to be some differences between BEF relationships in marine and terrestrial systems (Stachowicz *et al.* 2007). However, these comparisons may be hampered by differences in the approaches taken to terrestrial and marine BEF research (Stachowicz *et al.* 2008a). Recent re-analyses of two large meta-datasets of BEF studies did not identify an overall tendency for biodiversity effects to vary among ecosystems, and the findings suggested that similar mechanistic processes may underpin BEF relationships in terrestrial and aquatic systems (Schmid *et al.* 2009).

It is important to recognize that BEF research can be primarily motivated by different overall objectives: (a) pure ecological research, to understand something about how the natural world works, and (b) to generate results applicable to real-world management problems. Of course this distinction is not by any means exclusive, but it can influence decisions about approaches to research because each of these objectives has different requirements and constraints. Much of the research undertaken to date has been justified by the need to understand the consequences of changing biodiversity because of potential links between biodiversity loss, and the loss of ecosystem goods and services provided by organisms in intact ecosystems. However, much of it is more correctly classified as pure ecological research, based on abstracted systems (Duffy 2009). Particularly in marine systems, the majority of published studies have been undertaken in laboratory mesocosms (Figure 13.1), which potentially limits their applicability to real-world biodiversity declines. In pure ecological terms, a considerable degree of progress has been made, and some widely accepted generalities are emerging (Hooper *et al.* 2005; Stachowicz *et al.* 2007). However, the extent to which these findings can be applied to specific management or conservation issues is less clear (Srivastava and Vellend 2005).

In this chapter, we examine the nature of BEF research to date in marine systems, and specifically contrast it with that done in terrestrial systems. We then discuss limitations of our current knowledge, attributable in part to the approaches taken, before

Marine Biodiversity and Ecosystem Functioning. First Edition. Edited by Martin Solan, Rebecca J. Aspden, and David M. Paterson.
© Oxford University Press 2012. Published 2012 by Oxford University Press.

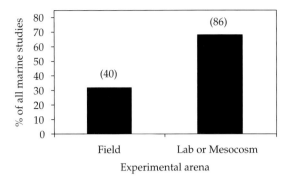

Figure 13.1 Experimental arenas of all marine BEF studies, including those evaluating both consumer and primary producer diversity. Data are based on papers summarized in Stachowicz *et al.* (2007). Relatively few studies (32%) have been conducted in the field; most (68%) have been done in the lab or in mesocosms. Parenthetical numbers over bars indicate the number of studies.

considering the relative merits of different approaches towards different objectives and making some suggestions for future research.

13.2 At which spatial and temporal scales have most biodiversity–ecosystem function (BEF) studies been conducted to date?

Two contemporary recent reviews, one summarizing BEF studies in marine systems (Stachowicz *et al.* 2007) and one meta-analysing the effects of producer diversity on biomass production in

terrestrial systems (Cardinale *et al.* 2007) allowed us to assess, compare, and contrast the spatial and temporal scales of BEF research. We added data on duration and area to the Supplementary Material on marine study characteristics in the Stachowicz *et al.* (2007) review and accessed the online material accompanying the Cardinale *et al.* (2007) analysis for information on duration, area, and effects in terrestrial studies (Table 13.1). Note that for the sake of comparison, a particular paper could contain more than one study if results of multiple experiments were reported.

On average, marine BEF studies ($N = 122$) have been run for 90 ± 14 (mean ±s.e.m.) days, and the average experimental arena is 0.29 ± 0.05 m². Most marine studies (69%) have been conducted in laboratory or mesocosm arenas (Figure 13.1). These patterns contrast sharply with the durations and spatial scales of terrestrial producer BEF studies ($N = 108$), where the average experiment has been run for 897 ± 75 days and the average experimental arena is 11 ± 2 m². The vast majority of terrestrial studies (89%) were conducted in the field. However, the Stachowicz *et al.* (2007) review includes studies evaluating not only the effects of producer diversity on plant community responses—e.g. biomass accumulation, resistance, recovery, or nutrient uptake; see Allison 2004; Bruno *et al.* 2005; Bracken and Stachowicz 2006 for examples—but also effects of infaunal invertebrates on sediment characteristics— e.g. Emmerson *et al.* 2001; Waldbusser *et al.* 2004),

Table 13.1 Attributes of marine and terrestrial BEF studies.

Attribute	Units	Values (means ± standard errors)			
		All marine studies	Marine producers	All terrestrial producers	Red. terrestrial producers[†]
Sample size	N	122	33	108	33
Area	m²	0.3 ± 0.1	0.8 ± 0.2	11.0 ± 2.2	3.0 ± 0.7
Duration	days	90 ± 14	158 ± 44	897 ± 75	552 ± 76
Expt. arena					
Field	%	32	65	89	85
Lab/mesocosm	%	68	35	11	15
Overyielding					
Non-transgr.	%	71	61	79	77
Transgressive	%	25	0	23	10

[†]'Reduced' terrestrial producer experiments are the earliest $N = 33$ studies reported in the Cardinale et al. (2007) database.

effects of producer or consumer diversity on invasibility (e.g. Stachowicz *et al.* 2002; Arenas *et al.* 2006), and effects of diversity at one trophic level on processes at other trophic levels (e.g. Duffy *et al.* 2003; O'Connor and Crowe 2005). In explicitly and statistically comparing marine and terrestrial studies, we therefore limited the marine database to only those studies ($N = 33$) that described effects of producer diversity on algal community responses.

Interestingly, the field–lab difference between marine and terrestrial studies narrowed greatly when only producers were considered. Whereas the majority of marine BEF studies have been conducted in lab and mesocosm settings (Figure 13.1), the majority of marine *producer* BEF studies (65%) have been conducted in the field (Figure 13.2a), and the overall pattern is not strikingly different from that in terrestrial systems. What is striking is the difference between the spatial and temporal scales over which marine and terrestrial experiments have been run (Figure 13.2b). Both the average duration (nearly 900 days) and area (11 m²) of terrestrial experiments are dramatically different from those in marine producer experiments, where the average duration is 158 ± 44 days and the average experimental arena measures only 0.8 ± 0.2 m².

There have been substantially more terrestrial producer BEF studies ($N = 108$) than marine producer studies ($N = 33$), which is due, in part, to a longer history of BEF experiments in terrestrial systems. The earliest experimental results reported in Cardinale *et al.* (2007) were published in 1985, whereas the earliest experiment reported in Stachowicz *et al.* (2007) was published in 1999, over a decade later. The longer time over which terrestrial producer BEF experiments have been evaluated could explain the longer duration of terrestrial studies, so we selected the $N = 33$ earliest terrestrial studies (published from 1985 to 2003) to balance sample sizes in marine and terrestrial systems. This reduced the average duration of terrestrial studies to 552 ± 76 days and the average experimental arena to 3.0 ± 0.7 m². Thus, balancing the number of studies and correcting for time partially, but not completely, compensated for differences in the duration ($t = 4.5, df = 64, P < 0.001$) and area ($t = 3.0, df = 64, P = 0.004$) of terrestrial versus marine pro-

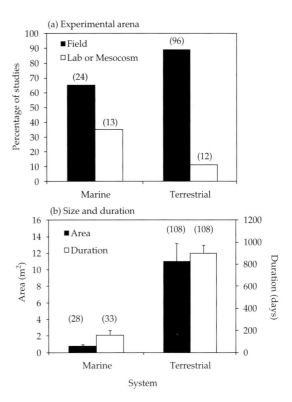

Figure 13.2 Characteristics of experiments evaluating the effects of primary producer diversity on plant community responses in marine and terrestrial systems. (a) Whereas overall patterns were similar—more studies were conducted in the field than in the lab in both systems—a relatively higher percentage of marine studies were conducted in lab and mesocosm arenas, whereas a higher percentage of terrestrial studies were conducted in the field($\chi^2 = 11.1, P < 0.001$). (b) Both the area ($t = 2.6$, df = 139, $P < 0.001$) and the duration ($t = 5.3$, df = 139, $P < 0.001$) of studies were higher in terrestrial systems. Parenthetical numbers over bars indicate the number of studies.

ducer BEF experiments. Similarly, the percentage of terrestrial studies conducted in the field declined slightly, to 85%, but that is still higher than the 65% of marine producer BEF studies conducted in the field ($\chi^2 = 10.7, P = 0.001$).

13.3 What important ecological processes or patterns may be lost in abstracting BEF experimental systems from natural ecosystems?

The legitimacy of all BEF research rests upon establishing a robust body of findings based on rigorous and realistic research (Naeem 2008). Therefore, comparison of different approaches and critical evaluation of their relative shortcomings and merits are essential. Although inferences from laboratory studies can be similar to those from comparable experiments done in the field, they can also differ in important respects (Carpenter 1996; Romanuk *et al.* 2009; Boyer *et al.* 2009). The magnitudes of effect sizes have also been shown to be inaccurately estimated by laboratory studies (Farrell and Crowe 2007), and a number of authors have argued that the domination of mesocosm studies in marine BEF research may have led to underestimation of the prevalence of diversity effects in marine systems (e.g. Stachowicz *et al.* 2008a; Duffy 2009). It is widely accepted that it is inappropriate to translate findings from laboratory or mesocosm studies directly into real-world applications, yet we face the challenge of designing more realistic empirical tests. Several innovative mesocosm-based studies have yielded insights into the mechanisms by which diversity may affect important patterns and processes (e.g. France and Duffy 2006; Cardinale 2011); however, scaling up from these experiments remains an issue. Here, we examine some of the specific deficiencies in our understanding of BEF relationships that may be arising from an overemphasis on lab- or mesocosm-based research.

The loss of natural patterns of dispersal is of particular concern. Mesocosm-based studies have shown that connectivity among habitat patches can modify the effect of species diversity on ecosystem properties (e.g. Matthiessen *et al.* 2007). A key advantage of most field experiments, including those in 'natural microcosms'—small, contained habitats that are naturally populated by minute organisms—is that they encompass movements of organisms between patches of varying degrees of suitability and connection—i.e. meta-community processes (Srivastava *et al.* 2004). Although the advantages of using natural microcosms for BEF studies have been identified (Srivastava *et al.* 2004), their potential remains underdeveloped. A drawback in their realistic interpretation, however, often revolves around the scale of the plots involved. Exchange of individuals between small patches on a shore—the scale at which most field experiments

are conducted—is a very different process from movements between whole shores or areas of coast—the scale at which local extinctions occur and at which managers are obliged to take action. The way that dispersal mediates the effects of biodiversity on ecosystem functioning and stability has only recently been investigated, and there remains a need to test experimentally how scales of resource heterogeneity and dispersal interact (Gonzalez *et al.* 2009). Similarly, ecosystem processes at small scales—in rock pools, for example—are likely to be quite different from those at larger scales and contingent on different limiting factors.

The expression of natural behavioural patterns is clearly compromised by confined laboratory settings—e.g. Thompson *et al.* 1998; Romanuk *et al.* 2009; Griffin *et al.* 2009). Impaired expression of natural behavioural patterns is more problematic in experiments involving consumers. Many species, especially intertidal animals, change their behaviour rapidly in response to variable environmental conditions (Chapman 2000). Thus, many of the problems related to highly controlled BEF experiments are even more challenging when multiple trophic levels are included (see below). In the field, intertidal grazers often reduce their foraging activity or shift foraging efforts to different habitats or times to avoid predators (Trussell *et al.* 2003). Such trait-dependent interactions may not be expressed or may be enhanced in the confines of an experimental system, particularly when predators are held in containers smaller than their normal territorial habitat, as is often the case. When considering questions that cannot be addressed easily in the field, it is therefore essential that such studies be accompanied by field tests of related hypotheses, so that differences between field-based behaviour and laboratory-based behaviour can be measured, and the relevance of the laboratory studies sensibly evaluated (Chapman 2000). Studies that compare the results of field- and mesocosm-based studies are rare but the few that exist have shown interesting and contrasting results (e.g., Stachowicz *et al.* 2008a; Boyer *et al.* 2009; O'Connor and Bruno 2009).

Several theories predict that effects of species richness should be stronger in heterogenous environments or landscapes (Tilman *et al.* 1997;

Cardinale *et al.* 2000; Loreau *et al.* 2003). Mathematical modelling (Cardinale *et al.* 2004), meta-analysis (Cardinale *et al.* 2007), and, more recently, direct experimental manipulations (Stachowicz *et al.* 2008a; Stachowicz *et al.* 2008b) have shown that the effect of diversity can increase with increasing experimental duration, and that increasing spatial scale should increase the strength of diversity effects due to the inclusion of a greater diversity of habitat patch types. Stachowicz *et al.* (2008a) emphasize the importance of structural heterogeneity, with a specific example that substrata used to 'cultivate' algae in assembly experiments may prevent natural recruitment or clonal spread and limit mechanisms to growth of individual ramets. The authors go further and suggest that by running analogous experimental manipulations in mesocosms and in the field over longer durations, it is possible to identify potential mechanisms that drive diversity effects. They argue that if mechanisms based on growth of adult thalli—such as complementarity of resource use—drive diversity effects, then we should expect to see similar responses from both types of experiments. If mechanisms that are more likely to operate in longer term studies and in systems open to processes such as recruitment—e.g. microhabitat differentiation, temporal complementarity, or facilitation—link diversity and function, then we would expect long-term field experiments to show strong diversity effects, and short-term mesocosm experiments to have weak or no effects. Stachowicz *et al.* (2008a) contend that short mesocosm experiments can only detect a subset of potential mechanisms, and conclude that the lack of species diversity effects reported to date should be interpreted with caution. They agree with others that call for longer experiments to test accurately for diversity effects (e.g. Cardinale *et al.* 2007).

Griffin *et al.* (2009) tested explicitly whether spatial heterogeneity of the physical environment can mediate effects of species diversity on ecosystem processes. They manipulated the diversity of intertidal molluscan grazers and the spatial heterogeneity of the substrata they grazed and measured algal consumption. They showed that on homogeneous substrata species identity had a strong effect, with the identity of best performing species dependent on the substratum. Heterogeneous substrata, containing suitable grazing conditions for all species, allowed the expression of spatial complementarity of resource use and thus enhanced total algal consumption, showing that spatial heterogeneity increases the importance of species richness for an ecosystem process.

As experimental systems become more complex and incorporate more environmental variability, the mechanisms by which species diversity stabilizes communities may also change because environmental control over populations may fluctuate (Ives and Carpenter 2007). Studies in a homogeneous lab environment may, therefore, be more likely to detect biotic interactions that may be far less significant under more natural conditions where abiotic or physical processes may dominate, and thus would not be detected in field experiments. Romanuk *et al.* (2009) tested for a relationship between species richness and community variability using similar communities in controlled laboratory microcosms, artificially constructed rock pools, and naturally occurring rock pools. The relationship was clear in laboratory microcosms, weaker in artificial rock pools, and absent in natural rock pools. The authors concluded that the effects of diversity may be difficult to detect in natural systems. It is also possible, however, that these effects are not as important as other abiotic or physical factors.

Mesocosm experiments, by their nature, limit temporal environmental heterogeneity and thereby reduce the need for multiple species capable of performing functional roles under different circumstances. Outbreaks of predators, storms, and other extreme events—heat waves, ice scour, etc.—can favour some species over others, such that diverse systems can continue to function effectively when depauperate systems may not. Homogeneous lab conditions are likely to favour one species, such that its monoculture is more likely to outperform mixtures than would be the case in a more variable natural environment. The lack of natural temporal variation in environmental conditions, coupled with the selection of convenient abundant taxa, also leads to underestimation of the importance of rare or temporarily redundant taxa, which may become abundant and/or influential under different environmental circumstances. This is a key omission, given that the majority of biodiversity in most ecosystems comprises rare taxa (de Aguiar *et al.* 2009). Many BEF experiments to date lack the temporal and spatial heterogeneity within replicates that may be key to species coexistence and enhance the likelihood that complementarity effects among species will be expressed (Stachowicz *et al.* 2007).

It may not be possible to accurately mimic multitrophic communities, including natural predator–prey interactions, using small-scale, short-term lab experiments. Moreover, the logistics involved in manipulating predator species diversity in the field are extremely difficult. This presents us with one the greatest challenges in BEF research because incorporating multitrophic-level effects into BEF theory, and increasing experimental realism have been identified as being among the next frontiers in BEF research (Srivastava and Vellend 2005; Stachowicz *et al.* 2007; Schmitz 2007; Bruno and Cardinale 2008; Duffy 2009). Paine (2002) suggested that the failure to incorporate consumers into BEF research has led to overly simplistic conclusions. In the quest to understand BEF relationships involving more realistic assemblages, it is important to consider appropriate densities of all species in experimental units. Many of the problems of extending model systems to incorporate multiple trophic levels are related to the problems associated with extending spatial and temporal scale (Bulling *et al.* 2006). The decision to include all species at similar densities or to create an assemblage that is representative of ambient densities in nature may affect the likelihood of identifying an effect, and should be based on the exact hypothesis being tested. For example, Bruno and O'Connor (2005) created tritrophic assemblages in mesocosms to test for effects of predator diversity. A similar biomass of each species of algae was added to each mesocosm at the start of the experiment, and a community-level trophic cascade effect was detected where the absence of carnivorous fish resulted in higher grazer abundance and lower algal biomass (Bruno and O'Connor 2005). O'Connor and Bruno (2007) used a similar experimental set-up but added algal species at different biomasses representative of each spe-

cies' natural abundances based on surveys at local field sites. In this experiment, the absence of fish also resulted in an increase in grazer abundance, but this did not lead to a reduction in total algal biomass. Instead, a shift in algal assemblage structure occurred, whereby red algae that were initially present in low proportions increased as the biomass of the initially dominant brown algae decreased (see also Jenkins *et al.* 1999; Benedetti-Cecchi *et al.* 2001). The more varied components of the initial algal assemblage in this experiment appear to have led to a different response of total algal mass, driven by varied responses of different algal taxa within the assemblage.

While it is clear that predator richness can influence the strength of trophic cascades by modifying indirect interactions, the traits of predators and their prey that influence the direction of these effects are not well understood (Schmitz 2007). Mixed results have been reported from the various experiments that have manipulated predator richness to date, most of which have been performed in featureless containers or homogenous landscapes (Bruno and Cardinale 2008, but see O'Connor *et al.* 2008).

Several authors have recently recommended that future studies should include biological and environmental contingencies that are likely to affect predator–prey interactions, and should explicitly test for effects of different spatial arrangements of habitats (e.g. Dyson *et al.* 2007; Griffin *et al.* 2009).

13.4   Does the reduced temporal/spatial scale or compromised ecological realism of marine BEF studies affect our ability to extrapolate results to other systems?

It is clear from our data and discussions above that the results of marine BEF studies, which have been largely conducted in laboratory or mesocosm arenas, in relatively small experimental units, and for short durations, may differ dramatically from those in other systems. What evidence exists to suggest that the functional consequences of marine biodiversity change are fundamentally different than those in terrestrial systems? Here, we turn again to data summarized in Stachowicz *et al.* (2007) and

Cardinale *et al.* (2007), and evaluate the frequency of mechanisms identified as links between biodiversity and ecosystem functioning.

A fundamental difference between marine and terrestrial results has been the relatively common finding that increases in terrestrial producer diversity result in *transgressive overyielding*, which occurs when a diverse assemblage performs better than the most effective component species by itself. This is the most conservative test for strong effects of changing biodiversity, and until recently, it had never been shown for marine producers (Figure 13.3). *Non-transgressive overyielding*, which occurs when a diverse assemblage performs better than the average of the component species, was a common finding in marine producer experiments; 61% of marine studies showed richness effects, compared to 79% of terrestrial studies. When we selected the 33 'earliest' terrestrial experiments to balance sample sizes and correct for duration effects, we still found slight differences in the percentage of non-transgressive overyielding effects in terrestrial (corrected to 77%) and marine studies ($\chi^2 = 5.9$, $P = 0.014$), but more significant differences in the percentage of experiments in terrestrial (corrected to 10%) and marine (0%) experiments exhibiting transgressive overyielding effects ($\chi^2 = 10.5$, $P = 0.001$). Note that limiting the potential duration of terrestrial producer experiments to balance the number of studies in marine and terrestrial systems more than halved (i.e. from 23% to 10%), the percentage of terrestrial studies exhibiting transgressive overyielding.

Why was transgressive overyielding—used synonymously with complementarity by many authors—so much more common in terrestrial systems? Two possibilities are suggested by the differences in the design of marine versus terrestrial experiments (Figure 13.2), and it seems likely that both play a role. Experimental duration has been linked to complementarity in both marine and terrestrial experiments (Cardinale *et al.* 2007; Fargione *et al.* 2007; Stachowicz *et al.* 2008a; Stachowicz *et al.* 2008b). All four of these studies demonstrate that as experimental duration increases, the importance of complementarity and

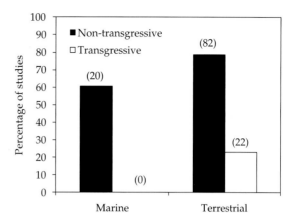

Figure 13.3 Overyielding in marine and terrestrial experiments evaluating the effects of primary producer diversity on plant community responses. Both transgressive and non-transgressive overyielding were more common in terrestrial systems ($\chi^2 = 5.1$, $P = 0.023$). In fact, Stachowicz *et al.* (2007) included no studies demonstrating transgressive overyielding in marine producer communities. Parenthetical numbers over bars indicate the number of studies.

richness effects increases, and the importance of selection and species identity effects declines. In the Cardinale *et al.* (2007) database, for example, the mean duration of experiments that demonstrated transgressive overyielding was 1592 days—over 4 years!—whereas the mean duration of experiments that showed non-transgressive overyielding was 665 days. In the one long-term marine study that has explicitly looked at how diversity effects on producer biomass change with time (Stachowicz *et al.* 2008b), the authors found that transgressive overyielding did not become apparent until the experiment had run for nine months, and the relative importance of diversity effects only consistently exceeded that of species identity effects after two years.

The area of the experimental arena also seems to be an important predictor of the potential for strong biodiversity effects. In the Cardinale *et al.* (2007) database, the experimental plots of terrestrial studies that demonstrated transgressive overyielding were considerably larger (31.9 ± 8.1 m²) than those that did not (5.1 ± 1.5 m²). The lack of transgressive overyielding in marine producer experiments precludes an identical comparison in marine systems, but marine experiments showing non-transgressive overyielding tended to be larger (1.0 ± 0.3 m^2) than those that did not (0.4 ± 0.1 m^2). In marine producer experiments, the effect of area, in and of itself, is difficult to separate from the effect of experimental arena—smaller experiments are usually conducted in lab or mesocosm settings—but the striking differences from terrestrial systems suggest that larger experiments are more likely to show strong effects of diversity.

It has been argued that complementarity effects are relatively rare, and identity effects—the key roles of particular species—are more important, in marine systems (see Crowe 2005; Gamfeldt and Bracken 2009). However, the comparisons we present above, as well as data in Cardinale *et al.* (2007) and Stachowicz *et al.* (2008a,b) suggest that most experiments evaluating the functional consequences of changes in marine producer diversity have not been big enough or long enough to detect some important mechanisms linking diversity and function.

Our discussion, thus far, has focused largely on producer diversity, as that is the common currency of the Stachowicz *et al.* (2007) and Cardinale *et al.* (2007) reviews. However, when the marine dataset was evaluated as a whole, transgressive overyielding became much more prevalent—it was found in 25% of all marine experiments—than in our analysis of producer experiments, despite the smaller area and shorter duration of experiments looking at sediment processes, invasibility, and trophic effects. Are primary producers anomalous in their requirements for space and time? The lack of a comparable dataset in terrestrial systems—where manipulations of consumer diversity are less common—makes it difficult to explicitly evaluate this possibility, but it seems likely that the benefits of conducting longer, larger experiments in the field would extend beyond primary producers. Some mechanisms linking diversity and function, such as seasonal complementarity in abundance or reproduction, will only become manifest over longer timescales, and larger experimental plots allow for more microhabitat heterogeneity, enhancing the potential for spatial complementarity (Stachowicz *et al.* 2008b).

13.5 Relative merits of different approaches to overcoming limitations of BEF studies

13.5.1 Empirical research to elucidate ecological concepts

Testing theoretical models of BEF relationships and characterizing their mechanistic underpinning requires complex experiments. Benedetti Cecchi (2004), for example, proposed a framework with an effective minimum of 16 treatments, and many more are required for experiments incorporating additional factors beyond diversity, density, and identity—e.g. environmental context, realistic versus random extinction, multitrophic effects, species versus functional group diversity. Each treatment must be replicated, and experiments with > 100 manipulated units may be needed to test complex models (e.g. Gamfeldt *et al.* 2005; Douglass *et al.* 2008; Nicol and Crowe, unpublished data). Designing experiments and selecting the combinations of species within treatments to examine BEF relationships requires that researchers make a series of decisions about how best to reduce the enormous numbers of possible species combinations to something manageable while not compromising realism, for example by mixing species that would not co-occur in nature (Srivastava *et al.* 2004; Naeem 2008). Tractable systems involving species that can be conveniently manipulated and functions that can be easily measured are required to run such experiments and statistical power is maximized if uncontrolled environmental variation is kept to a minimum.

It could be argued, therefore, that direct tests of theory are most easily achieved with laboratory or mesocosm experiments (Bulling *et al.* 2006), and that is certainly reflected in the decisions of many marine researchers to date (see above). Such experiments must inevitably focus on comparatively small organisms that are easily found in large numbers and can be conveniently kept in aquaria. Lab and mesocosm experiments can be well replicated and closely controlled, and offer a good opportunity to establish patterns and mechanisms that may arise in nature. Their main drawback is, of course, their artificiality, which potentially makes them irrelevant to the natural world (Carpenter 1996).

More realistic tests can be undertaken in manipulative field experiments, although procedural artifacts can certainly arise and careful design is necessary (e.g. Underwood, 1997; Benedetti-Cecchi 2004; Benedetti-Cecchi 2006). Field experiments are perhaps logistically more challenging and are certainly less controlled than lab experiments. The lack of control is, however, their greatest strength; they are able to test whether factors found to be important in the lab remain so against a background of natural variation. If, in a carefully designed field experiment, residual variation outweighs diversity effects, then diversity is a less important factor than other, uncontrolled factors in the system. In fact, such variation often drives diversity effects as different species perform optimally under different conditions. An important distinction has been drawn between Synthetic Assemblage Experiments (SAEs), in which comparatively small numbers of taxa are brought together artificially, and Removal Experiments (REs), in which taxa are removed from plots which initially contain complete natural assemblages (Diaz *et al.* 2003). SAEs offer a high degree of control, but may include combinations of species that do not occur in nature, which may severely impair their capacity to detect diversity effects (Bracken *et al.* 2008). REs can only be done in the field and benefit from involving assemblages that have developed naturally over time and have been structured by real interactions (Diaz *et al.* 2003). It is also worth distinguishing between experiments manipulating arbitrarily defined patches of habitat and those in which experimental units can more defensibly be defined as delimited ecosystems, such as rock pools, lagoons, or lakes, which again offer a higher degree of realism (Srivastava *et al.* 2004).

A compromise when field experimentation is not possible may be to use outdoor mesocosms that mimic a relatively open system, and should employ a through-flow raw seawater system that at least enables recruitment of ephemeral species during the experiment (e.g. O'Connor and Bruno 2007). The effect of propagule supply regulating marine benthic systems has recently been investigated (Lee and Bruno 2009), highlighting the importance of creating 'open systems' in BEF mesocosm studies.

Experiments must, however, be carried out over longer timescales to enhance our understanding of whether remaining species can numerically or behaviourally compensate for the loss of superficially similar species to assess the degree to which species are ecologically redundant (Stachowicz *et al.* 2007), and this can only be done accurately with field-based manipulations.

Scale is perhaps the biggest issue in extrapolating findings from complex manipulative experiments to real ecosystems (Symstad *et al.* 2003). There is a trade-off between spatial and temporal scale, and the degree of complexity and replication of studies (Raffaelli 2006). Experiments with large plot sizes tend to be poorly replicated or completely unreplicated (Raffaelli 2006). While we recognize the need for simpler experiments done at larger scales to bridge the gap between theory and practice, and to encompass whole ecosystems at scales relevant to management, unreplicated or pseudoreplicated comparisons between areas of differing diversity or condition—e.g. a marine protected area vs. a comparable area open to fishing—cannot be logically interpreted and should be avoided. We agree with Rafaelli (2006) that scales of experiments should be defined by features of the system, such as home ranges of organisms or generation times of their populations, rather than practical considerations, as is more commonly the case. If population processes are to be incorporated, using microbes as a model system offers one of the few opportunities for complex experiments over multi-generational time frames (e.g. Petchey *et al.* 2002). However, microbial microcosm experiments have been criticized for potentially low generality to larger fauna and flora, relying on oversimplified communities and the use of inappropriate scales to test theories based on the mechanisms driving ecosystem processes (Carpenter 1996; Srivastava *et al.* 2004; Jessup *et al.* 2004).

13.5.2 Empirical research for direct application to management/conservation

Research for direct application to specific management challenges requires a lesser degree of generality, but a greater degree of realism (Naeem, 2008). It is more likely to be focused on a particular ecosystem with particular species and stressors. Managers need to understand the potential consequences of a range of conservation strategies. Work with direct relevance to management often needs to be done at larger scales than work intended to explore nature for its own sake (see Chapter 11, this volume). Local extinctions and the actions of stressors tend to occur at the scale of whole shorelines or stretches of coast, and to span a number of years either in terms of the application of the stress—e.g. chronic sewage pollution or harvesting—or recovery—e.g. from an acute event such as an oil spill. It is not clear whether findings from small-scale studies can be scaled up, and doing so is likely to be misleading (Symstad *et al.* 2003; Cardinale *et al.* 2004; Srivastava *et al.* 2004; Bulling *et al.* 2006). For example, Byrnes and Stachowicz (2009) recently showed that the effects of changes in consumer species richness may manifest over longer time scales than permitted in short term lab-based studies.

It is rare that research projects receive funding to address processes over these time scales (but see Symstad *et al.* 2003), and, in many cases, it is logistically unfeasible and/or ethically indefensible to undertake experimental manipulations at large spatial scales. It is therefore necessary to adopt a more mensurative approach to characterizing patterns of impact. Peters (1991) argued eloquently that mechanistic understanding is not a prerequisite for management action; observed relationships between particular stressors and measured responses provide a valid basis for prediction. Evidence of real world benefits of biodiversity may be based on such predictive relationships (e.g. see Klein *et al.* 2003; Tylianakis *et al.* 2008, reviewed by Duffy 2009). The mechanistic understanding of cause and effect provided by experimental manipulations can greatly improve prediction, however. In this context, it is necessary to take advantage of opportunities presented by planned impacts or management interventions to design rudimentary large-scale experiments incorporating sampling at impacted sites and at carefully selected controls (e.g. Castilla and Fernandez 1998; Salomon *et al.* 2008; and see Walters 1986; Underwood 1995; Levin *et al.* 2009). Such experiments will rarely, if ever, be able to test complex BEF models, but they can be

valuable in assessing the impact of changing abundance of particular components of diversity. A few other opportunities are available. For example, Zedler *et al.* (2001) tested BEF hypotheses as a component of research into large-scale habitat restoration, an application for which SAEs are explicitly required. In their experiments, it was possible to use complex designs, replicated in comparatively large-scale plots, and to generate findings of direct relevance to management.

The information and mechanistic understanding gained from decades of pure research must also be made available to managers in a synthesized and accessible form ('decision support tools', Nobre and Ferreira 2009). Ideally, managers would have access to accurate models of the ecosystems for which they are responsible, which they could use to simulate disturbances or changes to management (e.g. reductions in fishing pressure or increases in nutrient input) and evaluate the consequences for the functioning of the system, its 'ecological quality' or 'conservation status', and its capacity to provide ecosystem goods and services to society. They could use such models to inform decisions about which combinations of activities to allow in a system, taking account of the modifying influence of forecast changes in climate. Such models would require a high degree of mechanistic understanding of specific ecosystems. We do not currently have a sufficient understanding to develop effective models for any but a few impacts on any but a few, comprehensively studied systems. Nevertheless, initial models could be constructed for a larger number of systems, and sensitivity analyses could help to identify priority areas for empirical research into key mechanistic links. The findings of such research could then be incorporated into the models such that theoretical and empirical research agendas are directly interlinked.

On a more immediately accessible level, existing empirical knowledge of the functional traits of the component taxa in systems can be used to help environmental managers predict ecosystem consequences of changes to the biota in a system by using tools such as Biololgical Traits Analysis (Frid *et al.* 2008). Biological Traits Analysis characterizes all relevant traits in all taxa present, and expresses these as a matrix of functionality in which values assigned for particular taxa are weighted by their abundance. Such matrices can be analysed using exactly the same methods as multivariate matrices of species' abundances or biomass, and are sensitive to functionally relevant changes in assemblage structure (Bremner *et al.* 2003). Biological traits composition has been shown to be responsive to human impacts (Bremner *et al.* 2003), permitting a degree of simulation/prediction of functional consequences of particular impacts (Frid *et al.* 2008).

13.6 Conclusions

There is a clear case for a combined approach to the empirical side of BEF research (Bulling *et al.* 2006; Naeem 2008). Laboratory experiments are of value because of the possibility to control the environment very precisely, measure a range of ecosystem functions quite easily, and replicate the sorts of complex designs necessary to disentangle richness, identity, and density effects. They give an idea of the sorts of BEF relationships that *may* occur in nature, but need to be supplemented by field-based experiments of some kind. Field-based studies of longer duration are more likely to reveal natural mechanisms by which biodiversity can affect the functioning of ecosystems (Stachowicz *et al.*, 2008a). A key point is to test whether potential effects observed in laboratories are actually significant against a background of natural variability, and with the possibility of dispersal into and out of the systems. To generate findings at the scales required by managers, larger, more opportunistic experiments that take advantage of perturbations or management interventions—e.g. fisheries or protected areas—must be done by sampling diversity and measuring functioning in those areas and making comparisons with carefully selected controls. It could be argued that if BEF researchers wish to make their work more directly relevant to society, their research should be driven by a focus on key ecosystem services and designed accordingly (Raffaelli 2006), rather than focusing on key theoretical questions and selecting the most tractable systems and response variables.

Arguably, the distinction we make between 'pure' and 'applied' research is arbitrary. Certainly, so-called 'pure' research should also aim to be as realistic as possible and develop meaningful insight into the functioning of real ecosystems, whereas applied research has no less a requirement for logical rigour (Underwood 1995). Whatever approach is taken, careful design and interpretation are essential (Huston 1997; Allison 1999; Benedetti-Cecchi 2004). This relies to a great extent on a clear statement of hypotheses at the outset, and due consideration of the ultimate rationale for the work.

In both marine and terrestrial BEF research, the need to work with tractable systems, either in the lab or the field, has inevitably led to a bias towards particular habitats and systems—e.g. intertidal and shallow subtidal benthic habitats as opposed to pelagic or deeper subtidal systems; see Chapter 9—which may lead to misleading generalizations. It is important, as we expand BEF research into more systems, that we use comparable designs to avoid comparisons confounded by the problems associated with scale and length of experiment highlighted above.

In this chapter, we have focused predominantly on experimental manipulations of organismal diversity. In closing, we wish to join other authors (e.g. Srivastava and Vellend 2005; Raffaelli 2006; Naeem 2006; Naeem 2008) in emphasizing that such experiments must fit within a broader framework if BEF research is to deliver genuine ecological insight and practical tools for environmental management. Other chapters in this volume emphasize the role of observational studies and modelling approaches, the importance of considering diversity at genetic and habitat levels, and the importance of linking research with societal needs. We fully agree with Naeem's (2008) assertion that pluralistic, synthetic approaches should be the dominant trend in BEF research as it seeks to provide real-world applications to the widespread problem of biodiversity loss.

Acknowledgements

We are grateful to Diane Srivastava for her input at an early stage in the preparation of this chapter. TPC was supported by Science Foundation Ireland,

MESB was supported by the National Science Foundation (OCE-0549944) and NEOC was supported by an Environmental Protection Agency Fellowship (STRIVE: 2007-FS-B-8-M5).

References

Allison, G. (2004) The influence of species diversity and stress intensity on community resistance and resilience. *Ecological Monographs*, **74**, 117–34.

Allison, G. W. (1999) The implications of experimental design for biodiversity manipulations. *American Naturalist* **153**, 26–45.

Arenas, F., Sanchez, I., Hawkins, S. J. *et al.* (2006) The invasibility of marine algal assemblages: Role of functional diversity and identity. *Ecology*, **87**, 2851–61.

Benedetti-Cecchi, L. (2004) Increasing accuracy of causal inference in experimental analyses of biodiversity. *Functional Ecology*, **18**, 761–68.

Benedetti-Cecchi, L. (2006) Understanding the consequences of changing biodiversity on rocky shores: How much have we learned from past experiments? *Journal of Experimental Marine Biology and Ecology*, **338**, 193–204.

Benedetti-Cecchi, L., Pannacciuli, F., Bulleri, F. *et al.* (2001) Predicting the consequences of anthropogenic disturbance: large-scale effects of loss of canopy algae on rocky shores. *Marine Ecology Progress Series* **214**, 137–50.

Boyer, K. E., Kertesz, J. S. and Bruno, J. F. (2009) Biodiversity effects on productivity and stability of marine macroalgal communities: the role of environmental context. *Oikos*, **118**, 1062–72.

Bracken, M. E. S., Friberg, S. E., Gonzalez-Dorantes, C. A. *et al.* (2008) Functional consequences of realistic biodiversity changes in a marine ecosystem. *Proceedings of the National Academy of Sciences of the United States of America*, **105**, 924–28.

Bracken, M. E. S. and Stachowicz, J. J. (2006) Seaweed diversity enhances nitrogen uptake via complementary use of nitrate and ammonium. *Ecology*, **87**, 2397–403.

Bremner, J., Rogers, S. I. and Frid, C. L. J. (2003) Assessing functional diversity in marine benthic ecosystems: A comparison of approaches. *Marine Ecology-Progress Series*, **254**, 11–25.

Bruno, J. F., Boyer, K. E., Duffy, J. E. *et al.* (2005) Effects of macroalgal species identity and richness on primary production in benthic marine communities. *Ecology Letters*, **8**, 1165–74.

Bruno, J. F. and Cardinale, B. J. (2008) Cascading effects of predator richness. *Frontiers in Ecology and the Environment*, **6**, 539–46.

Bruno, J. F. and O'Connor, M. I. (2005) Cascading effects of predator diversity and omnivory in a marine food web. *Ecology Letters*, **8**, 1048–56.

Bulling, M. T., White, P. C. L., Raffaelli, D. G. *et al.* (2006) Using model systems to address the biodiversity-ecosystem functioning process. *Marine Ecology-Progress Series*, **311**, 295–309.

Byrnes, J. and Stachowicz, J. J. (2009) Short and long term consequences of increases in exotic species richness on water filtration by marine invertebrates. *Ecology Letters*, **12**, 830–41.

Cardinale, B. J. (2011) Biodiversity improves water quality through niche partitioning. *Nature*, **472**, 86–9.

Cardinale, B. J., Ives, A. R. and Inchausti, P. (2004) Effects of species diversity on the primary productivity of ecosystems: extending our spatial and temporal scales of inference. *Oikos*, **104**, 437–50.

Cardinale, B. J., Nelson, K. and Palmer, M. A. (2000) Linking species diversity to the functioning of ecosystems: on the importance of environmental context. *Oikos*, **91**, 175–83.

Cardinale, B. J., Wright, J. P., Cadotte, M. W. *et al.* (2007) Impacts of plant diversity on biomass production increase through time because of species complementarity. *Proceedings of the National Academy of Sciences of the United States of America*, **104**, 18123–8.

Carpenter, S. R. (1996) Microcosm experiments have limited relevance for community and ecosystem ecology. *Ecology*, **77**, 677–80.

Castilla, J. C. and Fernandez, M. (1998) Small-scale benthic fisheries in Chile: On co-management and sustainable use of benthic invertebrates. *Ecological Applications*, **8**, S124–S132.

Chapman, M. G. (2000) Poor design of behavioural experiments gets poor results: examples from intertidal habitats. *Journal of Experimental Marine Biology and Ecology*, **250**, 77–95.

Crowe, T. P. (2005) What do species do in intertidal systems? In: Wilson, J. G. (Ed.) *The Intertidal Ecosystem: The Value of Ireland's Shores*. Dublin, Royal Irish Academy, pp. 115–33.

De Aguiar, M. A. M., Baranger, M., Baptestini, E. M. *et al.* (2009) Global patterns of speciation and diversity. *Nature*, **460**, 384–U98.

Diaz, S., Symstad, A. J., Chapin, F. S. *et al.* (2003) Functional diversity revealed by removal experiments. *Trends in Ecology and Evolution*, **18**, 140–6.

Douglass, J. G., Duffy, J. E. and Bruno, J. F. (2008) Herbivore and predator diversity interactively affect ecosystem properties in an experimental marine community. *Ecology Letters*, **11**, 598–608.

Duffy, J. E. (2009) Why biodiversity is important to the functioning of real-world ecosystems. *Frontiers in Ecology and the Environment,* **7**(8), 437–444.

Duffy, J. E., Richardson, J. P. and Canuel, E. A. (2003) Grazer diversity effects on ecosystem functioning in seagrass beds. *Ecology Letters,* **6**, 637–45.

Dyson, K. E., Bulling, M. T., Solan, M. *et al.* (2007) Influence of macrofaunal assemblages and environmental heterogeneity on microphytobenthic production in experimental systems. *Proceedings of the Royal Society B-Biological Sciences,* **274**, 2547–54.

Emmerson, M. C., Solan, M., Emes, C. *et al.* (2001) Consistent patterns and the idiosyncratic effects of biodiversity in marine ecosystems. *Nature,* **411**, 73–7.

Fargione, J., Tilman, D., Dybzinski, R. *et al.* (2007) From selection to complementarity: Shifts in the causes of biodiversity-productivity relationships in a long-term biodiversity experiment. *Proceedings of the Royal Society B: Biological Sciences,* **274**, 871–76.

Farrell, E. D. and Crowe, T. P. (2007) The use of byssus threads by Mytilus edulis as an active defence against Nucella lapillus. *Journal of the Marine Biological Association of the United Kingdom,* **87**, 559–64.

France, K. and Duffy, J. E. (2006) Diversity and dispersal interactively affect predictability of ecosystem function. *Nature,* **441**, 1139–43.

Frid, C. L. J., Paramor, O. A. L., Brockington, S. *et al.* (2008) Incorporating ecological functioning into the designation and management of marine protected areas. *Hydrobiologia,* **606**, 69–79.

Gamfeldt, L. and Bracken, M. E. S. (2009) The role of biodiversity for the functioning of rocky reef communities. In: Wahl, M. (ed), *Marine Hard Bottom Communities: Patterns, Dynamics, Diversity, and Change.* Heidelberg, Springer-Verlag, pp. 361–73.

Gamfeldt, L., Hillebrand, H. and Jonsson, P. R. (2005) Species richness changes across two trophic levels simultaneously affect prey and consumer biomass. *Ecology Letters,* **8**, 696–703.

Gonzalez, A., Mouquet, N. and Loreau, M. (2009) Biodiversity as spatial insurance: the effects of habitat fragmentation and dispersal on ecosystem functioning. In: Naeem, S., Bunker, D. E., Hector, A. *et al.* (eds), *Biodiversity, Ecosystem Functioning, and Human Wellbeing: an ecological perspective.* Oxford, Oxford University Press, pp. 134–46.

Griffin, J. N., Jenkins, S. R., Gamfeldt, L. *et al.* (2009) Spatial heterogeneity increases the importance of species richness for an ecosystem process. *Oikos,* **118**, 1335–42.

Hector, A., Schmid, B., Beierkuhnlein, C. *et al.* (1999) Plant diversity and productivity experiments in European grasslands. *Science,* **286**, 1123–7.

Hooper, D. U., Chapin, F. S., Ewel, J. J. *et al.* (2005) Effects of biodiversity on ecosystem functioning: a consensus of current knowledge. *Ecological Monographs,* **75**, 3–35.

Huston, M. A. (1997) Hidden treatments in ecological experiments: re-evaluating the ecosystem function of biodiversity. *Oecologia,* **110**, 449–60.

Ives, A. R. and Carpenter, S. R. (2007) Stability and diversity of ecosystems. *Science,* **317**, 58–62.

Jenkins, S. R., Hawkins, S. J., and Norton, T. A. (1999) Interaction between a fucoid canopy and limpet grazing in structuring a low shore intertidal community. *Journal of Experimental Marine Biology and Ecology* **233**, 41–63.

Jessup, C. M., Kassen, R., Forde, S. E., Kerr *et al.* (2004) Big questions, small worlds: microbial model systems in ecology. *Trends in Ecology and Evolution,* **19**, 189–97.

Klein, A. M., Steffan-Dewenter, I. and Tscharntke, T. (2003) Fruit set of highland coffee increases with the diversity of pollinating bees. *Proceedings of the Royal Society of London Series B-Biological Sciences,* **270**, 955–61.

Lee, S. C. and Bruno, J. F. (2009) Propagule supply controls grazer community structure and primary production in a benthic marine ecosystem. *Proceedings of the National Academy of Sciences of the United States of America,* **106**, 7052–7.

Levin, P. S., Kaplan, I., Grober-Dunsmore, R. *et al.* (2009) A framework for assessing the biodiversity and fishery aspects of marine reserves. *Journal of Applied Ecology,* **46**, 735–42.

Loreau, M., Mouquet, N. and Gonzalez, A. (2003) Biodiversity as spatial insurance in heterogeneous landscapes. *Proceedings of the National Academy of Sciences, USA,* **100**, 12765–70.

Matthiessen, B., Gamfeldt, L., Jonsson, P. R. *et al.* (2007) Effects of grazer richness and composition on algal biomass in a closed and open marine system. *Ecology,* **88**, 178–87.

Menge, B. A., Chan, F., Dudas, S. *et al.* (2009) Terrestrial ecologists ignore aquatic literature: Asymmetry in citation breadth in ecological publications and implications for generality and progress in ecology. *Journal of Experimental Marine Biology and Ecology,* **377**, 93–100.

Naeem, S. (2006) Expanding scales in biodiversity-based research: challenges and solutions for marine systems. *Marine Ecology-Progress Series,* **311**, 273–83.

Naeem, S. (2008) Advancing realism in biodiversity research. *Trends in Ecology and Evolution,* **23**, 414–16.

Naeem, S., Thompson, L. J., Lawler, S. P. *et al.* (1994) Declining biodiversity can alter the performance of ecosystems. *Nature,* **368**, 734–7.

Nobre, A. M. and Ferreira, J. G. (2009) Integration of ecosystem-based tools to support coastal zone management. *Journal of Coastal Research* **2**, 1676–80.

O'Connor, M. I. and Bruno, J. F. (2009) Predator richness has no effect in a diverse marine food web. *Journal of Animal Ecology*, **78**, 732–40.

O'Connor, N. E. and Bruno, J. F. (2007) Predatory fish loss affects the structure and functioning of a model marine food web. *Oikos*, **116**, 2027–38.

O'Connor, N. E. and Crowe, T. P. (2005) Biodiversity loss and ecosystem functioning: Distinguishing between number and identity of species. *Ecology*, **86**, 1783–96.

O'Connor, N. E., Grabowski, J. H., Ladwig, L. M. *et al.* (2008) Simulated predator extinctions: Predator identity affects survival and recruitment of oysters. *Ecology*, **89**, 428–38.

Paine, R. T. (2002) Trophic control of production in a rocky intertidal community. *Science*, **296**, 736–9.

Petchey, O.L., Casey, T., Jiang, L. *et al.* (2002) Species richness, environmental fluctuations, and temporal change in total community biomass. *Oikos* **99**, 231–40.

Peters, R. H. (1991) *A Critique for Ecology*, Cambridge, Cambridge University Press.

Raffaelli, D., Solan, M. and Webb, T. J. (2005) Do marine and terrestrial ecologists do it differently? *Marine Ecology-Progress Series*, **304**, 283–9.

Raffaelli, D. G. (2006) Biodiversity and ecosystem functioning: issues of scale and trophic complexity. *Marine Ecology-Progress Series*, **311**, 285–94.

Romanuk, T. N., Vogt, R. J. and Kolasa, J. (2009) Ecological realism and mechanisms by which diversity begets stability. *Oikos*, **118**, 819–28.

Salomon, A. K., Shears, N. T., Langlois, T. J. *et al.* (2008) Cascading effects of fishing can alter carbon flow through a temperate coastal system. *Ecological Applications*, **18**, 1874–87.

Schmid, B., Balvanera, P., Cardinale, B. J. *et al.* (2009) Consequences of species loss for ecosystem functioning: meta-analyses of data from biodiversity experiments. In: Naeem, S., Bunker, D. E., Hector, A. *et al.* (eds), *Biodiversity, Ecosystem Functioning, and Human Wellbeing: an ecological perspective.* Oxford, Oxford University Press.

Schmitz, O. J. (2007) Predator diversity and trophic interactions. *Ecology*, **88**, 2415–26.

Solan, M., Cardinale, B. J., Downing, A. L. *et al.* (2004) Extinction and ecosystem function in the marine benthos. *Science*, **306**, 1177–80.

Srivastava, D. S., Kolasa, J., Bengtsson, J. *et al.* (2004) Are natural microcosms useful model systems for ecology? *Trends in Ecology and Evolution*, **19**, 379–84.

Srivastava, D. S. and Vellend, M. (2005) Biodiversity-ecosystem function research: is it relevant to conservation? *Annual Review of Ecology Evolution and Systematics*, **36**, 267–94.

Stachowicz, J. J., Best, R. J., Bracken, M. E. S. *et al.* (2008a) Complementarity in marine biodiversity manipulations: Reconciling divergent evidence from field and mesocosm experiments. *Proceedings of the National Academy of Sciences of the United States of America*, **105**, 18842–7.

Stachowicz, J. J., Bruno, J. F. and Duffy, J. E. (2007) Understanding the effects of marine biodiversity on communities and ecosystems. *Annual Review of Ecology Evolution and Systematics*, **38**, 739–66.

Stachowicz, J. J., Fried, H., Osman, R. W. *et al.* (2002) Biodiversity, invasion resistance, and marine ecosystem function: Reconciling pattern and process. *Ecology*, **83**, 2575–90.

Stachowicz, J. J., Graham, M., Bracken, M. E. S. *et al.* (2008b) Diversity enhances cover and stability of seaweed assemblages: The role of heterogeneity and time. *Ecology*, **89**, 3008–19.

Stachowicz, J. J., Whitlatch, R. B. and Osman, R. W. (1999) Species diversity and invasion resistance in a marine ecosystem. *Science*, **286**, 1577–9.

Symstad, A. J., Chapin Iii, F. S., Wall, D. H. *et al.* (2003) Long-term and large-scale perspectives on the relationship between biodiversity and ecosystem functioning. *Bioscience*, **53**, 89–98.

Thompson, R. C., Norton, T. A. and Hawkins, S. J. (1998) The influence of epilithic microbial films on the settlement of Semibalanus balanoides cyprids—a comparison between laboratory and field experiments. *Hydrobiologia*, 375–376, 203–16.

Tilman, D., Naeem, S., Knops, J. *et al.* (1997) Biodiversity and ecosystem properties. *Science*, **278**, 1865–6.

Tilman, D., Wedin, D. and Knops, J. (1996) Productivity and sustainability influenced by biodiversity in grassland systems. *Nature*, **379**, 718–20.

Trussell, G. C., Ewanchuk, P. J. and Bertness, M. D. (2003) Trait-mediated effects in rocky intertidal food chains: Predator risk cues alter prey feeding rates. *Ecology*, **84**, 629–40.

Tylianakis, J. M., Rand, T. A., Kahmen, A. *et al.* (2008) Resource heterogeneity moderates the biodiversity-function relationship in real world ecosystems. *PLoS Biology*, **6**, 947–56.

Underwood, A. J. (1995) Ecological research (and research into) environmental management. *Ecological Applications*, **5**, 232–47.

Underwood, A. J. (1997) *Experiments in ecology: their logistical design and interpretation using analysis of variance.*, Cambridge, Cambridge University Press.

Waldbusser, G. G., Marinelli, R. L., Whitlatch, R. B. and et al. (2004) The effects of infaunal biodiversity on bio-geochemistry of coastal marine sediments. *Limnology and Oceanography*, **49**, 1482–92.

Walker, B. H. (1992) Biodiversity and ecological redundancy. *Conservation Biology*, **6**, 18–23.

Walters, C. (1986) *Adaptive management of renewable resources*. New York, McGraw Hill.

Zedler, J. B., Callaway, J. C. and Sullivan, G. (2001) Declining biodiversity: Why species matter and how their functions might be restored in Californian tidal marshes. *Bioscience*, **51**, 1005–17.

CHAPTER 14

Why bother going outside: the role of observational studies in understanding biodiversity–ecosystem function relationships

Simon F. Thrush and Andrew M. Lohrer

14.1 The role of observation in the design, execution, and interpretation of BEF relationships

Observation of nature is a fundamental element of ecology and environmental science that provides inspiration, fires our imagination, challenges us to define useful experiments, and helps us select relevant theory. However, there is no single natural scale at which ecological phenomena can be observed and our own perceptual filters influence the patterns we observe (Levin 1992). Furthermore, mechanisms and patterns do not necessarily occur at the same scales. Patterns can be constrained by large-scale environmental forcing—e.g. wave energy can influence the strength of adult/juvenile interactions in bivalves (Thrush *et al.* 2000)—or emerge as a result of the collective activities of organisms interacting on fine scales—e.g. the hummock and hollow patterns on some mudflats generated by the influence of microphytobenthos on sediment stability (Weerman *et al.* 2010). Nevertheless, it is important to try to understand the mechanisms that underlie observable patterns. This is a major theme in the biodiversity–ecosystem functioning (BEF) literature, where ecologists are seeking to clarify the contribution of biodiversity to the delivery of ecosystem goods and services.

Biodiversity loss is a systemic response to stress or disturbance and not simply a random process in a homogeneous landscape, as is assumed in many models and controlled experiments. In this chapter, we discuss the role of observational studies; we classify observational studies as empirical research that does not involve manipulative experimentation. This can include a broad range of approaches from observations of natural history, behaviour, and bio-physical interactions, to the descriptions of patterns in space and time, to correlative studies and explicit broad-scale hypothesis tests of patterns and relationships. Our intention is to encourage a more integrated empirical approach to BEF research by emphasizing what is gained by field observation compared with controlled experiments in studying these patterns and processes. We seek to emphasize the merits of combined and iterative approaches using both experimental and observational studies. We draw on a long history of experimental marine ecology to demonstrate the value of observational studies as a route to providing context, generality, and mechanisms to scale up our results. Where possible we illustrate with examples from the BEF literature, but we also draw on key species–ecosystem function studies to illustrate the value of more integrative approaches.

Observational studies illustrate how function can be driven by changes in size, density and spatial pattern of organisms. Marine soft-sediment habitats deliver a number of ecosystem functions associated with nutrient and carbon cycling, sediment trapping/transport/stability, habitat

Marine Biodiversity and Ecosystem Functioning. First Edition. Edited by Martin Solan, Rebecca J. Aspden, and David M. Paterson.
© Oxford University Press 2012. Published 2012 by Oxford University Press.

formation, provision of settlement sites and refugia, productivity, and resilience. We discuss the heterogeneous nature of seafloor landscapes and the potential influences of this variability on BEF relationships. Next, we discuss scaling issues for experiments, in particular the issue of generality in determining the significance of biodiversity–ecosystem function relations across seafloor landscapes. Finally, we emphasize the importance of observational studies in providing a context for model development, verification, and testing.

The very nature of BEF experiments makes them complicated and difficult to perform because they involve measuring interactions between biodiversity and various bio-geo-physico-chemical processes. When designing an experiment that seeks to identify the role of particular combinations of species in affecting an ecosystem function, it does not take more than a few species before the number of possible combinations swamps the most ardent experimentalist (Naeem 2008). In aquatic studies, this complexity is often reduced using aquarium experiments or mesocosms as 'model systems'. Mesocosm studies simplify nature by using a limited suit of species in a controlled environment—e.g. buckets or plastic containers. Often the 'ecosystem response' is limited to one or a few related functions. These model system studies can be practical, tractable, and rigorously designed, and can be useful for isolating some of the mechanisms underpinning BEF relationships (Benton *et al.* 2007). However, the process of simplification can hide valuable information about the workings of natural systems and the applicability of mechanisms at varying scales. These implications must be carefully considered when we assess the ecological relevance of these model systems (Petersen *et al.* 2009).

There are two basic, but strongly overlapping, reasons to improve our understanding of BEF relationships: to understand how ecosystems work, and to apply this knowledge to resource management, conservation, and environmental economics (Stachowicz *et al.* 2007). When scientific information is applied to environmental issues, it is imperative that we understand its relevance and generality. In fact, Srivastava and Vellend (2005) have questioned the value of justifying the conservation of biodiver-

sity on the basis of maintaining ecosystem functions because of the continued uncertainty over the relevance and generality of BEF relationships. The weight of evidence for biodiversity to be important in defining ecosystem function, at least for some functions and in some locations, is growing (Caliman *et al.* 2010), but Srivastava and Vellend (2005) highlight the need for rigorous and integrative science if general principles are to be found. There is the potential for tight coupling between biodiversity and ecosystem function, though these relationships are not necessarily uni-directional (Figure 14.1). BEF studies have tended to emphasize the way some particular element of biodiversity—e.g. genotypic, or phenotypic diversity, species richness, or some measure of functional diversity—influences some measure(s) of ecosystem functional performance as a cause and effect relationship, rather than as a feedback loop. When feedback process are important, designing cause and effect experiments requires careful consideration of the processes involved and the scales over which they interact. In terms of recognizing the value of observational studies, BEF studies have much to learn from past research focused on the role of key species or functional groups in influencing ecosystem functions.

To understand the generality and limits of extrapolation of model systems, and to build a more integrative science, observations from the real world are essential (Table 14.1). For example, many experiments lack the spatial and temporal heterogeneity within replicates that can vary functional responses and allow the complementarity among species to be expressed as increased functional performance. We know from observational studies and experiments that heterogeneity is an important element of the functionality of ecosystems (Legendre and Fortin 1989; Legendre 1993; Seitz *et al.* 2001; Huffaker, 1958). Yet heterogeneity occurs across a range of space and time scales. The importance of specific scales of variability to individual species will depend on the traits of those species—e.g. size, mobility and resource specificity (Wiens, 1989; Wiens *et al.* 1995; Hewitt *et al.* 1996). However, only small scales of spatial heterogeneity have been directly

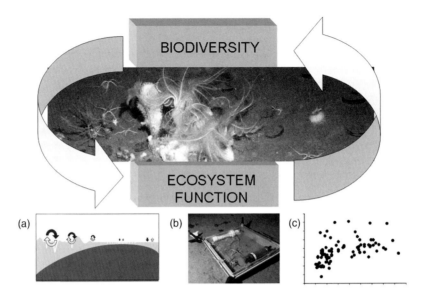

Figure 14.1 BEF relationships are not simple cause–effect relationships. They should be viewed as feedback loop relationships. This chapter argues for developing a process of iterating between (a) observation of patterns, (b) local measurement and manipulation, and (c) formulation of general relationships. See also Plate 6.

Table 14.1 The merits of mesocosm experiments versus observation versus an integrative approach

Atttributes	Mesocosm experiments	Observational studies	An integrative approach
Defining changes with spatial or temporal scale	+	+ + +	+ + +
Defining changes in rates and processes across locations	+	+ +	+ + +
Control of experimental treatments/conditions	+ + +	+	+ + +
Low variability in replicates within treatments/conditions	+ + +	+	+
Inference of mechanisms	+ +	+	+ +
Assessing generality	+	+ +	+ + +
Potential for spurious correlation	+	+ +	+
Inclusion of potentially relevant environmental or biotic factors as co-variables	+	+ + +	+ + +
Reality checking of models and model systems	+	+ +	+ + +
Ease of conducting over large space and time scales	+	+ + +	+ +
Multi-scale analysis is possible	+	+ + +	+ + +
Potential to test theoretical predictions	+ +	+ +	+ + +

manipulated in BEF experiments (see Bulling *et al.* 2008; Dyson *et al.* 2007; Jones and Frid 2009). Just as spatial heterogeneity and extent can be an important element in defining function, temporal variability and extent can influence the strength of diversity effects and allow complex interactions to emerge (Stachowicz *et al.*; 2008, Suttle *et al.* 2007). Many processes in ecology are not directly amenable to traditional randomized, replicated, small-scale experiments (Carpenter 1996; Doak *et al.* 2008). Observation can help us design and interpret our experimental studies and test the generality of our new findings and theories. Integration can be achieved by testing predictions based on theory or small-scale experiments with broad-scale observational studies.

14.2   The heterogeneous nature of seafloor landscapes

Aside from highly stressed and disturbed systems that are practically devoid of macrofauna, the seafloor is not composed of homogenous plains of mud and sand (Levin and Dayton 2009; Thrush, 1991; Zajac 2008). These ecosystems are capable of supporting high diversity (Coleman *et al.* 1997; Gray 1997; Snelgrove 1999). Heterogeneity in these systems is generated by the interaction of a range of biotic and abiotic processes, with organisms that modify sediment stability, boundary flows, and below-ground sediment structure being important in defining and creating spatial patterns that occur over a wide range of spatial scales, from millimetres to ocean basins. Patterns apparent at one scale can collapse to noise when viewed from other scales, indicating that patterns, processes, and our perceptions vary in a scale-dependent manner (Thrush *et al.* 1997). Spatial pattern and habitat variation can affect the strength and direction of processes that influence ecosystem function especially within landscapes of habitat-altering species (Norkko *et al.* 2006). In estuarine and marine ecosystems, many of these patterns and associated processes are influenced by hydrodynamics (Barry and Dayton 1991).

The contribution of biogenic processes in the creation and maintenance of seafloor heterogeneity emphasizes the role of context dependence in species interactions and diversity effects as both a driver of functioning and as a response variable. Heterogeneity is not noise obscuring an averaged relationship that may be detected in a controlled experiment; it is an important and functional attribute of natural ecosystems that can be quantitatively addressed. Through time-series or spatial analysis, we can describe spatial and temporal patterns that occur across a range of scales, we can measure the direction and strength of relationships between different but co-occurring processes, and we can determine how the strength and direction of focal variables change with biotic or environmental factors, thus providing a more realistic analysis of BEF relationships.

Gradient studies have emphasized how environmental context can directly influence the effects of species diversity on ecosystem functioning. Godbold and Solan (2009) studied the strong but short organic enrichment gradients that typically exist under salmon farms to assess the relative importance of macrobenthic species richness and sediment characteristics in driving changes in the mixing depth within the sediment, as revealed by sediment profile imaging. They found that species richness was much more important in explaining mixing depth—accounting for 34% of the total variability—than was the most important sediment characteristic—total sediment organic carbon content—which only accounted for 5% of the total variability. This emphasizes the important role of macrofauna in mediating eutrophication effects (Conely *et al.* 2007), and further illustrates the non-uni-directional coupling between biodiversity and ecosystem function (Figure 14. 1). A much broader scale analysis of BEF relationships along a gradient in wave energy in the Irish Sea showed that biodiversity–ecosystem function relationships can be influenced by environmental conditions in different ways depending on which ecosystem function is studied (Hiddink *et al.* 2009). This is an excellent example of how broad-scale observational studies can help us to recognize how the strength of BEF relationships can depend on environmental context and identify the important environmental drivers of change. The strength and the direction of biotic interactions often changes along environmental gradients, consequently defining important environmental gradients can help interpret changes in the strength of ecosystem functions (Bruno *et al.* 2003).

A variety of techniques has been developed to describe spatial heterogeneity, summarized by four theories that have been developed to link observed pattern to process. These theories are not mutually exclusive but can be represented by (1) patch analysis—linked to island biogeography, meta-population, and meta-community theories; (2) gradient analysis—linked to large-scale environmental gradients and niche theory; (3) hierarchy theory—linked to allocating different processes to different scales, and (4) multi-scale theory—linked to identifying how the role of different processes and functions vary across scale. An excellent recent analysis of the relative merits of these different approaches emphasizes the importance of integration in untangling the complexity of natural ecosystems (Talley 2007). This, after all, is the challenge that ecologists set themselves. A narrow view of the system, such as a focus on the wrong scales or types of heterogeneity, can lead to erroneous or incomplete conclusions and can obscure generalities.

Descriptions of spatial heterogeneity can provide important context for the interpretation of experiments, provide a reality check on the scales and complexity encompassed in laboratory trials and help to improve the design and analysis of field experiments by converting spatial 'noise' into information (Ellis and Schneider 2008; Hewitt *et al.* 2007a). Gibbs *et al.* (2005) provided an example of this from the sea floor. Fluxes of inorganic nitrogen from the benthos to the water column—which fuel primary production and serve as an index of functioning in many BEF studies—were shown to be substantially greater near pinnid bivalves than they were in adjacent bare sediments. Interestingly, however, the functional importance of the bivalves—the production of organic-rich biodeposits—changed dramatically along an estuarine turbidity gradient, with biodeposition-related organic enrichment having a stronger influence in organic-limited parts of the system. In the upper estuary, light-limited microphytobenthos were being smothered by biodeposits, and in the lower estuary, nutrient-limited microphytobenthos were benefitting from biodeposition-related organic enrichment and associated nutrient regeneration. Norkko *et al.* (2006) also documented conditional facilitation by the same bivalves along the same gradient, though the response in this latter paper was macrofaunal richness and diversity. So, again, it is important to recognize that biodiversity can be both a cause and a consequence of large changes in functioning (Figure 14.1).

Observation of temporal variability can also be used to improve experiments and better assess their generality. Time-series data allow us to define temporal variability and the dynamic nature of populations and communities. This information also provides insights into the interactions between species and environmental factors that influence time-

dependent functions such as stability and resilience (Ives *et al.* 2003; Gaston and McArdle 1994). Finally, temporal information on the dynamics of populations, functional groups, and communities is critical to our understanding of the magnitudes of ecological response to anthropogenic stressors. Thus time-series data may be especially useful in elucidating relationships between diversity and functional interactions within communities, even though the increased temporal replication typically limits spatial replication (Thrush *et al.* 1994; Thrush *et al.* 2008). Short-term experiments can produce results that do not reveal the complexity of ecological responses as sufficient time has not accrued in experiment plots for relationships to emerge (Stachowicz *et al.* 2008). In the laboratory, extending the time of experiments can also be problematic potentially introducing artifacts that confound the interpretation of experimental treatments.

In natural ecosystems, spatial and temporal variability can interact influencing the rigor of BEF measurements in two ways. Firstly, observations of ecological time-series have shown that the dynamics of coastal marine populations can be driven by interactions between broad-scale climatic factors, such as the El Niño-Southern Oscillation (ENSO), and local-scale environmental and biological factors (Hewitt and Thrush 2009). ENSO offers many superb examples of temporal variability, though humans may have difficulty perceiving ENSO-driven variability with their own senses without the help of long-term datasets. For example, several macrofaunal groups in sandflat communities—20-year datasets from New Zealand—show multi-year aperiodic cycles that correlate to ENSO (Thrush *et al.* 2008). ENSO-associated changes in the oxygen minimum zone on the central Chile margin results in the dominance of different functional groups with different ecosystem level consequences (Gallardo *et al.* 2004).

Understanding the nature of such broad-scale patterns requires observation. Although there is a robust tradition in ecology of using experiments to provide strong tests of hypotheses, experimental results often emerge from a bewildering number of processes that are encompassed even by the simplest ecological experiment (Figure 14. 2). Thus

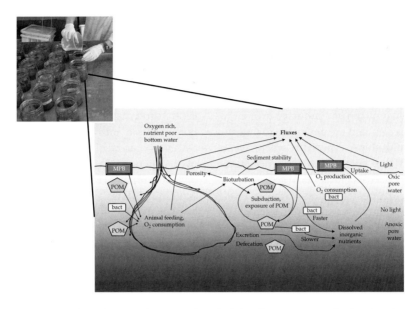

Figure 14.2   BEF processes are complex, and it may not be easy to isolate individual mechanisms even in small-scale enclosure experiments. There are many mechanisms involved in the way in which animals, such as burrowing urchins, interact with biogeochemical processes to influence nutrient flux and productivity. See also Plate 7.

even the most rigorous experiment may not actually tease apart all of the mechanisms that influence a BEF relationship. Interactions between processes occurring on different temporal and spatial scales produce the complex and interesting dynamics that characterize many benthic ecosystems.

14.3   Observing the nature of functions

Communities and ecosystems have many different functional attributes derived from a variety of processes. Another important scaling challenge for BEF research is the development of our understanding of these multiple functions rather than consideration of only one function (Hector and Bagchi 2007). Observational studies are useful in determining the suite of functions that may be delivered from a particular system, and assessing which ones are likely to dominate. Observations can also help to identify the kinds of functions that are especially important due to surrounding environmental conditions and context (Hiddink *et al.* 2009).

One approach that has proved useful in linking the functional attributes of species across broad space or time scales has been the aggregations of species into functional groups or the application of biological traits analysis. The formulation and analysis of functional groups has a long history in benthic ecology (Norling *et al.* 2007; Posey 1990; Pearson 2001). Functional groups are consortia of species that share some common attribute that is likely to influence function in a specific way. For example, surface-deposit feeders and suspension feeders should generate strong differences in sediment biogeochemistry, irrespective of species-specific attributes. In BEF relationships, the manipulation of functional groups has been recommended as a way of identifying general interactions (Balvanera *et al.* 2006). Biological traits analysis provides a more multidimensional analysis of species attributes that affect function (Bremner *et al.* 2006a; Bremner *et al.* 2006b). For example, filter-feeding, attached epifaunal organisms, and large organisms tend to show negative correlations with trawling intensity on the seafloor, while small infauna and scavengers tend to become more abundant (Tillin *et al.* 2006; Robinson and Frid 2008). Such broad-scale description of functional attributes provides insight into how BEF relationships can change. A prototype trait-based approach to assessing how ecosystem functions are modified by stress has recently been proposed for terrestrial plant communities to overcome the context specificity that often dominates community ecology (Suding *et al.* 2008). Functional groups and biological trait analysis can be conducted over large space and time-scales, and are amenable to investigating BEF relationships, especially where weak interactions and rare species can influence functional performance across gradients in species composition or environmental factors (Walker *et al.* 1999).

Few studies have addressed how diversity relates to functional linkages in community dynamics (Cottingham *et al.* 2001). One example is an analysis of the connection strength between functional groups using time-series data collected from three different sandflat communities (Thrush *et al.* 2008). This analysis indicates that the greater number of connections existing in diverse sites could play an important role in community persistence in the face of temporal variability of populations. Analyses of the temporal dynamics of functional group interactions indicated that more connections can occur with high temporal variability in species richness, demonstrating the role of species substitutions in functional groups in dynamic ecosystems. Thus declines in overall diversity will lead to a loss of intrinsic stabilizing mechanisms and accelerated simplification of multitrophic communities.

Many functions derived from soft-sediment communities are not the product of species abundance or presence/absence alone, yet this is typically what is manipulated in BEF experiments. Organism size, spatial distribution, physical structure, and behaviour matter too. The mass ratio theory of Grime (1998) contends that the immediate effects of biodiversity on ecosystem function in terrestrial plants can be predicted by the individual species' contribution to the total plant biomass. Similarly, in marine soft-sediments, large organisms are particularly important in influencing processes that affect the fluxes of energy and matter (Green *et al.* 1998; Sandnes *et al.* 2000; Thayer, 1983). Observations of the loss of biodiversity tend to emphasize the loss of larger organisms (Beukema *et al.* 1999; Thrush *et al.*

2003; Thrush and Whitlatch 2001; McConnaughey *et al.* 2000).

Many functions associated with soft-sediment habitats are influenced by 'patch-scale' processes rather than individual-scale processes. This is often apparent in functions influenced by organism-hydrodynamic interactions; for example, flows around individual pinnid bivalves (*Atrina zealandica*) can be quite different to those that occur over patches. The spacing of the individual bivalves in patches will determine whether or not 'skimming flows' develop. Thus, dependent on bivalve density, the patches have differing flow dynamics that influence stress responses and ultimately population stability (Coco *et al.* 2006). Furthermore, the type of flow across a patch can profoundly modify connections between the water column and seabed, thus affecting macrofaunal community composition and important transfers of energy and matter (Hewitt *et al.* 2002; Norkko *et al.* 2006; Gibbs *et al.* 2005).

There are many examples of how the behaviour of organisms is modified by their surrounding environment. In the context of BEF relationships, this forms a link between trophic relationships and landscape structure. Examples of behaviour-dependent functions arise from studies of mobile predators and their interactions with variations in food resources or the provision of refugia. An understanding of trophic functions is particularly important in marine ecosystems because there is indirect evidence that the functional extinction of large, high trophic level predators has profoundly changed community composition and functional relationships (Duffy 2003; Dayton *et al.* 1998; Jackson, 2001). Trophic interactions are strongly influenced by the spatial configuration of food resources, as consumers may only be able to detect or utilize resource patches above a specific size (Thrush 1999; Schneider 1978; Van de Koppel *et al.* 2005). Interactions between consumers that limit their ability to exploit prey patches can also be dependent on the spatial configuration of prey patches (Hines *et al.* 2009). Complex habitat patches can provide meso-consumers with refugia from roaming predators, or conversely can harbour ambush predators. These variations in risk can influence predator–prey interactions and associated functions. Heithaus *et al.* (2008) highlighted the importance of such interactions when predicting the community and ecosystem consequences of predator removal. Detailed observations of the behaviors of predators and prey—including the risk associated with feeding in the presence of predators—are a crucial first step. These results emphasize the importance of understanding how environmental processes operating over different spatial and temporal scales influence the strengths and directions of interactions between species associated with changes in species richness. Observation and an understanding of the natural history of resident species can highlight such context dependency.

14.4 Scaling laws and relevance to BEF

Resource managers are generally concerned with the 'big picture'—sustainability in the long-term, suitability/efficacy of protected areas, and ecosystem-based management. To ecologists, the 'big picture' means the net effects of drivers occurring over long periods of time and at large spatial scales, over which it is impossible to conduct manipulative experiments. Thus, if BEF studies are to be used for management purposes, the development of scaling laws to link small-scale experimental results to ecosystem-scale processes is essential. Although this is a daunting task due to the complicated and idiosyncratic relationships between organisms and their environments, there are aspects of organisms and ecosystems that are self-similar across a wide range of spatial and temporal scales that offer promise of scaling relationships (Brown *et al.* 2002; Schneider 2001a; Schneider 2001b).

Scale transition models (Melbourne and Chesson 2005) are one approach designed to understand the overall dynamics of networks of local communities. The argument is that spatial variation *between* local communities interacts with nonlinearities *within* local communities to determine the dynamics of the overall network. Rather than avoiding or imposing artificial controls on variation, generality is sought by building on and measuring variation (see also Hewitt *et al.* 2007b). To their credit, Melbourne and

colleagues have applied the theory using concrete examples with biological data (Melbourne *et al.* 2005; Melbourne and Chesson 2006). Crucial to the process of developing scale transition models is the identification of 'key quantities', which represent spatial mechanisms that contribute to change with spatial scale. Identification of key quantities is likely to be a product of careful and repeated observations of natural systems in multiple places (Hewitt *et al.* 2007b; Urban, 2005).

The approach of Melbourne and colleagues is a step forward in understanding functional scaling when functioning is determined largely by single 'key' species. When this is not the case, the approach of Suding *et al.* (2008) can be adapted. Suding *et al.* (2008) discuss a framework that integrates two main components: how a community responds to change, and how the changed community affects ecosystem processes. The species in a community that are responding to environmental change may, or may not, be important in a functional sense. So it is important to consider traits that determine responses to change (response traits), traits that determine levels of functioning (effects traits), and correlations between the two. Understanding the impacts of particular anthropogenic activities on community responses may be informed by sampling and experimentation across well-defined environmental gradients. It can also be done theoretically when computer models are informed by observational data. A marine example is provided by Solan *et al.* (2004), who discussed how changes in bioturbation rate and sediment mixed depth were heavily dependent on the presence of a single key species, as opposed to species richness per se, due to correlations between extinction risk and functional traits. Model results differed dramatically when the extinction of benthic species was random, as opposed to when extinction was modelled under more realistic scenarios. A full accounting of the traits of all the species in natural communities will be dependent upon excellent natural history information and numerous real-world observations.

Much of the work conducted by marine ecologists aimed at determining effects traits—i.e. the relative importance of different species to functioning—has been done in enclosed model experimental ecosystems—'mesocosms', 'micro-cosms'. Obviously, enclosed micro/mesocosms are reduced in both spatial and temporal scale and remain a limited representation of natural systems. However, they can be useful tools for comprehending BEF relationships (Petersen *et al.* 2009; Benton *et al.* 2007). Mesocosm results may reflect null models or define hypotheses that can be tested against nature to indicate the effects of scale and natural variability. There are ways to scale up mesocosm results so long as the enclosed model ecosystem captures critical elements of the systems of interest (Schneider 1994; Schneider 2001a; Rastetter *et al.* 1992).

BEF researchers have a considerable amount of influence over the extent of scale distortion in meso-cosm experiments, which allows them to strike a balance between experimental control, realism, and scale (Petersen and Hastings 2001). We emphasize the importance of realism and minimizing scale distortions, even if it means sacrificing some experimental control. For example, several studies have suggested that large mobile animals such as spatan-goid urchins are functionally important species in marine soft-sediment systems. These species have movement rates that can exceed 100 cm d^{-1}, and high densities of urchins can rework the entire upper 3–5 cm of sediment once every three days (Lohrer *et al.* 2005). Limiting their movements, by placing them in undersized containers, negates much of their functional importance. Although the inclusion of single individuals in small containers may be equivalent to realistic densities observed in the field—in terms of numbers per m^2—sediment characteristics in the field are often the product of integrated movements of numerous urchin individuals over long periods of time. Thus the type of information needed to design realistic laboratory experiments comes from familiarity with the natural conditions in which the organisms live, and the degree to which their behaviours change across naturally heterogeneous landscapes.

Cardinale *et al.* (2004) suggest that small-scale experiments can provide qualitatively robust insights into the regional consequences of species loss for communities structured by local processes, so long as the experiments are conducted across a range of heterogeneity that is relevant to species coexistence. Unfortunately, heterogeneity is not well incorporated into most small enclosure experiments; experimental control is generally the dominant feature of enclosure designs. This suggests a need for multiple iterations of experiments under different sets of conditions to create the necessary range of heterogeneity that would increase applicability at ecosystem scales. On the bright side, incorporating heterogeneity and realism can actually be easier than controlling for it. For example, instead of sieving and homogenizing sediments to create identical replicates—to which animals are added and from which functional variables such as fluxes are later measured—animals can be added to intact cores of sediment collected from the field (Giblin *et al.* 1997; Sundback *et al.* 2000; Eyre and Ferguson 2002). With cores collected from defined areas along measured gradients, the range of extrapolation becomes more transparent and easier to justify. For example, Levin *et al.* (2009) sampled at 50 m intervals along an oxygen gradient between 700—1100 m depth off the coast of Pakistan. This unusually high spatial resolution sampling revealed dramatic, threshold changes in bioturbation and community structure that could not be simply related to oxygen concentration. Finally, benthic chambers (Hughes *et al.* 2000; Webb and Eyre 2004; Janssen *et al.* 2005) and field flumes (Asmus *et al.* 1998; Cornelisen and Thomas 2006) can be positioned in experimentally manipulated field plots or across larger-scale environmental gradients. Although control of all potentially important variables is difficult or impossible in most field experiments, researchers with a detailed understanding of the system should be able to identify and measure such variables. There are numerous statistical techniques that can be applied to datasets with continuous co-variables, and these are likely to be more powerful than categorical analyses in heterogeneous landscapes (McCullagh and Nelder 1989; Hastie and Tibshirani 1990; Cressie *et al.* 2009).

Meta-analyses of multiple BEF experiments have shown a reasonable amount of consistency, with a generally positive but saturating effect of richness on functioning (Balvanera *et al.* 2006; Cardinale *et al.* 2006; Worm *et al.* 2006). However, Loreau (2008) cautions that the relationships emerging from small-scale experiments may still not match up with broad-scale patterns. For example, Danovaro *et al.* (2008) demonstrate a non-saturating, exponential increase in functions with increasing species richness in deep sea sediments. Deep sea sediments are one of the most widespread, globally important, yet unexplored habitat types on Earth. Danovaro *et al.* (2008) present convincing evidence, despite conducting an observational study, that environmental factors are unlikely to explain fully the observed exponential relationships between biodiversity and ecosystem properties in the deep sea. Positive species interactions are known to lead to accelerating relationships (Gross and Cardinale 2005). This could be important in the deep sea, where we have very limited knowledge of species interactions or even species identities and levels of diversity (Levin and Dayton 2009).

14.5   A more integrative approach to empirical research in biodiversity–ecosystem function studies

Observational studies have great potential to add insight, relevance and generality to experimental studies of BEF (Figure 14.3). The scientific merits of enclosure experiments and surveys are different, but the two approaches can be complementary when integrated. In Table 14.1, we illustrate the merits of small-scale model experiments, observations, and a more integrative approach. Regardless of whether small-scale model experiments are conducted in the laboratory or the field, we still think an integrative approach is best (Figure 14. 4). Experiments are often defended because they

provide tests of mechanisms—cause-and-effect relationships—but even the simplest laboratory experiments can have high biological and geochemical complexities that are difficult to fully elucidate (Figure 14. 2). The reality is that experiments do not always provide a definitive mechanistic understanding. They are important and very useful, but they need context. To develop a mechanistic understanding of how ecological communities are structured, it is important to recognize how processes change from place to place, and time to time. That means recognizing the existence of natural variance and designing studies to explain why the processes of interest may vary. We are beginning to see some illustrations of integrative frameworks as illustrated by combinations of enclosure experiments with numerical simulations and small-scale field experiments (Naeem 2008). Such approaches have added much needed heuristic value.

Ecology is full of surprises (Doak *et al.* 2008). This fact emphasizes the importance of strong empirical research to develop, parameterize, and ground truth theory. Observations on spatial patterns and broader scale gradients can be used as a basis for embedding experimental sites into landscapes, facilitating multi-scale analysis. This can strengthen the potential contrasts between replicates within treatments when sites are arrayed across broader gradients of biotic or environmental factors. When processes can be statistically modelled as a monotonic gradient, their role in affecting local ecological processes can be understood. If the effect of the broad, gradient-scale processes modifies the intensity rather than the direction of the local processes, then multiple regressions can be used to analyse effects across sites. Meta-analysis can be used in this circumstance, and also when broad-scale effects change the direction of the local processes (Rosenthal 1991; Gurevitch *et al.* 2001). Some basic observation of how the system is likely to function is required. This offers a way forward in terms of heuristically developing models and determining ecological generalities.

Observational studies are essential for improving the design of field experiments, and as part of series of planned iterations of testing BEF relationships across multiple studies as well as for building up natural history information (Figure 14. 4). This

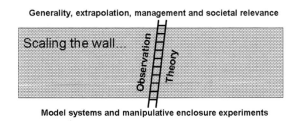

Figure 14.3   Scaling up is essential to address important scientific and societal issues.

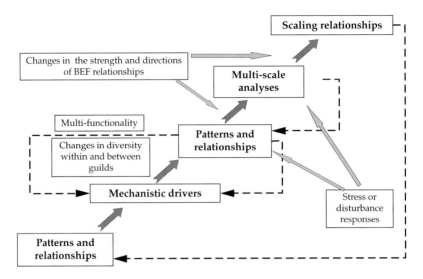

Figure 14.4 The integrative process in understanding the generality of BEF relationships involves linking pattern, mechanistic process, and theoretical studies by observation.

framework calls for an iterative process of prediction and testing, which remains one of the most promising ways for scaling up (Schneider 1994; Schneider 2001a; Rastetter *et al.* 1992).

Observational studies can assist in developing a more complete and relevant understanding of BEF relationships (Table 14.1, Figure 14.4). Simple dichotomous views that experiments are rigorous and reveal mechanism, while surveys are correlative and spuriously confounded, are overly simplistic and will not advance our science—as noted by many, e.g. Eberhardt and Thomas 1991; Legendre and Legendre 1998; Hewitt *et al.* 2007a). We do not argue for some hierarchy of approaches based on narrow views of rigor and inference, but think the best approach is an integrative one that seeks to combine the merits of different approaches and iterate between observation, theory and testing to assess the generality of BEF relationships across scales. Analysis of the trends and biases in BEF research publications has recently emphasized the need to expand the diversity of approaches, specifically to include more long-term and trophically complex experiments, to employ large scale observational studies and increase integration (Caliman *et al.* 2010). This does not mean that individual stud-

ies that take one approach are wrong. Some will find it easier to conduct experiments in mesocosms, while others will prefer field experiments and surveys. The key is having the wisdom to recognize the value of different approaches. An integrative framework in which individual studies can be incorporated is needed and should be a priority in the evolution of BEF research.

Box 14.1 Take-home message

- Individual species, communities, and biodiversity exhibit variation across scales of space and time
- Observation of natural ecosystems is essential to advance our understanding of BEF relationships.
- Processes that drive ecosystem functions vary across scales of space and time because of interactions between chemical, physical and biological processes that operate on differing spatio-temporal scales.
- While this leads to potential problems in the generality and extrapolation of small-scale model system experiments, observations of nature can help in all phases of an integrated BEF research agenda.

Acknowledgements

This work was supported by FRST C01X0501 and a Marie Curie International Incoming Fellowship to SFT. AML benefited from travel funds from MarBEF. We thank Judi Hewitt, Lisa Levin, Theresa Talley, and Hazel Needham and an anonymous reviewer for many helpful comments on the manuscript.

References

Asmus, R. M., Jensen, M. H., Jensen, K. M. *et al.* (1998) The role of water movement and spatial scaling for measurement of dissolved inorganic nitrogen fluxes in intertidal sediments. *Estuarine Coastal and Shelf Science,* **46**, 221–32.

Balvanera, P., Pfisterer, A. B., Buchmann, N. *et al.* (2006) Quantifying the evidence for biodiversity effects on ecosystem functioning and services. *Ecology Letters,* **9**, 1146–56.

Barry, J. P. and Dayton, P. K. (1991) Physical heterogeneity and the organisation of marine communities. In: Kolasa, K. and Pickett, S. T. A. (eds), *Ecological heterogeneity.* New York, Springer-Verlag, pp. 207–320.

Benton, T. G., Solan, M., Travis, J. M. J. *et al.* (2007) Microcosm experiments can inform global ecological problems. *Trends in Ecology and Evolution,* **22**, 516–21.

Beukema, J. J., Flach, E. C., Dekker, R. *et al.* (1999) A long-term study of the recovery of the macrozoobenthos on large defaunated plots on a tidal flat in the Wadden Sea. *Journal of Sea Research,* **42**, 235–54.

Bremner, J., Rogers, S. I. and Frid, C. L. J. (2006a) Matching biological traits to environmental conditions in marine benthic ecosystems. *Journal of Marine Systems,* **60**, 302–16.

Bremner, J., Rogers, S. I. and Frid, C. L. J. (2006b) Methods for describing ecological functioning of marine benthic assemblages using biological traits analysis (BTA). *Ecological Indicators,* **6**, 609–22.

Brown, J. H., Gupta, V. K., Li, B.-L. *et al.* (2002) The fractal nature of nature: power laws, ecological complexity and biodiversity. *Philosophical Transactions of the Royal Society of London B,* **357**, 619–26.

Bruno, J.F., Stachowicz, J.J., and Bertness, M.D. (2003) Inclusion of facilitation into ecological theory. *Trends in Ecology & Evolution,* **18**, 119–25.

Bulling, M. T., Solan, M., Dyson, K. E. *et al.* (2008) Species effects on ecosystem processes are modified by faunal responses to habitat composition. *Oecologia,* **158**, 511–20.

Caliman, A., Pires, A.F., Esteves, F.A. *et al.* (2010) The prominance of and biases in biodiversity and ecosystem functioning research. *Biodiversity and Conservation,* **19**, 651–64.

Cardinale, B. J., Ives, A. R. and Inchausti, P. (2004) Effects of species diversity on the primary productivity of ecosystems: extending our spatial and temporal scales of inference. *Oikos,* **104**, 437–50.

Cardinale, B. J., Srivastava, D. S., Duffy, E. J. *et al.* (2006) Effects of biodiversity on the functioning of trophic groups and ecosystems. *Nature,* **443**, 989–92.

Carpenter, S. R. (1996) Microcosm experiments have limited relevance for community and ecosystem ecology. *Ecology,* **77**, 677–80.

Coco, G., Thrush, S. F., Green, M. O. *et al.* (2006) Feedbacks between bivalve density, flow, suspended sediment concentration on patch stable states. *Ecology,* **87**, 2862–70.

Coleman, N., Gason, A. S. H. and Poore, G. C. B. (1997) High species richness in the shallow marine waters of south east Australia. *Marine Ecology Progress Series,* **154**, 17–26.

Conely, D.J., Carstensen, J., Aertebjerg, G. *et al.* (2007) Long-term changes and impacts of hypoxia in Danish coastal waters. *Ecological Applications,* **17**, S165–S184.

Cornelisen, C. D. and Thomas, F. I. M. (2006) Water flow enhances ammonium and nitrate uptake in a seagrass community. *Marine Ecology Progress Series,* **312**, 1–13.

Cottingham, K. L., Brown, B. L. and Lennon, J. T. (2001) Biodiversity may regulate the temporal variability of ecological systems. *Ecology Letters,* **4**, 72–85.

Cressie, N., Calder, C. A., Clark, J. S. *et al.* (2009) Accounting for uncertainty in ecological analysis: the strengths and limitations of hierarchical statistical modelling. *Ecological Applications,* **19**, 553–70.

Danovaro, R., Gambi, C., Dell'anno, A. *et al.* (2008) Exponential decline of deep-sea ecosystem functioning linked to benthic biodiversity loss. *Current Biology,* **18**, 1–8.

Dayton, P. K., Tegner, M. J., Edwards, P. B. *et al.* (1998) Sliding baselines, ghosts, and reduced expectations in kelp forest communities. *Ecological Applications,* **8**, 309–22.

Doak, D. F., Estes, J. A., Halpern, B. S. *et al.* (2008) Understanding and predicting ecological dynamics: Are major surprises inevitable? *Ecology,* **89**, 952–61.

Duffy, J. E. (2003) Biodiversity loss, trophic skew and ecosystem functioning. *Ecology Letters,* **6**, 680–7.

Dyson, K. E., Bulling, M. T., Solan, M. *et al.* (2007) Influence of macrofaunal assemblages and environmental heterogeneity on microphytobenthic production in experimental systems. *Proceedings of the Royal Society B-Biological Sciences*, **274**, 2547–54.

Eberhardt, L. L. and Thomas, J. M. (1991) Designing environmental field studies. *Ecological Monographs*, **61**, 53–73.

Ellis, J. and Schneider, D. C. (2008) Spatial and temporal scaling in benthic ecology. *Journal of Experimental Marine Biology and Ecology*, **366**, 92–8.

Eyre, B. D. and Ferguson, A. J. P. (2002) Comparison of carbon production and decomposition, benthic nutrient fluxes and denitrification in seagrass, phytoplankton, benthic micro- and macroalgae-dominated warm-temperate Australian lagoons. *Marine Ecology Progress Series*, **229**, 43–59.

Gallardo, V. A., Palma, M., Carrasco, F. D. *et al.* (2004) Macrobenthic zonation caused by the oxygen minimum zone on the shelf and slope off central Chile. *Deep Sea Research II*, **51**, 2475–90.

Gaston, K. J. and McCardle, B. H. (1994) The temporal variability of animal abundances: measures, and patterns. *Philosophical Transactions of the Royal Society of London Series B Biological Sciences*, **345**, 335–58.

Gibbs, M., Funnell, G., Pickmere, S. *et al.* (2005) Benthic nutrient fluxes along an estuarine gradient: influence of the pinnid bivalve Atrina zelandica in summer. *Marine Ecology-Progress Series*, **288**, 151–64.

Giblin, A. E., Hopkinson, C. S. and Tucker, J. (1997) Benthic metabolism and nutrient cycling in Boston Harbor, Massachusetts. *Estuaries*, **20**, 346–64.

Godbold, J.A. and Solan, M. (2009) Relative importance of biodiversity and the abiotic environment in mediating an ecosystem process. *Marine Ecology Progress Series*, **396**, 273–82.

Gray, J. S. (1997) Marine biodiversity: patterns, threats and conservation needs. *Biodiversity and Conservation*, **6**, 153–75.

Green, M. O., Hewitt, J. E. and Thrush, S. F. (1998) Seabed drag coefficients over natural beds of horse mussels (*Atrina zelandica*). *Journal of Marine Research*, **56**, 613–37.

Grime, J. P. (1998) Benefits for plant diversity to ecosystems; immediate, filter and founder effects. *Journal of Ecology*, **86**, 902–10.

Gross, K. and Cardinale, B. J. (2005) The functional consequences of random vs. ordered species extinctions. *Ecology Letters*, **8**, 409–18.

Gurevitch, J., Curtis, P. S. and Jones, M. H. (2001) *Meta-analysis in ecology*.

Hastie, T. and Tibshirani, R. J. (1990) *Generalized additive models*, London, Chapman and Hall.

Hector, A. and Bagchi, R. (2007) Biodiversity and ecosystem multifunctionality. *Nature*, **448**, 188–91.

Heithaus, M. R., Frid, A., Wirsing, A. J. *et al.* (2008) Predicting ecological consequences of marine top predator declines. *Trends in Ecology & Evolution*, **23**, 202–10.

Hewitt, J. E., Legendre, P., Thrush, S. F. *et al.* (2002) Integrating heterogeneity across spatial scales: interactions between *Atrina zelandica* and benthic macrofauna. *Marine Ecology Progress Series*, **239**, 115–28.

Hewitt, J. E. and Thrush, S. F. (2009) Reconciling the influence of global climate phenomena on macrofaunal temporal dynamics at a variety of spatial scales. *Global Change Biology*, **15**, 1911–29.

Hewitt, J. E., Thrush, S. F., Cummings, V. J. *et al.* (1996) Matching patterns with processes: predicting the effect of size and mobility on the spatial distributions of the bivalves *Macomona liliana* and *Austrovenus stutchburyi*. *Marine Ecology Progress Series*, **135**, 57–67.

Hewitt, J. E., Thrush, S. F., Dayton, P. K. *et al.* (2007a) The effect of scale on empirical studies of ecology. *American Naturalist*, **169**, 398–408.

Hewitt, J. E., Thrush, S. F., Dayton, P. K. and *et al.* (2007b) The effect of spatial and temporal heterogeneity on the design and analysis of empirical studies of scale-dependent systems. *The American Naturalist*, **169**, 398–408.

Hiddink, J.G., Davies, T.W., Perkins, M. *et al.* (2009) Context dependency of relationships between biodiversity and ecosystem functioning is different for multiple ecosystem functions. *Oikos*, **118**, 1892–900.

Hines, A. H., Long, C. W., Terwin, J. R. *et al.* (2009) Facilitation, interference, and scale: the spatial distribution of prey patches affects predation rates in an estuarine benthic community. *Marine Ecology Progress Series* **385**, 127–35.

Huffaker, C. B. (1958) Experimental studies on predation: dispersion factors and predator-prey oscillations. *Hilgardia*, **27**, 343–83.

Hughes, D. J., Atkinson, R. J. A. and Ansell, A. D. (2000) A field test of the effects of megafaunal burrows on benthic chamber measurements of sediment-water solute fluxes. *Marine Ecology Progress Series*, **195**, 189–99.

Ives, A. R., Dennis, B., Cottingham, K. L. *et al.* (2003) Estimating community stability and ecological interactions from time-series data. *Ecological Monographs*, **73**, 301–30.

Jackson, J. B. C. (2001) What was natural in the coastal oceans? *Proceedings of the National Academy of Science*, **98**, 5411–18.

Janssen, F., Faerber, P., Huettel, M. *et al.* (2005) Porewater advection and solute fluxes in permeable marine sediments (I): Calibration and performance of the novel benthic chamber system Sandy. *Limnology and Oceanography*, **50**, 768–78.

Jones, D. and Frid, C.L.J. (2009) Altering intertidal sediment topography: effects on biodiversity and ecosystem functioning. *Marine Ecology*, **30**, 83–96.

Legendre, P. (1993) Spatial autocorrelation: trouble or new paradigm? *Ecology*, **74**, 1659–73.

Legendre, P. and Fortin, M. J. (1989) Spatial pattern and ecological analysis. *Vegetato*, **80**, 107–38.

Legendre, P. and Legendre, L. (1998) *Numerical Ecology*, Amsterdam, Elsevier.

Levin, L. A. and Dayton, P. K. (2009) Integration and application of ecological theory on continental margins. *Trends in Ecology & Evolution*, **24**, 606–17.

Levin, L. A., Whitcraft, C. R., Mendoza, G. F. *et al.* (2009) Oxygen and organic matter thresholds for benthic faunal activity on the Pakistan margin oxygen minimum zone (700–1100 m) *Deep Sea Research Part I Oceanographic Research Papers*, **56**, 449–71.

Levin, S. A. (1992) The problem of pattern and scale in ecology. *Ecology*, **73**, 1943–67.

Lohrer, A. M., Thrush, S. F., Hunt, L. *et al.* (2005) Rapid reworking of subtidal sediments by burrowing spatangoid urchins. *Journal of Experimental Marine Biology and Ecology*, **321**, 155–69.

Loreau, M. (2008) Biodiversity and ecosystem functioning: the mystery of the deep sea. *Current Biology*, **18**, 126–7.

Mcconnaughey, R. A., Mier, K. L. and Dew, C. B. (2000) An examination of chronic trawling effects on soft-bottom benthos of the eastern Bering Sea. *Ices Journal of Marine Science*, **57**, 1377–88.

Mccullagh, P. and Nelder, J. A. (1989) *Generalised Linear Models*, London, Chapman and Hall.

Melbourne, B. A. and Chesson, P. (2005) Scaling up population dynamics: integrating theory and data. *Oecologia*, **145**, 179–87.

Melbourne, B. A. and Chesson, P. (2006) The scale transition: Scaling up population dynamics with field data. *Ecology*, **87**, 1478–88.

Melbourne, B. A., Sears, A. L., Donahue, M. J. *et al.* (2005) Applying scale transition theory to metacommunities in the field. In: *Metacommunities: spatial dynamics and ecological communities.* Holyoak, M., Leibold, M. A. and Holt, R. D. (eds), Chicago, University of Chicago Press.

Naeem, S. (2008) Advancing realism in biodiversity research. *Trends in Ecology & Evolution*, **23**, 414–16.

Norkko, A., Hewitt, J. E., Thrush, S. F. *et al.* (2006) Conditional outcomes of facilitation by a habitat-modifying subtidal bivalve. *Ecology*, **87**, 226–34.

Norling, K., Rosenberg, R., Hulth, S. *et al.* (2007) Importance of functional biodiversity and species-specific traits of benthic fauna for ecosystem functions in marine sediment. *Marine Ecology-Progress Series*, **332**, 11–23.

Pearson, T. H. (2001) Functional group ecology in soft-sediment marine benthos: The role of bioturbation. *Oceanography and Marine Biology*, **39**, 233–67.

Petersen, J. E. and Hastings, A. (2001) Dimensional approaches to scaling experimental ecosystems: Designing mousetraps to catch elephants. *American Naturalist*, **157**, 324–33.

Petersen, J. E., Kennedy, V. S., Dennison, W. C. *et al.* (eds) (2009) *Enclosed experimental ecosystems and scale: tools for understanding and managing coastal ecosystems,* New York, Springer.

Posey, M. H. (1990) Functional approaches to soft-substrate communities: How useful are they? *Review of Aquatic Sciences*, **2**, 343–56.

Rastetter, E. B., King, A. W., Cosby, B. J. *et al.* (1992) Aggregating fine-scale ecological knowledge to model coarser-scale attributes of ecosystems. *Ecological Applications*, **2**, 55–70.

Robinson, L. A. and Frid, C. L. J. (2008) Historical marine ecology: Examining the role of fisheries in changes in North Sea benthos. *Ambio*, **37**, 362–71.

Rosenthal, R. (1991) *Meta-analytic procedures for social research*, Newbury Park, USA, Sage Publications.

Sandnes, J., Forbes, T., Hansen, R. *et al.* (2000) Bioturbation and irrigation in natural sediments, described by animal-community parameters. *Marine Ecology Progress Series*, **197**, 169–79.

Schneider, D. (1978) Equalisation of prey numbers of migratory shorebirds. *Nature*, **271**, 353–54.

Schneider, D. C. (1994) *Quantitative Ecology: Spatial and Temporal Scaling*, San Diego, USA, Academic Press.

Schneider, D. C. (2001a) The rise of the concept of scale in ecology. *Bioscience*, **51**, 545–53.

Schneider, D. C. (2001b) Spatial allometry; theory and application to experimental and natural aquatic ecosystems. In: *Scaling relationships in experimental ecology.* Gardener, R. H., Kemp, W. M., Kennedy, V. S. *et al.* (eds), New York, Columbia University Press, pp. 113–53.

Seitz, R. D., Lipcius, R. N., Hines, A. H. *et al.* (2001) Density-dependent predation, habitat variation, and the persistence of marine bivalve prey. *Ecology*, **82**, 2435–51.

Snelgrove, P. V. R. (1999) Getting to the bottom of marine biodiversity: Sedimentary habitats—Ocean bottoms are the most widespread habitat on Earth and support high biodiversity and key ecosystem services. *Bioscience*, **49**, 129–38.

Solan, M., Cardinale, B. J., Downing, A. L. *et al.* (2004) Extinction and ecosystem function in the marine benthos. *Science*, **306**, 1177–80.

Srivastava, D. S. and Vellend, M. (2005) Biodiversity-ecosystem function research: Is It Relevant to Conservation? *Annual Review of Ecology and Systematics*, **36**, 267–94.

Stachowicz, J. J., Best, R. J., Bracken, M. E. S. *et al.* (2008) Complementarity in marine biodiversity manipulations: Reconciling divergent evidence from field and mesocosm experiments. *Procedings of the National Academy of Science*, **105**, 18842–7.

Stachowicz, J. J., Bruno, J. F. and Duffy, J. E. (2007) Understanding the Effects of Marine Biodiversity on Communities and Ecosystems. *Annual Review of Ecology and Systematics*, **38**, 739–66.

Suding, K. N., Lavorel, S., Chapin, F. S. *et al.* (2008) Scaling environmental change through the community-level: a trait-based response-and-effect framework for plants. *Global Change Biology*, **14**, 1125–40.

Sundback, K., Miles, A. and Goransson, E. (2000) Nitrogen fluxes, denitrification and the role of microphytobenthos in microtidal shallow-water sediments: an annual study. *Marine Ecology Progress Series*, **200**, 59–76.

Suttle, K. B., Thomsen, M. A. and Power, M. E. (2007) Species interactions reverse grassland responses to climate change. *Science*, **315**, 640–42.

Talley, T. S. (2007) Which spatial heterogeneity framework? Consequences for conclusions about patchy population distributions. *Ecology*, **88**, 1476–89.

Thayer, C. W. (1983) Sediment-mediated biological disturbance and the evolution of marine benthos. In: *Biotic Interactions in Recent and Fossil Benthic Communities*. Tevesz, M. J. S. (ed.), New York, London, Plenum Press.

Thrush, S. F. (1991) Spatial patterns in soft-bottom communities. *Trends in Ecology and Evolution*, **6**, 75–9.

Thrush, S. F. (1999) Complex role of predators in structuring soft-sediment macrobenthic communities: Implications of changes in spatial scale of experimental studies. *Australian Journal of Ecology*, **24**, 344–54.

Thrush, S. F., Coco, G. and Hewitt, J. E. (2008) Complex positive connections between functional groups are revealed by neural network analysis of ecological time-series *American Naturalist*, **171**, 669–77.

Thrush, S. F., Hewitt, J. E., Norkko, A. *et al.* (2003) Catastrophic sedimentation on estuarine sandflats: recovery of macrobenthic communities is influenced by a variety of environmental factors. *Ecological Applications*, **13**, 1433–55.

Thrush, S. F., Pridmore, R. D., Bell, R. G. *et al.* (1997) The sandflat habitat: Scaling from experiments to conclusions. *Journal of Experimental Marine Biology and Ecology*, **216**, 1–10.

Thrush, S. F., Pridmore, R. D. and Hewitt, J. E. (1994) Impacts on soft-sediment macrofauna: The effects of spatial variation on temporal trends. *Ecological Applications*, **4**, 31–41.

Thrush, S. F. and Whitlatch, R. B. (2001) Recovery dynamics in benthic communities: Balancing detail with simplification. IN Reise, K. (Ed.) *Ecological Comparisons of Sedimentary Shores*. Berlin, Springer-Verlag.

Tillin, H. M., Hiddink, J. G., Jennings, S. *et al.* (2006) Chronic bottom trawling alters the functional composition of benthic invertebrate communities on a sea-basin scale. *Marine Ecology Progress Series*, **318**, 31–45.

Urban, D. L. (2005) Modeling ecological processes across scales. *Ecology*, **86**, 1996–2006.

Van De Koppel, J., Bardgett, R. D., Bengtsson, J. *et al.* (2005) The effects of spatial scale on trophic interactions. *Ecosystems*, **8**, 801–7.

Walker, B., Kinzig, A. and Langridge, J. (1999) Plant attribute diversity, resilience and ecosystem function: The nature and significance of dominant and minor species. *Ecosystems*, **2**, 95–113.

Webb, A. P. and Eyre, B. D. (2004) Effect of natural populations of burrowing thalassinidean shrimp on sediment irrigation, benthic metabolism, nutrient fluxes and denitrification. *Marine Ecology Progress Series*, **268**, 205–20.

Wiens, J. A. (1989) Spatial scaling in ecology. *Functional Ecology*, **3**, 385–97.

Wiens, J. A., Crist, T. O., With, K. A. *et al.* (1995) Fractal patterns of insect movement in microlandscape mosaics. *Ecology*, **76**, 663–6.

Worm, B., Barbier, E. B., Beaumont, N. *et al.* (2006) Impacts of biodiversity loss on ocean ecosystem services. *Science*, **314**, 787–90.

Zajac, R. N. (2008) Macrobenthic biodiversity and sea floor landscape structure. *Journal of Experimental Marine Biology and Ecology*, **366**, 198–203.

CHAPTER 15

Implementing an ecosystem approach: predicting and safeguarding marine biodiversity futures

Alison R. Holt, Caroline Hattam, Stephen Mangi, Anton Edwards, and Scot Mathieson

15.1 Introduction

Evidence of the links between marine biodiversity and ecosystem function, as highlighted in previous chapters, indicates that the loss of biodiversity is likely to have serious consequences for human well-being (Diaz *et al.* 2006). Humans derive a variety of important goods and services from marine systems (see Covich *et al.* 2004; MA 2005), for instance, the provision of food to millions (Worm *et al.* 2006), and the protection of coastal populations from flooding (Adger *et al.* 2005). The underlying causes of marine degradation stem from the unsustainable use of marine resources; for example, over-fishing, and other activities that may directly or indirectly destroy habitat, pollute, warm, and acidify the ocean (Dulvy *et al.* 2003; Royal Society 2005; Lotze *et al.* 2006). Maintaining human well-being into the future requires finding ways of sustaining ecosystems that are resilient to change, and have the ability to continue to provide important services (Levin and Lubchenco 2008). It means rethinking the way that natural resources are managed, taking a whole-ecosystem perspective, understanding the interactions between the human and ecological systems, and employing sophisticated and intelligent monitoring, management, and governance practices.

In the first section of this chapter, we discuss the emergence of what has been termed the 'ecosystem approach' to environmental management, central to which is the concept of ecosystem services. In the second, we show the importance of biodiversity and ecosystem function research in implementing an ecosystem approach to marine management. We outline the perspectives, approaches, and evidence that are required to provide a base for marine biodiversity valuation and decision-making, highlighting the gaps in knowledge. In the third, we link ecology and economics to focus on the importance of an economic perspective and outline tools that can be used to value the social benefits arising from ecosystem services and biodiversity. This has become a popular area of investigation across aquatic and terrestrial systems, in an attempt to bring environmental values into decision-making and to create societal understanding of the value of nature in providing ecosystem services (Daily *et al.* 2009). In the fourth section, we suggest a general framework for the application of the ecosystem approach that integrates ecological and socio-economic knowledge and approaches to achieve suitable monitoring, management , and governance, to ensure sustainable biodiversity futures and human well-being. We conclude with a discussion of the future challenges of integrating across disciplines, science, and policy to achieve the implementation of an ecosystem approach.

15.1.1 Taking an ecosystem approach

To address the challenges posed by biodiversity loss and the degradation of ecosystem services, an integrated approach is required to achieve sustainable biodiversity futures. The ecosystem approach was first defined as a strategy for the integrated management of land, water, and living resources that promotes conservation in an equitable way (CBD 2000). However, definitions of what the ecosystem approach could be have evolved since, and the same philosophy is alluded to using different terminology depending on which country, institution, or discipline is using it. For example, an ecosystem approach is analogous to ecosystem-based management (EBM), terminology used in marine management in the US (see Leslie and McLeod 2007; Palumbi *et al.* 2009), and the Adaptive Ecosystem Approach outlined in Kay *et al.* (2001). Common to all definitions is that the approach identifies humans as an integral part of ecosystems, recognizing that social, cultural, and economic factors affect ecosystems, drive change, and feed back to affect human well-being. The ecosystem approach was first applied in a policy context by the Convention on Biological Diversity (CBD) to achieve their three primary objectives: the conservation of biodiversity, sustainable use of natural resources, and the equitable sharing of benefits from natural resources (CBD 2000). As outlined by the CBD, the ecosystem approach requires an understanding of how the components of biodiversity that create ecosystems resilient to biodiversity loss and environmental change function; how the benefits that flow from these functions can be equitably shared; how adaptive management and policy can address the uncertainties in both ecological and social systems; and how to include all relevant stakeholders in decision-making processes. The ecosystem approach was endorsed by the 2002 World Summit on Sustainable Development in Johannesburg. It is implicit in the EU Water Framework Directive, the 2010 target to halt biodiversity, the Ramsar convention (Laffoley *et al.* 2004), and it has been suggested that the Common Fisheries Policy and European Marine Strategy follow suit.

The ecosystem approach is increasingly part of the vocabulary used across many sectors, government, NGOs, and business (Holt and Hattam 2009), and is seen as a way of achieving a shared vision for sustainable environmental management. Among institutions that are adopting this approach is the Department for Environment, Food and Rural Affairs (Defra) in the UK, where it is used as a coherent framework to integrate across environmental challenges, to ensure that we can live within our environmental limits and deal with the pressures of environmental change (Defra 2007). Also gaining in popularity is the concept of ecosystem services, which is now seen as an integral part of the ecosystem approach as a way of understanding and conserving the benefits that ecosystems provide to society.

15.2 Ecosystem services, function, and biodiversity

The concept of ecosystem services has gained popularity since the completion of the Millennium Ecosystem Assessment (MA 2005). It offers an anthropocentric view of ecosystem management that aims to understand ecosystem service provision and how humans value the benefits that arise from these services. Although there is no one standardized definition of ecosystem services, they are broadly defined as the benefits that humans derive from the processes and functions of ecosystems (MA 2005; Costanza *et al.* 1997 and Daily 1997). The MA categorized the services provided by ecosystems into: *provisioning services* (e.g. food and fibre), *supporting services* (e.g. primary production, nutrient cycling), *regulating services* (e.g. waste processing, flood alleviation), and *cultural services* (e.g. recreation and cultural heritage). However, a number of other typologies for ecosystem services exist (de Groot *et al.* 2002; Boyd and Banzhaf 2007; Wallace 2007; Fisher *et al.* 2009). These have evolved from the need for a logical classification system that avoids double counting when valuing ecosystem services, and have caused considerable debate (e.g. Costanza 2008). Drawing on the work of Boyd and Banzhaf (2007), Fisher *et al.* (2009, p. 645) defined ecosystem services as 'aspects of ecosystems utilized (actively or passively) to produce human well-being', and consider this to be a more operational definition. Although the concept of ecosystem services has received much discussion, it remains in the early stages of translation into practical application.

Taking an ecosystem approach requires decision-making that balances maintaining well functioning, resilient ecosystems with the needs of humans. Decision- and policy-makers need to be able to make informed choices about the inevitable trade-offs between ecosystem services that will occur when choosing different management alternatives. This requires an integration of ecological, economic, and social sciences together with policy. Particular attention also needs to be given to certain knowledge gaps, with the development of tools that aid decision-making in the face of uncertainty in knowledge of the systems with which we are dealing.

Despite the support for an ecosystem service perspective to be integrated within an ecosystem approach, the contribution that ecosystem service management (management of specific use values) can make to the conservation of biodiversity (management of non-use or intrinsic value) is not yet clear. Recent mapping studies by Chan *et al.* (2006), Naidoo and Ricketts (2006), and Anderson *et al.* (2009) have shown that there may be circumstances where there will be little compatibility between management for specific ecosystem services and what is optimal for the conservation of biodiversity. To make informed trade-off decisions, we need to understand how biodiversity is related to the provision of ecosystem services and to maintain the ability of a system to provide suites of services now and in the future. This will only come from knowledge of the species and units that directly and indirectly supply ecosystem services (ecological production functions). This is where the importance of biodiversity and ecosystem functioning research lies.

In employing an ecosystem approach and the concept of ecosystem services, it is vital to understand the links between biodiversity and ecosystem function, and how these are related to the variability in ecosystem service provision in space and time. Kremen (2005) identified a gap that needs to be bridged between the approach of mapping the provision (and demand) of services with no assessment of how biodiversity is providing them, and the biodiversity-function approach that examines the structure of experimental communities that are much less complex than real landscapes. Raffaelli (2006) shows how making the leap between biodiversity, function, and services is challenging for

several reasons: multiple processes may give rise to one service; there may be weak relationships between biodiversity and a service if there are a number of direct and indirect processes involved; and processes and services may be differently affected depending on the biodiversity loss scenario. To relate biodiversity and ecosystem functioning to ecosystem services, a different approach is required. As many of the processes that have been investigated may themselves be services or support important services, there is a temptation to continue to link effects from biodiversity to processes and then to services. However, a service-led approach (Figure 15.1) may be more suitable (Raffaelli 2006) that first identifies the service of interest (for example, shellfish provision) then explores the processes involved in its delivery followed by focusing on how change in biodiversity (functional or species richness depending on which is the most

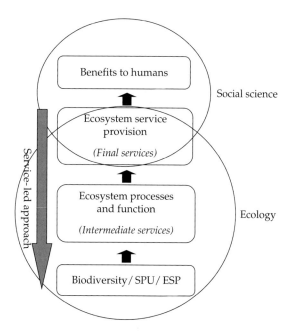

Figure 15.1   The steps involved in understanding the relationships between ecosystem services and biodiversity, the service providing unit (SPU), and ecosystem service provider (ESP). Intermediate and final services are indicated, integrating the typology of Fisher *et al.* (2009). The left side arrow shows the service-led approach to understanding the components of biodiversity that may supply ecosystem services. This approach requires the integration of natural and social sciences.

appropriate) effects those processes. This approach fits neatly into the classification suggested by Fisher *et al.* (2009) for the valuation of the benefits derived from ecosystem services. Identifying the ecosystem services is the suggested first step, to be followed by an understanding of the characteristics of the ecosystems that produce them.

With the continuing rise in popularity of the concept of ecosystem services within ecology, studies have aimed to understand how the biodiversity and ecosystem functioning literature to date links to ecosystem services. Balvanera *et al.* (2006) researched the link between species, functional diversity, and ecosystem productivity, using meta-analyses, and categorizing the ecosystem processes and functions that underpin certain ecosystem services such as primary productivity, erosion control, nutrient cycling, and regulation of biodiversity. To create a theoretical framework to link biodiversity and ecosystem services, Luck *et al.* (2009) integrated the concepts of service providing units (the collection of individuals from a species and their characteristics necessary to deliver an ecosystem service at the desired level, Luck *et al.* (2003)) and ecosystem service providers (the component of populations, communities, functional groups, interaction networks, or habitat types that provide ecosystem services (Kremen 2005)) into the service providing unit–ecosystem service providers continuum. They define it as an approach that quantifies the organism, community, or habitat characteristics needed to provide an ecosystem service given beneficiary demands and ecosystem dynamics (Luck *et al.* 2009). They provide a list of some of the studies that have focused on this continuum, which range from services such as biological control, pollination, waste decomposition, water regulation, filtration, and seed dispersal.

These examples have been largely terrestrial; however, marine studies are emerging. The use of the term ecosystem services in the marine literature began around 1992 but became more widely used from 2003 onwards (Figure 15.2). One such example is Worm *et al.* (2006). They studied the effects of changes in biodiversity on ecosystem function and services using meta-analyses of data from small-scale experiments, and investigated whether these

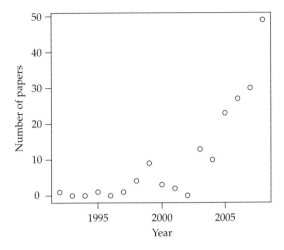

Figure 15.2 The number of publications between 1992 and 2008 that used the term 'ecosystem services' in a marine context, from ISI Web of Knowledge.

trends scaled up to the ecosystem level using global fisheries data. Positive relationships were found and systems with higher regional species richness showed greater stability with lower collapse rates of taxa over time. They also showed that regional biodiversity losses from coastal and estuarine systems impaired three important ecosystem services: the number of viable fisheries, provision of nursery habitats (such as seagrass beds and wetlands) and filtering and detoxification services. Losses may also have had an indirect effect on other services such as water quality and tourism.

These studies demonstrate a single-service-orientated approach. For example, Luck *et al.* (2009) explain service providing units and ecosystem service providers in terms of single services, although they recognize that these concepts can be extended to multiple services. The challenge for scientists, policy, and decision-making lies in understanding how marine system change affects multiple services and how to use this to predict future scenarios. This may be particularly difficult given that ecosystem services are underpinned by multiple and non-independent ecosystem processes and are supported by multifaceted service providing units and ecosystem service providers. While it may be impossible to gain a full understanding of the relationships between services, processes, and biodiversity, general rules are vital to help inform decisions. Such complexity is not possible to capture using traditional population or community approaches alone, but it is this very complexity that needs to tackled in order to successfully manage for sustainable marine biodiversity futures.

15.2.1   Taking a systems perspective

Ecosystems are complex adaptive systems (Gunderson and Holling 2002; Levin and Lubchenco 2008). The dynamics of interactions at small scales shape macroscopic system properties (e.g. trophic structure or patterns of nutrient flux) that in turn feed back and influence the smaller scales (Levin 1998). These interactions are non-linear and generate multiple stable states (Gunderson and Holling 2002). Understanding the interactions across scales is key to understanding biodiversity and ecosystem functioning (Levin 1998). In addition, to achieve an ecosystem approach to management, it is important to understand the consequences for ecosystems of human behaviour, institutional arrangements (systems of established and embedded social rules that structure social interaction) and economic activity. Social systems are also complex systems that do not operate in isolation from natural ones; they are highly connected and co-evolve at a range of spatial and temporal scales (Folke 2007). Management failure is common because actions are not based on an understanding that these social-ecological systems are linked. This is often the result of so-called information asymmetries or failures, resulting in decision-makers being ill-equipped to make decisions that avoid biodiversity loss. As a result, management practices may be myopic in nature, have difficulties in incorporating complex information, and are consequently narrowly focused (e.g. on the sustainability of one fish stock) resulting in the erosion of the ability of ecosystems to withstand perturbations, create inflexible institutions, and produce a society that becomes ever more dependent on the specific resource being managed (Gunderson and Holling 2002). Understanding this so-called 'pathology' (Holling and Meffe 1996) is key to improving envi-

ronmental protection and the sustainability of ecosystem service provision.

Multiple stable states that exist in natural and social systems (e.g. coral reefs or democracies) may suddenly shift into another state that is very different to the original, dominated by unfamiliar processes, that may be irreversible and undesirable to humans (Scheffer *et al.* 2001; Kinzig *et al.* 2006). These have been termed 'regime shifts' (Carpenter 2003). These shifts may have large impacts on the provision of ecosystem services and consequently human well-being. Regime shifts have been observed in a number of systems and are caused by a range of factors that occur within and outside the system; they may be driven by human activity. For example, a study of Caribbean coral reefs shows evidence of a shift to a system dominated by algae due to factors such as high nutrient loads, intensive fishing, and pathogens increasing their vulnerability to hurricanes (Nyström *et al.* 2000). One of the properties of systems that have received much discussion in this context is resilience. Resilience is defined as 'the capacity of a system to absorb disturbance and reorganize while undergoing change so as to retain essentially the same function, structure, identity, and feedbacks' (Walker *et al.* 2004). Elucidating the complex dynamics of these linked social-ecological systems requires an understanding of their ability to recover from perturbations (Gunderson and Pritchard 2002). This is a point of contact with the BEF community through the idea that more diverse systems provide resilience, robustness, and stability (see Levin and Lubchenco 2008, Palumbi *et al.* 2008) for discussion in marine systems) and understanding the implication for systems of the manipulation of service providing units and ecosystem service providers. Indeed, Vandewalle *et al.* (2008) have placed the service providing units concept (Luck *et al.* 2003) within that of linked social-ecological systems.

Resilience also allows systems to withstand management failure (Gunderson and Pritchard 2002). As we are unlikely to ever fully understand such complex systems, resilience allows society time to adapt to change. This is why adaptive management of systems is crucial. It stresses that each management action should be seen as an opportunity to learn about the system under consideration, and that the information derived from these management actions should be used to update understanding of the system continuously (Carpenter and Gunderson 2001). Marine managers need to understand how to maintain the resilience of systems to sustain ecosystem services into the future (Levin and Lubchenco 2008).

15.2.2   Linking ecology and economics

If an ecosystem approach is to be successfully applied to marine management, it is important to explore ecosystems and ecosystem services in an economic context. In so doing, it is argued that the full costs and benefits of human activities can be identified, that multiple objectives can be better balanced, and that greater incentives will be created for the protection and sustainable use of biodiversity and the ecosystem services it provides. It also contributes to our understanding of the consequence for ecosystems of economic activity.

As mentioned earlier, a number of typologies for ecosystem services have been put forward. Whatever the classification framework for ecosystem services chosen, it is important to examine the characteristics of the ecosystem services of interest. The characteristics will influence the appropriate scale at which the services are measured and valued, and will help identify who benefits from the services, where, and when. Scale is of interest because ecosystem services are not static or constant across a seascape or landscape. Some services will be utilized when and where they are provided (e.g. flood defence by coastal wetlands) while with others, the benefit will be generated in one location, but used in another at a different time (e.g. the consumption of fish). Who benefits from the service is particularly important as it shapes what is understood to be an ecosystem service (Boyd 2007; Boyd and Banzhaf 2007; Fisher *et al.* 2009). Different stakeholder groups can be expected to derive different benefits from the same ecosystem processes, and these benefits may be conflicting. The complexity of ecosystem services also needs to be taken into account. As already mentioned, services are generated from complex systems, and our knowledge of

the dynamics of these systems is limited. Furthermore, one service may provide multiple benefits, while multiple services may contribute to the same benefit.

15.3   An economic framework for ecosystem services

Economics views ecosystems as assets from which flow a variety of goods and services. They therefore have an economic value (Pearce 2007). These goods and services may be physical in nature, but also aesthetic, intrinsic, and moral, and humankind is dependent upon them as a life support system as well as for enhancing well-being (Turner *et al.* 2003). Pearce (2007) linked ecosystem services to human welfare through a stylized supply-and-demand model (Figure 15.3), and demonstrated why society currently undervalues the environment and the services obtained from it. The model shows that the demand for ecosystem services is only seen through the services that are traded on the market ($D_{ES(M)}$, e.g. fish and timber). Not all ecosystem services, however, are traded through markets (e.g. climate regulation and coastal protection). The total of the ecosystem services which are actually used is represented by $D_{ES(MNM)}$, the demand for both marketed (M), and non-marketed (NM) services. We do not know the actual shape of these demand curves, but both curves are downward sloping (representing the decreasing value placed on an additional unit of ecosystem service). We also know that the demand for marketed and non-marketed ecosystem services combined ($D_{ES(MNM)}$) always lies above the demand for marketed ecosystem services alone ($D_{ES(M)}$); the gap between them reflects the additional demand for the non-marketed ecosystem services. The cost of supplying additional units of ecosystem service is shown by MC_{ES}, the marginal (or incremental) cost curve. The point where the marginal cost curve crosses the demand curve for marketed ecosystem services (shown in Figure 15.3 by the dotted line ES_M) is the point where the cost of a unit of ecosystem service demanded is considered equal to the benefits gained from that unit. As a consequence of missing markets for non-marketed ecosystem services, the point where cost is equal to benefit is below

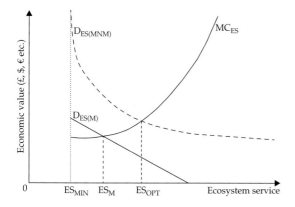

Figure 15.3 Supply and demand model for ecosystem service provision. From Pearce (2007), with kind permission from Springer Science and Business Media.

the optimal level (shown by ES_{OPT}). This means that we are using more ecosystem services than we realizes, and that we pay less for them than we should; a situation which can result in their over-exploitation. It also shows why the market will not be able to provide for the optimal level of ecosystem services unless new markets for the ecosystem services that are currently untraded are developed or mechanisms are used to correct the market (e.g. through subsidies or taxes). Another important feature is what lies to the left of ES_{MIN}. ES_{MIN} indicates the minimum level of ecosystem structure required to maintain a functional ecosystem capable of providing services (i.e. some threshold level of ecosystem services). Where exactly this level lies is uncertain and can be expected to be different for different ecosystem services (Dobson *et al.* 2006).

Fisher *et al.* (2008) highlight three key issues arising from this model that they suggest should help make operational ecosystem services research. First, ecosystem services should be studied in terms of marginal changes (i.e. unit changes in quantity, and hence value, of ecosystem services) rather than total values; second, the minimum level of ecosystem structure that is required to maintain a functional ecosystem needs to be understood better (i.e. where are the 'tipping points' that lead to regime shifts) and third, measures need to be put in place to capture the benefits arising from non-marketed ecosystem services (e.g. taxes and levies or payments for ecosystem services).

15.3.1 Valuation of ecosystem services

Valuation can be defined as the process of assessing the socio-economic importance of an ecosystem service by determining its monetary or non-monetary value. It is a way of capturing the value of ecosystem services that are not traded on markets. A distinction can be made between instrumental (or utilitarian) values, which are derived from some objective function, goal or purpose that is being sought (e.g. maximizing welfare) and intrinsic values, which are non-derivative, in that they are independent of their utility to humans. Valuation methods are only designed to capture instrumental values.

There are two classes of instrumental values: use and non-use values. Use value is a measure of the relative satisfaction or desirability of the consumption derived from of a good or service. Non-use values (also known as passive values) are values that are not associated with actual use of a good or service, but potential future use; or the value derived from knowing that a good or service exists. These use and non-use values have been encapsulated in the total economic value (TEV) framework. This provides a structure for valuing a diverse range of ecosystem services. If the individual services can be quantified and valued in monetary terms, then theoretically, an aggregated figure for TEV could be estimated. Most studies have managed to provide monetary values for some of the services but not for others. They have therefore not stated a single figure for TEV of biodiversity (e.g. Beaumont *et al.* 2008). In practice, however, it has been recognized that a single figure for TEV is undesirable because it represents a snapshot in time, rather than an understanding of change in value over time (i.e. value at the margin). It is also likely that a single figure for TEV would represent double counting as the different methods use to calculate the different use values often overlap and capture more than one aspect of use.

Not all commentators accept the necessity of valuation. Heal (2000) states that 'Valuation is neither necessary nor sufficient for conservation. We conserve much that we do not value, and do not conserve much that we value' (p. 29). Others argue from a rights-based perspective, that nature has a right to exist irrespective of human needs and that economic valuation is morally suspect (e.g. Sagoff (2002) and Maguire and Justus (2008), in reference to the stance of preservationists). Heal goes on to say that what is important is not to value ecosystem services, but to demonstrate the incentives for their conservation. While valuation is not a panacea, it can do just that. The use of a common language (i.e. money and markets) provides a yardstick to measure the benefits humans derive from ecosystems. These benefits can then be compared more easily against alternative uses of the ecosystems and their assets (Turner *et al.* 2003), and thus help persuade decision-makers to incur the opportunity cost of the protection of such ecosystem services (i.e. the benefits foregone if ecosystem services are not conserved).

15.3.2 Valuation methods

A number of methods can be used to value the ecosystem services provided by biodiversity. The variety of methods reflects, to a certain extent, the different perspectives among the population about the meaning of value. People from different disciplines, cultures, ethical, and philosophical view points express and understand value in different ways (Goulder and Kennedy 1997).

At a general level, valuation methods fall into two broad groups: monetary and non-monetary. Monetary methods result in an explicit statement of value and are quantitative in nature. They are used to represent utilitarian (or instrumental) values of direct and indirect use, and non-use values (bequest, existence, and option use). While most biodiversity valuations are based on monetary methods, it has been recognized that these methods are unable to capture all aspects of value. Alternatively, non-monetary methods have been proposed that tend to be more qualitative in nature, assigning values to biodiversity according to its qualities or attributes. They are often techniques that are not specific to valuation studies, and value is usually implied rather than explicitly stated. Table 15.1 summarizes a number of commonly applied methods for both monetary and non-monetary valuation.

Decision-support tools are also commonly used in valuation exercises. Although they are not valuation methods as such, they can incorporate the findings from valuation exercises to support the decision-maker in making trade-offs and choices between resources and their use. Some of the common decision support tools that draw from valuation studies include Cost–Benefit Analysis (CBA), Bayesian Belief Networks (BBN), and Multi-Criteria Analysis (MCA) (see Table 15.1).

Uncertainty is inherent in ecosystem services valuation studies and at all stages of the valuation process. There is uncertainty surrounding the ecosystem processes, uncertainty regarding the data collected (both social and natural science) and uncertainty in the management decision made as a result of the valuation research. The values estimated should therefore be used with caution and not taken as absolute. Communicating, and potentially minimizing uncertainty, should therefore be a priority and an integral part of valuation research.

Valuation studies also face a number of additional challenges. This is partly due to the varying characteristics of ecosystem services and debate over what constitutes an ecosystem service. It is also a consequence of our limited understanding of ecosystem services and the role of biodiversity within them (see section 15.2); the availability of appropriate data (both biological and socio-economic) at suitable temporal and spatial scales; difficulties in avoiding double counting given that different stakeholders may derive different benefits from the same ecosystem services (Turner *et al.* 2003); arguments of incommensurability, that valuation is not possible because ecosystem services cannot be measured using the same scale (for a full review, see Aldred 2006); and the incompatibility of value estimates from different valuation techniques. Overcoming these difficulties is now a major theme within environmental and ecological economic research.

15.4 A framework for implementing an ecosystem approach

Crucial to the implementation of an ecosystem approach to management, and to the prediction of marine biodiversity futures, is a framework to guide

Table 15.1 Summary of monetary and non-monetary methods for biodiversity and ecosystem service valuation

	Approach/method		Characteristics
Monetary methods	Market based		Actual market values used and income induced multiplier effects along the supply chain
	Revealed preference	Travel costs Hedonic pricing Averting behaviour Production function	A relationship is assumed between observed behaviour and associated environmental attributes to identify how individuals value changes in those attributes
	Stated preference (willingness-to-pay or willingness-to-accept)	Contingent valuation Choice experiments	Values are elicited as responses to a set of hypothetical questions
	Replacement and damage cost avoided		Values are based on the costs of replacing or restoring the ecosystem services of interest
	Deliberative approaches		Novel approach that use small-scale group deliberation to generate consensus about the social value of changes in some attribute of biodiversity
	Benefit or value transfer		Estimates of the value of some aspect of biodiversity are transferred from one place, time or population to another setting
Non-monetary methods	Emergy		Uses the amount of energy needed to produce a good or services as a unit of account
	Q-methodology		Allows the values people place on biodiversity to be inferred from their subjective preferences
	Citizens' juries		Similar to deliberative approaches, it allows value to be inferred from the verdict reached about a policy or project
	Delphi method		Produces a synthesis of expert opinion and inferred value through an iterative, anonymous deliberative process among experts
	Scoring, ranking and weighting methods		Stakeholders are asked to score, rank of weight natural resources and biodiversity according to their attributes and hence implied value
	Biological methods	Phylogenetic methods Species richness indices Other biological indicators	Integrates available biological information for an area into an indicator of biological value
Decision support systems	Cost benefit analysis		Explores all costs and benefits of proposed interventions in financial terms over time.
	Bayesian belief networks		These are graphical tools that can facilitate decision-making under uncertain conditions; they use conditional probabilities to explore the impact of environmental change on each variable of interest.
	Multi-criteria analysis		Involves the weighting of different scenarios or interventions according to economic, social and environmental criteria.

monitoring, analysis, decision- and policy-making. A number of frameworks have recently emerged as scientists, policy makers, and managers grapple with how it might be possible to deal with trade-offs between ecosystem services, understand the role of and conserve biodiversity, encourage stakeholder participation, deal with uncertainties, monitor appropriate indicators, guide policy, and create the appropriate governance institutions (Turner and Daily 2008; Vandewalle *et al.* 2008; Daily *et al.* 2009; Levin *et al.* 2009; Surridge *et al.* 2010). Here, we present a framework for the implementation of an ecosystem approach (Figure 15.4) that is centred on adaptive management and based on the understanding that ecological and social systems are linked. The intent is for stakeholder participation

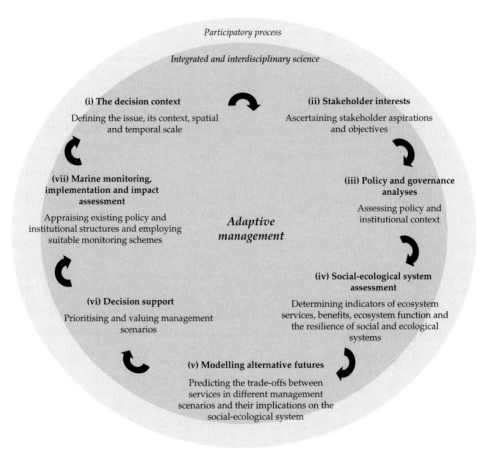

Figure 15.4   Framework for the implementation of an ecosystem approach. The cyclical nature of the framework illustrates how knowledge is accumulated through each of its steps. Adaptive management can occur through iterations of the framework, with alterations to approaches within each step depending on what has been learnt (see section *(vi)*). Participation of stakeholders (which includes scientists and policy makers) is assumed throughout, as is the need for interdisciplinary and integrated science. Adapted from Surridge *et al.* (2010). Reproduced by permission of The Royal Society of Chemistry.

throughout the process, including scientists and policy-makers, therefore creating much needed science-policy interfaces. The cyclical nature of the framework illustrates how knowledge is accumulated through each of its steps. Adaptive management can occur through iterations of the framework, with alterations to approaches within each step, depending on what has been learnt. Indeed, in following iterations it may not be necessary to complete some of the steps. For example, monitoring *(vii)* may directly inform uncertainty estimates in the modelling phase *(v)* without having the need to update the policy analysis *(iii)*. However, due to the

dynamic nature of social-ecological systems, repeating each step of the framework is recommended. We discuss each stage including examples of the tools and approaches that will help with its implementation.

(i) The decision context

This stage is a scoping exercise to define the issue or decision context, and the appropriate spatial and temporal scales at which it is to be tackled. The driver for employing an ecosystem approach to marine management may be defined by the need to manage a site; for example, a marine reserve, estuary, or Large Marine Ecosystem, and/or for a specific issue (e.g. over-fishing, pollution event). Regardless of the driver, identification of the scales at which the related suite of ecosystem services operate, and the likely ecosystem providing unit, needs to be established. This will also facilitate the choice of appropriate scales for any valuation of the identified ecosystem services. Related policies and institutions involved also need to be identified, and these are likely to be at nested scales from local to global. Taking an ecosystem approach means dealing with multi-scale issues that will inevitably create challenges, no matter where and how the framework is applied.

For integrated environmental management and spatial planning, it may be appropriate to fix a reference context in which to deal with all issues. For example, the catchment scale is a logical scale at which to manage land, freshwater, and marine coastal areas, promoting cross-sector links to deal with the effects of land management on marine ecosystems (Surridge *et al.* 2010). However, there will be a need for different reference scales for the management of the open oceans; for example, using the concept of Large Marine Ecosystems (Sherman *et al.* 2005; Rosenberg and McLeod 2005).

(ii) Stakeholder interests

The identification of the issue and scale in *(i)* serves as a guide for the identification of stakeholder groups that need to be included in the process. Stakeholder participation throughout the framework process is important to ensure that the management outcome is credible and enforceable (Leslie and McLeod 2007). Stakeholders may include fisheries managers, fishermen (local or commercial), government bodies, conservation NGOs, and the general public. Choosing who should be involved in the participatory process is important, as the legitimacy and success of subsequent stages of the framework may be influenced by the composition of stakeholders. Stakeholder assessment techniques such as social network and stakeholder analysis can be useful here (Grimble and Wellard 1997; Prell *et al.* 2009). The scale at which participation is achieved is crucial, and may need to extend from the local to the national scale, depending on the ecosystem

services and the nature of the issues being managed.

Stakeholder participation is essential to understanding which ecosystem services are of interest to them, what their perspectives are on the significant management issues, their objectives and aspirations for the ecosystem. Stakeholder knowledge is crucial to create a focus in other stages in the framework, most importantly in the development and modelling of alternative scenarios *(v)*. Stakeholders can be brought together in focus groups or meetings using techniques such as participatory mapping (Cinderby 2007), to establish how stakeholders use and value ecosystem services, and through participatory modelling (Prell *et al.* 2007), where cognitive maps of issues and objectives can be constructed. Holding group meetings may not be the most efficient way of understanding stakeholder interests and values, and indeed may not be possible at the range of scales necessary. Here, surveys may be a more appropriate method. Participation is designed to foster social learning (Pahl-Wostl 2007) about the complexity of the system being managed, and the likely causes of, and possible solutions to management issues. Social learning will promote understanding of the multiple standpoints of stakeholders and their perceptions of marine issues, which in turn can help to build sustainable relationships between stakeholders, reduce conflicts, and improve management capacity.

(iii) Policy and governance analyses

Analysis of existing policy and governance of marine systems is essential for understanding what has and has not worked, and what policy mechanisms may be most appropriate for the management of the issue in question. A multitude of international and national legislation operates in marine systems. For example, regulations exist in the UK under the Common Fisheries Policy, European Marine Strategy Framework Directive, Water Framework Directive, Habitats Directive, Birds Directive, the OSPAR Convention, the Ramsar Convention, the forthcoming Marine and Coastal Access Bill, and Marine Biodiversity Action Plans for species and habitats. However, a key barrier to more integrated management is that individual regulations are often directed towards specific sectors (e.g. fisheries, water quality, conservation). Overcoming this barrier is dependent on understanding how the existing regulatory frameworks interact, and identifying how management can be integrated across multiple sectors. Ignoring such interactions compromises the ability of the system to provide a full suite of ecosystem services and policies to meet individual sector goals (Rosenberg and McLeod 2005). In the US, there have been urgent calls for a single administrative and legislative structure for integrated oceans management (Palumbi *et al.* 2009 and references therein). Such a structure has been achieved in the UK through the Marine and Coastal Access Act (2009), and the Marine (Scotland) Act (2010). Both have been developed with the ecosystem approach in mind and provide a framework for the management of the competing demands placed on the UK's seas. How they will operate in practice, however, remains to be seen.

Centralized (top-down) resource management does not always facilitate important stakeholder participation. In addition, yield orientated 'command and control' management is unsuitable in systems that are dynamic and non-linear (Levin 1998). Consequently co-management arrangements (e.g. networks of partnerships between state, public and private stakeholders) are emerging (Armitage, *et al.* 2007), and are encouraged for effective ecosystem management. These types of arrangements harness the activity of local-scale user groups that would not usually be supported or engaged by top-down governance structures. Bottom-up approaches, therefore, need to be balanced with carefully selected top-down mechanisms. For example, taxes, and levies on economic activities will work more effectively if introduced centrally, rather than by local stakeholders, and enforcement of environmental regulations may be better achieved if implemented from above, although it will require the buy-in of local stakeholders. Thus, top-down approaches may be particularly important in situations where property rights are poorly defined (e.g. throughout many coastal areas and the open ocean). Where the individual does not have ownership over a predefined

area, the incentive to protect its resources and ecosystem services may be lost. In such cases, there is a need to build the capacity of local stakeholders to self-govern the management of the resources, while at the same time developing appropriate state intervention to ensure that the environmental agenda is met. An example of where this has happened is in the collection of shellfish by women in Galicia, Spain (Frangoudes *et al.* 2008).

Understanding the web of governance of marine systems is a significant challenge. This understanding is crucial if revised, multi-scale, and polycentric governance structures (stakeholders in different institutional settings cooperating and collaborating in the formulation and implementation of public policy) that facilitate adaptive co-management and governance (see Folke *et al.* 2005) are to be developed as part of an ecosystem approach. Mapping policy networks and governance structures can be very useful for visualization and understanding at this stage (e.g. Ekstrom *et al.* 2009).

(iv) Assessing social-ecological systems

To set reasonable management objectives with respect to the decision context identified in (*i*), the ability of the ecosystem to provide ecosystem services sustainably and equitably into the future needs to be assessed. Assessment depends on measuring social, economic, ecological, and institutional characteristics that reflect the provision of services (i.e. through the assessment of indicators). Indicators need to encapsulate the values of the benefits derived from ecosystem services, the services that ecosystems provide (intermediate and final services) the functional aspects of the ecosystem that supply them (if different from the intermediate services, including information on service providing units and ecosystem service providers) and the ability of the social-ecological systems to withstand perturbation (resilience measures that may be related to biodiversity). Measuring the flexibility/resilience of social systems may mean understanding the structure of social networks of governance, social capital, trust, and leadership (see Folke *et al.* 2005). It may also require an understanding of how society responds to risk, as this may influence societal response to ecosystem change.

It is important in the context of the management goals and stakeholder values to also develop indicators of what causes shifts in the state of systems, and how human-environment interactions facilitate or undermine sustainable management (Leslie and McLeod 2007; Dietz *et al.* 2003). For example, institutional indicators might show how the social system reacts to policy change and biological shifts in ecosystems. Indicators of ecosystem services and processes may show how these change in response to shifts in societal values or policy change. Change in the status of services (for example, due to the implementation of new management actions) will provide important information to feed back into decision-making as part of the adaptive management process.

Indicators need to be based on good theory or evidence, be measureable, appropriate for the system at hand, understandable to the general public where appropriate and cost-effective (see Levin *et al.* 2009; Rice and Rouchet 2005). Where ecosystem service valuation is also an objective, these indicators must also be suitable for the valuation methods chosen. Developing a series of indicators in collaboration with the stakeholders together with a number that are scientifically derived may be a way of achieving this (Reed *et al.* 2008). However, deriving a suite of indicators is not without its challenges. One of the most significant from an ecological perspective is that our understanding of how ecosystem services are provided is currently limited (as discussed earlier in the chapter). Knowledge of the ecological units and functions that underpin resilient systems across a range of spatial and temporal scales is essential to configure appropriate indicators for system assessment. When establishing social and economic indicators that reflect the value of the benefits received from ecosystem services, market values for certain commodities (e.g. fish stocks) may be appropriate but less tangible values (e.g. spiritual or cultural values) may be much harder to capture (see section 15.3).

Novel approaches exist that can be used to assess the health or quality of the social-ecological systems and their ecosystem services. The Monitoring Ecosystem Health through Trends Analysis (MEHTA) (Raffaelli *et al.* 2004) is a good example of a method that develops indicators of the social and ecological components of ecosystems and the interaction between them. Indicators of ecological, financial, human, and social capital stocks of the ecosystem are analysed in relation to the thresholds necessary to maintain the delivery of a suite of ecosystem services, and the rate of depletion of the stocks. The participatory approach allows stakeholders to express their preferred values attached to the ecosystem services, and allows expert knowledge of ecosystem function to be incorporated into the overall assessment of system health (see Raffaelli *et al.* 2004 for case studies). Other frameworks exist, for example, the Integrated Ecosystem Assessment (IEA) (Levin *et al.* 2009) that emerges from decision theory, and was developed in the context of ocean ecosystem-based management. This is a comprehensive and rigorous way of testing and establishing appropriate ecosystem indicators (and modelling scenarios, see next section). Fulton *et al.* (2005) illustrate how the robustness of these indicators can be tested using computer simulation. Spatial mapping of provision and demand of ecosystem services, and comparison with governance and policy mapping of *(iii)* is useful to visualize spatial and temporal mismatches, and an excellent tool to discuss management with stakeholders.

A major drawback is that data on which to base system assessment, valuation, and modelling are not always available, and where they do exist may not be at the required spatial or temporal scales. When it is not possible to collect further data, this issue can be overcome in early iterations of the framework by using tools that can deal with uncertainty and incorporate non-quantitative expert knowledge (e.g. Bayesian Belief Networks). The framework can then be used to guide monitoring programmes for the collection of appropriate indicator or valuation data for future use.

(v) Modelling alternative futures

Once an assessment of the social-ecological system and its associated ecosystem services has been established, a number of different management scenarios can be developed. The range of possible scenarios draws on the objectives and knowledge of the stakeholders established in step *(ii)* of the framework, and also the potential futures possible given the existing policy, legislation, and institutional structures *(iii)*. The impacts of these scenarios on the social-ecological systems indicators developed are predicted through the use of modelling. An important reason for undertaking this modelling stage is to highlight the trade-offs that emerge as a result of management scenarios, and what the implications are for social-ecological systems and the provision of ecosystem services. Ecosystem services are not independent of one another, and management that affects one particular service may impact on the viability of another in a synergistic or competitive way. Models allow decisions about preferred management options to be more informed and used in combination with stakeholder participation supports knowledge exchange and learning.

Models of ecological systems can be used to understand their response to particular management practices. Kaplan and Levin (2009) use Atlantis, a process-based ecosystem model developed by Fulton *et al.* (2004), to evaluate a range of management strategies, based around fishing intensity, and how the outcome effects conservation and economic goals in the California Current Ecosystem of the US West Coast. However, a number of techniques can model both ecological and social dynamics. For example, Bayesian Belief Network models (Castelletti and Soncini-Sessa 2007), agent-based modelling (Moss *et al.* 2001), and integrated assessment models (Warren *et al.* 2008). For example, McDonald *et al.* (2008) demonstrate the use of an integrated agent-based modelling system to evaluate both the social and ecological impacts of management strategies, that combine uses such as fisheries and conservation, natural resource extraction and coastal development, in the north-west shelf coastal region of Australia. Currently being developed is the technology for integrated modelling systems. Such systems allow the use of a combination of the different sources of knowledge (e.g. quantitative data and opinions) and the integration of different models focused on different components of social-ecological systems, which enables a more realistic incorporation of the human dimensions (Surridge *et al.* 2010). In the absence of suitable models, or the capacity to undertake formal model-

ling, it may be possible to undertake a more qualitative assessment based on expert knowledge of potentially expected outcomes.

(vi) Decision support

The alternative management scenarios produced in step *(v)* of the framework need to be prioritized. Deciding which management option to choose should be made within a framework that enables a range of scenarios, including business as usual, to be compared against each other. This step should include the estimation of ecosystem services values and involve stakeholder participation.

Choices can be made based on a range of criteria that describe each scenario including traditional economic metrics such as cost effectiveness. However, they should also be based on other criteria, such as how the estimated values of ecosystem services are likely to change following an intervention, and the degree to which benefits are equitably distributed across different groups of stakeholders (for a discussion of decision-support tools see section 15.3). Multi-Criteria Analysis can provide a framework within which decisions about the alternative scenarios can be made. This approach enables individual scenarios to be presented in a number of ways, for example, through the use of monetary and qualitative values. Cost Benefit Analysis could also be applied to provide a comparative evaluation of the different scenarios. Sophisticated Cost Benefit Analysis techniques have been developed in the context of wetlands (Luisetti *et al.* 2008). The modelling in step *(v)* can be used as a basis of decision-support systems, along with other data information, and a user-friendly graphical interface (e.g. OCEAN, <http:www.ecotrust. org/ocean/>). Bayesian Belief Networks have also been applied to environmental decision-making. They are particularly useful where data are scarce and uncertainty is high. Langmead *et al.* (2007) used Bayesian Belief Networks to explore the impacts of habitat change, eutrophication, chemical pollution, and fishing on four European seas (the Baltic Sea, Black Sea, Mediterranean Sea, and North East Atlantic). Where data permitted, they were used to simulate the consequences of a 'business-as-usual' management scenario with four alternatives for economic and social development over the next two to three decades. The analysis of the different scenarios, however, was limited by a shortage of historic time-series data.

(vii) Marine monitoring, implementation, and impact assessment

Once a preferred management scenario has been identified, its compatibility with existing institutional, policy, and regulatory structures must be appraised. In many cases, these structures are unlikely to be able to deliver the desired management objectives by themselves. This may require the development of new structures; for example, financial incentives that help overcome the problem of missing markets for some ecosystem services and reward stewardship (Turner and Daily 2008). Collaborations between stakeholder groups may also be necessary to implement the chosen strategy e.g. the development of co-management networks and partnerships, which includes science–policy interfaces.

Monitoring and evaluation are vital aspects of this framework, and are necessary to supply evidence on which decisions about the sustainable management of ecosystem services can be based and modified. To ensure that monitoring is targeted appropriately at any one time, it should be part of an adaptive management approach. Monitoring needs to cover the indicators established in step *(iv)* generating spatial and temporal data on the health of different aspects of the social-ecological systems at a number of scales. The evidence collated from monitoring activities need to be carefully evaluated against management objectives, and the social, economic, and environmental impacts of management must be assessed. Adaptive management stresses that each management action is an opportunity to learn about the system under consideration, and that information derived from such management actions should be used to continuously update understanding of the system (Carpenter and Gunderson 2001). Updated understanding can subsequently be used to alter management objectives, system assessment, integrated models, and reduce uncertainties when these models are used to support decision-making. Such learning and adapting

necessitates the formulation of flexible policies and institutions (for adaptive governance, see Folke *et al.* (2005)). An important part of monitoring relates to the collection of data from monitoring networks in space and time. SMART (Self-Monitoring, Analysis, and Reporting Technology) can facilitate the collection of indicator data. Marine biosensors, photography, aerial, surface, and underwater monitoring and sampling will be necessary to capture information on social-ecological systems, their interactions and resilience, and ecosystem service provision over a range of spatial and temporal scales in real time. The social and economic aspects of the system can also be monitored using a variety of methods; for example, through the monitoring of market values for certain commodities. It may also be possible to regularly sample the values held by the general public to different ecosystem services through the inclusion of suitable questions into routine surveys such as the British Household Panel Survey or tourism surveys.

15.5   Challenges for the future

The integrative nature of the ecosystem approach makes it well placed for monitoring, predicting, and safeguarding marine biodiversity futures given the multiple pressures that marine systems experience. Here, we have highlighted the importance of the role of biodiversity and ecosystem functioning research in developing the ecosystem approach, and the links between ecosystems and socio-economic systems. The framework described outlines how such an approach might be implemented; however, there are some real challenges to be overcome to make this framework operational. It requires integration across ecological, economic, and social sciences, co-operation between scientists, policy-makers, managers, and other stakeholders, and coherence in the policy process which may require major organizational and institutional change.

15.5.1   Science needs

Developments are needed both within and between science disciplines. In ecology, employing a service-led approach (section 15.2, Figure 15.1) may accelerate understanding about the relationships between biodiversity, ecosystem function, and services, a vital evidence base for management. At the same time such research needs to be focussed on larger scales, those appropriate to the management issues faced in marine systems. Validating trends in data at large scales with what has been learnt in small-scale experiments is an important avenue for future biodiversity and ecosystem function research, following Worm *et al.* (2006). Description of the form of such relationships may lead to predictions of critical ecosystem thresholds; at what level of use of ecosystem services are ecosystem processes and functions compromised? In addition, the scientific basis of resilience requires further investigation (Palumbi *et al.* 2008). Studying the response of ecosystems to change and of their ability to recover is possible in dynamic ecosystems and can therefore inform ecosystem management. How this ability differs across different types of disturbances, that originate from natural and social systems, and from their different combinations, needs special attention.

This ecological knowledge is important to inform accurate valuation of marine ecosystem services. There is increasing pressure on the scientific community to support environmental management decisions with estimates of the value of ecosystem services. Information on the likely consequence of medium to long-term environmental change and the sustained provision of ecological services is often absent from these studies, limiting their current usefulness. Uncertainties regarding the appropriate spatial and temporal scales at which valuation studies should be carried out, therefore, need to be addressed, especially given that there are likely to be linkages between different service-providing units and their connectivity across different scales. We also need greater understanding of the dynamics of biophysical processes across different scales. Inter-disciplinary research focusing on the natural and social science components of biodiversity valuations needs to be encouraged. Many valuations are dominated by either natural or social science, with little appreciation of what the other discipline can bring in terms of understanding. There is also a need to further develop methods, explore compara-

bility of methods and the ability to combine the findings from the different methods used in valuation research.

Integrating across these disciplines is essential for understanding the links and feedbacks between social and ecological systems, and requires much greater investigation in marine systems (Leslie and McLeod 2007). It is important to decipher the response of social systems to change in ecological ones, and *vice versa*, in order to build in flexibility and resilience. Whilst waiting for these new science developments there is some urgency to research that develops models and decision-support tools that can deal with the uncertainty in systems, and to deal with the gaps in our current knowledge.

15.5.2 Policy needs

Building interfaces between scientists and policy-makers is essential to implementing an ecosystem approach. However, this interaction has never been easy (Houck 2003). Creating interfaces between scientists and policy-makers through the ecosystem approach framework can aid the two-way communication of information, at the same time as building trust and understanding.

Another issue is the mismatch between science and policy production. By the time that legislation, based on current science, has been developed and has been through the political and administrative processes and government consultation (which may take years) the resultant policy is no longer up-to-date as the science has moved on. Policy makers are becoming increasingly aware of this issue, as illustrated by the new Marine and Coastal Access Act in the UK. The Act requires the designation of a network of marine conservation zones, but states that this network will need to be reviewed with site boundaries being altered, new sites designated and some sites de-designated based upon the best available scientific and socio-economic data. However, current documentation is unclear about how exactly this will be achieved.

Over recent decades, there has been a move away from rigid prescription of science in policy, and there has been a tendency for policies and their implementation to express the science in an increasingly flexible manner, more subjectively, and in terms that permit implementation to reflect up-to-date science. This use of flexibly expressed science within policy and legislation is a way of improving the match of scientific and policy timetables that will be necessary for promoting the ecosystem approach as it develops in the future. There is no doubt that to put in place appropriate governance and management structures that are flexible and adaptive to change will mean major organizational and institutional alteration. However, it is possible, for example Canada and Australia have operational governance frameworks that provide integrated management of human impacts on coastal and ocean ecosystems (Leslie and McLeod 2007).

15.5.3 Conclusions

There may be major challenges in the science, social science, economics, and implementation required to take an ecosystem approach, but to address the environmental issues in marine systems, frameworks such as the one presented here must be applied as soon as possible. It is important to accept the uncertainties and the knowledge gaps in the short-term, develop tools to deal with the uncertainties, and put in place programmes of monitoring and research that will increase knowledge in the longer term. We have to accept that such an approach is an 'experiment', and that policy and society will have to be flexible enough to adapt to necessary change. The alternative is to persist in undervaluing the role of nature in the provision of ecosystem services, in the process continuing to jeopardize the life-support system of humans and the future of marine biodiversity.

References

Adger, W.N., Hughes, T. P., Folke, C. et al. (2005) Social-ecological resilience to coastal disasters. *Science*, **309**, 1036–9.

Aldred, J. (2006) Incommensurability and monetary valuation. *Land economics*, **82**, 141–61.

Anderson, B.J., Armsworth, P.R., Eigenbrod, F. et al. (2009) Spatial covariance between biodiversity and other ecosystem service priorities. *Journal of Applied Ecology*, **46**, 888–96.

Armitage, D., Berkes, F. and Doubleday, N. (2007) *Adaptive co-management: Collaboration, Learning and multi-level governance*. UBC Press, Vancouver.

Balvanera, P., Pfisterer, A.B., Buchmann, N. et al. (2006) Quantifying the evidence for biodiversity effects on ecosystem functioning and services. *Ecology Letters*, **9**, 1146–56.

Beaumont, N. J., Austen, M. C., Mangi, S. C. et al. (2008) Economic valuation for the conservation of marine biodiversity. *Marine Pollution Bulletin*, **56**, 386–96.

Boyd, J. (2007) Nonmarket benefits of nature: What should be counted in green GDP? *Ecological Economics*, **61**, 716–23.

Boyd, J. and Banzhaf, S. (2007) What are ecosystem services? The need for standardized environmental accounting units. *Ecological Economics*, **63**, 616–26.

Carpenter, S. R. (2003) *Regime shifts in lake ecosystems: pattern and variation*. Excellence in Ecology Series Volume 15. Ecology Institute, Oldendorf/Luhe, Germany.

Carpenter, S.R. and Gunderson, L.H. (2001) Coping with collapse: ecological and social dynamics in ecosystem management. *BioScience*, **6**, 451–7.

Castelletti, A. and Soncini-Sessa, R. (2007) Bayesian Networks and participatory modelling in water resource management. *Environmental Modelling and Software*, **22**, 1075–88.

Convention on Biological Diversity (CBD) (2000) SBSTTA 5 Recommendations adopted by the subsidiary body on scientific, technical and technological advice at its fifth meeting (SBSTTA 5). Recommendation V/10, Ecosystem approach: further conceptual elaboration, 78–84. <http://www.cbd.int/recommendation/sbstta/?id = 7027>.

Chan, K.M.A., Shaw, M.R., Cameron, D.R., et al. (2006) Conservation planning for ecosystem services. *PLoS Biology*, **4**, e379. DOI: 10.1371/journal.pbio.0040379.

Cinderby, S. (2007) How communities can use Geographical Information Systems. In: C Clay, M Madden and L Potts, (eds), *Towards Understanding Community: people and places*. Palgrave-Macmillan, Basingstoke.

Costanza, R. (2008) Ecosystem services: Multiple classification systems are needed. *Biological Conservation*, **141**, 350–2.

Costanza, R., d'Arge, R., de Groot, R. et al. (1997) The value of the world's ecosystem services and natural capital. *Nature*, **387**, 253–60.

Covich A.P., Austen M.C., Barlocher F. et al. (2004) The role of biodiversity in the functioning of freshwater and marine benthic ecosystems. *BioScience*, **54**, 767–75.

Daily, G.C. (1997) *Nature's Services. Societal Dependence on Natural Ecosystems*, Island Press, Washington DC.

Daily, G. C., Polasky, S., Goldstein, J. *et al.* (2009) Ecosystem services in decision making: time to deliver. *Frontiers in Ecology and the Environment*, **7**, 21–8.

De Groot, R.S., Wilson, M.A. and Boumans, R.M.J. (2002) A typology for the classification, description and valuation of ecosystem functions, goods and services. *Ecological Economics*, **41**, 393–408.

Department for Environment, Food and Rural Affairs (Defra) (2007) *Securing a healthy natural environment: An action plan for embedding an ecosystems, approach.* <http://www.defra.gov.uk/wildlife-countryside/natural-environ/eco-actionplan.htm>.

Diaz, S., Fargione, J., Chapin III, F.S. *et al.* (2006) Biodiversity loss threatens human well-being. *PLoS Biology*, **4**, e277. DOI: 10.1371/journal. pbio.0040277.

Dietz, T., Ostrom, E. and Stern, P.C. (2003) The struggle to govern the commons. *Nature*, **302**, 1907–12.

Dobson, A., Lodge, D., Alder, J. *et al.* (2006) Habitat loss, trophic collapse, and the decline of ecosystem services. *Ecology*, **87**, 1915–24.

Dulvy, N.K., Sadovy, Y., Reynolds, J.D. (2003) Extinction vulnerability in marine populations. *Fish and fisheries*, **4**, 25–64.

Ekstrom, J.A., Young, Or., Gaines, S.D. *et al.* (2009) A tool to navigate overlaps in fragmented ocean governance. *Marine Policy*, **33**, 532–5.

Fisher, B., Turner, K., Zylstra, M. *et al.* (2008) Ecosystem services and economic theory: integration for policy-relevant research. *Ecological Applications*, **18**, 2050–67.

Fisher, B., Turner, R. K. and Morling, P. (2009) Defining and classifying ecosystem services for decision making. *Ecological Economics*, **68**, 643–53.

Folke, C. (2007) Social-ecological systems and adaptive governance of the commons. *Ecological Research*, **22**, 14–5.

Folke, C., Hahn, T., Olsson, P. *et al.* (2005) Adaptive governance of social-ecological systems. *Annual Review of Environment and Resources*, **30**, 441–73.

Frangoudes, K., Marugán-Pintos, B. and Pascual-Fernández, J.J. (2008) From open access to co-governance and conservation: The case of women shellfish collectors in Galicia (Spain). *Marine Policy*, **32**, 223–32.

Fulton, E.A., Parslow, J.S., Smith, A.D.M. *et al.* (2004) Biogeochemical marine ecosystem models II: the effect of physiological detail on model performance. *Ecological Modelling*, **173**, 371–406.

Fulton, E.A., Smith, A.D.M., Punt, A.E. (2005) Which ecological indicators can robustly detect effects of fishing. *ICES Journal of Marine Science*, **62**, 540–51.

Goulder, L. H. and Kennedy, D. (1997) Valuing ecosystem services: philosophical bases and empirical methods. In:

Nature's Services. Societal Dependence on Natural Ecosystems. Daily, G. (ed.), Washington DC: Island Press, 23–48.

Grimble, R. and Wellard, K. (1997) Stakeholder methodologies in natural resources management: a review of principles, contexts, experiences and opportunities. *Agricultural Systems*, **55**, 173–93.

Gunderson, L.H. and Holling, C.S. (2002) *Panarchy: Understanding transformations in human and natural systems.* Island Press, Washington DC.

Gunderson, L.H. and Pritchard Jr, L. (2002) *Resilience and the behaviour of large scale systems.* Island Press, Washington DC.

Heal, G. (2000) Valuing Ecosystem Services. *Ecosystems*, **3**, 24–30.

Holling, C.S. and Meffe, G.K. (1996) Command and control and the pathology of natural resource management. *Conservation Biology*, **10**, 328–37.

Holt, A.R. and Hattam, C. (2009) Capitalizing on nature: how to implement and ecosystem approach. *Biology Letters*, 5(5): 580–82.

Houck, O. (2003) Tales from a troubled marriage: science and law in environmental policy. *Science*, **302**, 1926–9.

Kaplan, I.C. and Levin, P. (2009) Ecosystem-based management of what? An emerging approach for balancing conflicting objectives in marine resource management. In: *The future of fisheries science in North America.* Beamish, R. J. and Rothschild, B. J. (eds), Springer, New York, 77–96.

Kay, J.J., Boyle, M. and Pond, B. (2001) Monitoring in support of policy: an adaptive ecosystem approach. In: *Encyclopedia of Global Environmental Change* (4). Munn, T. (ed.) Wiley, New York, 116–37.

Kinzig, A., Ryan, P., Etienne, M. *et al.* (2006) Reslience and regime shifts: assessing cascading effects. *Ecology and Society*, **11**, 20. <http://www.ecologyandsociety.org/vol11/iss1/art20/>.

Kremen, C. (2005) Managing ecosystem services: what do we need to know about their ecology? *Ecology Letters*, **8**, 468–79.

Laffoley, D., Maltby, E., Vincent, M.A. *et al.* (2004) *The Ecosystem Approach. Coherent actions for marine and coastal environments.* A report to the UK Government. Peterborough, English Nature. P

Langmead, O., McQuatters-Gallop, A. and Mee, L.D. (eds) (2007) *European Lifestyles and Marine Ecosystems: Exploring the Challenges for managing Europe's Seas.* 43 pp. University of Plymouth Marine Institute, Plymouth, UK.

Leslie, H.M. and McLeod, K. (2007) Confronting the challenges of implementing marine ecosystem-based management. *Frontiers in Ecology and the Environment*, **5**, 540–8.

Levin, P.S., Fogarty, M.J., Murawski, S.A. *et al.* (2009) Integrated ecosystem assessments: Developing the scientific basis for ecosystem-based management of the ocean. *PloS Biology*, **7**, e1000014.

Levin, S.A. (1998) Ecosystem and the biosphere as complex adaptive systems. *Ecosystems*, **1**, 431–6.

Levin, S. A. and Lubchenco, J. (2008) Resilience, robustness and marine ecosystem-based management. *Bioscience*, **58**, 27–32.

Lotze, H.K., Lenihan, H.S., Bourque, B.J. *et al.* (2006) Depletion, degradation and recovery potential of estuaries and coastal seas. *Science*, **312**, 1806–9.

Luck, G.W., Daily, G.C. and Ehrlich, P.R. (2003) Population diversity and ecosystem services. *Trends in Ecology and Evolution*, **18**, 331–6.

Luck, G.W., Harrington, R., Harrison, P.A. *et al.* (2009) Quantifying the contribution of organisms to the provision of ecosystem services. *BioScience*, **59**, 223–35.

Maguire, L. A. and Justus, J. (2008) Why intrinsic value is a poor basis for conservation decisions. *BioScience*, **58**, 910–11.

McDonald, A.D., Little, L.R., Gray, R. *et al.* (2008) An agent-based modelling approach to evaluation of multiple-use management strategies for coastal marine ecosystems. *Mathematics and Computers in Simulation*, **78**, 401–11.

Millennium Ecosystem Assessment (MA) (2005) *Ecosystems and human well-being* Synthesis report. Washington, DC: World Resources Institute, Island Press.

Moss, S., Pahl-Wostl, C. and Downing, T. (2001) Agent based integrated assessment modelling: the example of climate change. *Integrated Assessment*, **2**, 17–30.

Naidoo, R. and Ricketts, T.H. (2006) Mapping the economic costs and benefits of conservation. *PLoS Biology*, **4**, e360.

Nyström, M., Folke, C. and Moberg, F. (2000) Coral reef disturbance and resilience in a human-dominated environment. *Trends in Ecology and Evolution*, **15**, 413–17.

Palumbi, S.R., McLeod, K.L., Grünbaum, D. (2008) Ecosystems in action: lessons from marine ecology about recovery, resistance and reversibility. *BioScience*, **58**, 33–42.

Palumbi, S.R., Sandifer, P.A., Allan, J.D. *et al.* (2009). Managing for ocean biodiversity to sustain marine ecosystem services. *Frontiers in Ecology and the Environment*, **7**, 204–11.

Pahl-Wostl, C. (2007) The implications of complexity for integrated resources management. *Environmental Modeling and Software*, **22**, 561–9.

Pearce, D. (2007) Do we really care about biodiversity? *Environmental and Resource Economics*, **37**, 313–33.

Prell, C., Hubacek, K., Reed, M. *et al.* (2007) If you have a hammer everything looks like a nail: 'traditional' versus participatory model building. *Interdisciplinary Science Review*, **32**, 263–82.

Prell C., Hubacek, K. and Reed, M. (2009) Stakeholder analysis and social network analysis in natural resource management. *Society and Natural Resources*, **22**, 501–18.

Raffaelli, D. (2006) Biodiversity and ecosystem functioning: issues of scale and trophic complexity. *Marine Ecology Progress Series*, **311**, 285–94.

Raffaelli, D.G., White, P.C.L., Renwick, A. *et al.* (2004) The Health of Ecosystems: the Ythan estuary case study. In: *Handbook of indicators for assessment of ecosystem health*. Jørgensen, S.E., Costanza, R. and Xu, F. L. (eds), CRC Press, Florida, 379–394.

Reed, M.S., Doughill, A.J. and Baker, T.R. (2008) Participatory indicator development: what can ecologists and local communities learn from each other? *Ecological Applications*, **18**, 1253–69.

Rice, J.C. and Rouchet, M.J. (2005) A framework for selecting a suite of indicators for fisheries management. *ICES Journal of Marine Science*, **62**, 516–27.

Rosenberg, A.A. and McLeod, K.L. (2005) Implementing ecosystem-based approaches to management for the conservation of ecosystem services. *Marine Ecology Progress Series*, **300**, 270–4.

Royal Society (2005) *Ocean acidification due to increasing atmospheric carbon dioxide*. Policy document 12/05. The Royal Society <http://royalsociety.org/displaypagedoc.asp?id = 13539>.

Sagoff, M. (2002) On the value of natural ecosystems. The Catskills parable. *Politics and the Life Sciences*, **21**, 19–25.

Scheffer, M., Carpenter, S., Foley, J.A. *et al.* (2001) Catastrophic shifts in ecosystems. *Nature*, **413**, 591–3.

Sherman, K., Sissenwine, M, Christensen, V. *et al.* (2005) A global movement towards an ecosystem approach to management of marine resources. *Marine Ecology Progress Series*, **300**, 275–9.

Surridge, B., Holt, A.R. and Harris, R. (2010) Developing the evidence base for Integrated Catchment Management: challenges and opportunities. In: *Water System Science and Policy Interfacing*. RSC Publishing. Quevauviller, P. (ed.), Cambridge, 63–100.

Turner, R. K., Paavola, J., Cooper, P. *et al.* (2003) Valuing nature: lessons learned and future research directions. *Ecological Economics*, **46**, 493–510.

Turner, R.K. and Daily, G.C. (2008) The ecosystem services framework and natural capital conservation. *Environmental Resource Economics*, **39**, 25–35.

Vandewalle, M., Sykes, M.T., Harrison, P.A., *et al.* (2008) Review paper on concepts of dynamic ecosystems and their services. <http://www.rubicode.net/rubicode/RUBICODE_Review_on_Ecosystem_Services.pdf>

Walker, B. H., Holling, C.S., Carpenter, S. *et al.* (2004) Resilience, adaptability, and transformability in social-ecological systems. *Ecology and Society*, **9**, 5. http://www.ecologyandsociety.org/vol9/iss2/art5/.

Wallace, K.J. (2007) Classification of ecosystem services: problems and solutions. *Biological Conservation*, **139**, 235–46.

Warren, R., de la Nava Santos, S., Arnell, N.W. *et al.* (2008) Development and illustrative outputs of the Community Integrated Assessment System (CIAS), a multi-institutional modular integrated assessment approach for modelling climate change. *Environmental Modelling and Software*, **23**, 592–610.

Worm B, Barbier E.B., Beaumont N. *et al.* (2006) Impacts of biodiversity loss on ocean ecosystem services. *Science*, **314**, 787–90.

Index